SCIENCE EDUCATION IN THE 21ST CENTURY

SCIENCE EDUCATION IN THE 21ST CENTURY

INGRID V. ERIKSSON
EDITOR

Nova Science Publishers, Inc.
New York

For permission to use material from this book please contact us:
Telephone 631-231-7269; Fax 631-231-8175
Web Site: http://www.novapublishers.com

LIBRARY OF CONGRESS CATALOGING-IN-PUBLICATION DATA
Science education in the 21st century / Ingrid V. Eriksson (editor).
p. cm.
Includes index.
ISBN-13: 978-1-60021-951-1 (hardcover)
ISBN-10: 1-60021-951-9 (hardcover)
1. Science--Study and teaching--Forecasting. 2. Education--Forecasting. I. Eriksson, Ingrid V. II. Title: Science education in the twenty-first century.
Q181.S373413 2008
507.1--dc22 2007034040

Published by Nova Science Publishers, Inc. ✢ New York

CONTENTS

PREFACE

This is hardly another field in education which is more important for a country's future than science education. Yet more and more students elect to concentrate on other fields to the exclusion of science for a variety of reasons:

1. The perception of degree of difficulty,
2. The actual degree of difficulty,
3. The lack of perceived prestige and earnings associated with the field.
4. The dearth of good and easy to use texts.
5. The lack of society in comprehending the significance of science and creating attractive incentives for those who enter the field.

This book presents new issues and challenges for the field.

Modern science has departed radically from the foundations laid in the renaissance period. As is well known through documented history, religious dogma hindered the advance of physics early in that era. Although some of its characterists are antithetical to the dogma of the 17th century, modern scientific hegemony has formed its own sytle of “religion”. The scientific dogma that grew to maturity by the end of the 20th century has now with its prejudices also become a source of unhealthy resistance. Through oversimplifications of theoretical type, often due to overly specialized education, too few scientists are open to the new ways of thinking required of the 21st century. Chapter 1 attempts to validate this claim, based on things learned in a case-study. This ongoing research has involved a multitude of differing mechanical oscillator types, studied intensely over a period of approximately two decades. Some of the discoveries from this work are far-reaching, impacting fundamental physics on the one hand and applied physics and engineering on the other. The study became possible for reason of a triad of modern technological tools. These comprise (i) new forms of an old instrument, the pendulum, (ii) modern electronics allowing remarkably better measurements than were possible just two decades ago, and (iii) personal computers outfitted with powerful, yet user-friendly acquisition and analysis tools. The pendulum was a key instrument in the early development of physics. Unfortunately, its continued positive influence is seriously hindered by those who view it only from a historical perspective. They fail to recognize the continued central role that it deserves for the study of material complexities. Those who like the author have been trying to improve the present state of science education, recognize a great need. The textbook descriptions of mathematically elegant models, that often miss the mark when it comes to predicting the behavior of real systems, need to be labeled as such and not viewed as sacrosanct. Material properties regulated by defect structures at the mesoscale are ubiquitous. Our understanding of internal friction from first principles is at best severely limited and at worst non-existent. One of the last frontiers of physics may prove to be one that was highly accessible and yet which lay fallow before us for the greater part of a century.

Research into student ideas and learning difficulties in science education is undertaken to inform developments in pedagogy. Complex phenomena may be best understood by exploring them from a range of viewpoints, and it is argued here that using a battery of 'analytical lenses' to illuminate research data may be an appropriate strategy in researching student learning. The approach is considered methodologically sound provided the 'lenses' derive from perspectives that are congruous with the theoretical framework underpinning the research. Chapter 2 illustrates such 'analytical pluralism' by applying three 'analytical lenses' ('modelling mentality', 'learning impediments' and 'student ontologies') to illuminate data from research into student understanding of the orbital model of atoms and molecules. The three 'lenses' focus on different features of the data in order to offer possible explanations for student learning difficulties and recommendations for improving teaching.

In recent times a considerable interest has been developed, in science education community, in fostering model-based reasoning at all level of schooling. Furthermore, research reports and projects recommended shifting the focus of science education from facts to overarching themes and toward the increasing relevance of models in teaching and learning science. Chapter 3 describes an approach to teaching/learning physics focused at developing pupil reasoning aimed at making sense of their own physical experiences and at evaluating information reported by others. The authors analyze how modeling procedures are relevant to the process of knowledge construction as well as to the epistemology of physics. Moreover, since models are representations of reality, they describe how to project and develop pedagogical tools able to scaffold the process of constructing different kinds of models: descriptive as well as interpretative and explicative models that concretize abstract ideas, simplify and clarify complex phenomena, predict trends and explain mechanism and processes. The authors also report the design process and the experimentation of two teaching-learning sequences focusing on the central issue of the relationships between observable phenomena, like macroscopic behaviors of matter, and their interpretation and/or explanation in terms of microscopic structure models. The design process with respect to the general framework of the *Educational Reconstruction Model* and the used pedagogical tools are described. Results, coming from teaching/learning experiments involving high school pupils, are also reported. The main focus is on the pupils' representations of phenomena and on the cognitive strategies put in action in order to modify or support their descriptive and interpretative mental models. Results are discussed from two different points of views: 1) by pointing out the efficacy of strategies focusing on the process of constructing predictive conceptual models and by identifying the concept of 'level of analysis' as different ways to look at the same phenomenon; 2) by analyzing how different kinds of modeling environments can stimulate pupil abilities in building mental models about the dynamical evolution of simple physical systems and improve their formalization.

The work described in Chapter 4 is a case study of a secondary education science teacher, whom shall be refered to as Ana, about the interaction between action-oriented reflection and action itself, and their influence on professional development. The study was carried out from two different perspectives —one with a qualitative orientation using diverse data collection and analysis instruments, and the other collaborative action-research to form the backbone of Ana's professional development. In the theoretical outline, the authors stress two concepts—that of reflection which sustains the theoretical-practical dialectic, and that of complexity as a progression hypothesis of central importance in which the authors distinguish the three dimensions of technique, practice, and criticism.

The results showed Ana to be in transition from a technical to a practical dimension, with her reflection and her classroom practice being closely integrated and in the process of becoming more complex. A central core of obstacles was observed that impeded her professional development.

In the 1990's two movements began to challenge the practice of science education from an historical and philosophical perspective. The International History and Philosophy of Science and Science Teaching Group (IHPST) maintains that not enough attention is being given to the intellectual heritage of many of our concepts in science with the result that, while students might be able to solve some quantitative problems in an algorithmic fashion, they have little understanding of the meaning or significance of the concepts they are using. The International Society for the Philosophy of Chemistry (ISPC) focuses its attention on the discipline of chemistry and has challenged practising chemists and chemistry educators to address, amongst other things, such fundamental questions as the meaning of the terms *atom*, *element*, *law*, and *theory* as they are used in chemistry. Is it possible to seek and accomplish conceptual depth, a desire of both international groups, without compromising conceptual usefulness, or must one be compromised in the pursuit of the other? In Chapter 5, this question is addressed in relation to the thermodynamic concepts of *energy*, *heat*, and *work*; the foundational concepts of *element* and the *mole*; and *quantum mechanical* concepts in chemistry. A textbook analysis of the definitions and use of these concepts in a tertiary level general chemistry course is given as is the implications of the study for chemistry education.

Newton's discovery of the laws of motion changed the way we view our universe. They are a standard part of any physicist's toolbox. Still, there is an overwhelming body of evidence that students often fail to master even the most basic aspects and everyday experience often appears to contradict the laws of motion.

Chapter 6 focuses on acceleration, which, unlike velocity, can be detected by a body in motion and measured within the moving system. Acceleration is also more fundamental; it is absolute, whereas velocity is relative and depends on the frame of reference chosen. This makes it possible to build on, rather than reject, the experience by learners and start in situations which where everyday observations do not contradict Newtonian concepts. The vector character of acceleration does not need to be seen as an abstract concept, but is evident in our daily three-dimensional motions, without the need to invoke a formal mathematical treatment.

After an introduction presenting different aspects of acceleration, including different ways of measuring and experiencing acceleration in different types of motion, follows a presentation of an investigation of new physics students' understanding of acceleration. Consequences of introducing acceleration through the mathematical definition, $\mathbf{a}=d\mathbf{v}/dt$, or through Newton's second law, $\mathbf{a}=\mathbf{F}/m$ are discussed and related textbook presentations are analysed. Which aspects of acceleration are discerned and which are not? Can teaching be modified to make critical aspects of acceleration accessible to the students?

The last part of this chapter describes pilot projects with 10-11 year olds who were invited to guided physics experiments in an amusement park or playground. Is it possible and meaningful for them to investigate acceleration in three dimensions?

Achieving scientific literacy for all citizens has become a national goal in many countries. Although admirable, it represents a challenge for science teachers and for those responsible for professional development. To achieve this goal, there must be a reform in the way science is taught in schools and in the strategies used to professionalize science teachers in general as well as the development of leadership among science teachers in particular.

One of the more effective methods to attain the goals of reform and its proper implementation is to actively involve the teachers in the curricular processes. Action research is only one of many effective strategies that could be used in order to effectively prepare science teachers to cope with new content and pedagogical standards. Having a long and varied history, action research was first introduced by Kurt Lewin in the 1940s. However, only within the last 15 years have we discovered the potential of action research to bring about lasting change in the science classroom.

It has been suggested that long-term professional development activities could create a clearing-house for pedagogical ideas derived from the results of action research. Chapter 7 presents four case studies, two from the USA and two from Israel, in which action research was used as a key strategy for professional development. For each case study, both the action research components and the professional development components are described. The chapter concludes with insights from these cases on issues and challenges involved in using action research as a professional development strategy.

There is a great need for science-based methods to estimate and rank the strength of different threats to the environment and Man. This demand has increased during the last decade, driven principally by the challenge to society of embracing sustainable development. In the case of the university sector, the inclusion in programmes and courses of facts and tools of general and subject-specific relevance for sustainable development is an appropriate educational response..

In working with environmental management and protection, science-based methods for risk-assessment and the ranking of environmental threats are of the utmost importance. Therefore, in environmental science education, it is necessary to include in course syllabi, a presentation, use and evaluation of relevant, available methodologies. The Potential Ecological Risk method (PER) represents one practicable approach. Whilst the PER-method was initially developed for ranking chemical threats to Swedish water ecosystems, it was subsequently extended for use not only in terrestrial ecosystems but also to characterise biological and physical threats having a global perspective.

The PER-methodology has been installed as a course module (value 2 ECTS credits) in an Erasmus Mundus Masters programme entitled "Coastal and Water Management". To date, the course module has been taught twice to international students at the University of Algarve, Faro, Portugal and once at University of Bergen, Bergen, Norway. The module has also been integrated into a general Ecological Risk Assessment course that is regularly delivered at Mälardalen University, Västerås, Sweden. Additionally, it was included in the mandatory course "Ecological Risk Assessment" in the joint Masters programme entitled "Master of Inland Water Quality Assessment" which commenced in 2007.

Chapter 8 presents the methodology, educational results, including student responses, reflections and chosen fields of assignment in relation to PER. Student-feedback concerning the methodology, learning outcomes and applicability of the PER method for managing threats to the environment are detailed. Potential enhancement of the course-model for integration into education for sustainable development is discussed.

A subject matter with a long history of development, such as science and mathematics, has a cultural notion of what counts as the practice of the disciplinary inquiry. Earth science, integrating geology, atmospheric and oceanic sciences, and astronomy into a whole, has a distinctive methodology in its intellectual and cultural tradition. Many authors agree that abduction is the scientific method characterizing the backward reasoning process of earth science. They also contend that abduction should be applied to the school curriculum so that students can be familiar with the discipline-specific inquiry method. Considering this position, the authors have adopted the Abductive Inquiry Model (AIM) into their earth science laboratory course for prospective secondary science teachers. In Chapter 9, the authors will describe what abduction is and what the AIM looks like, and then present some features of the undergraduate laboratories in which the AIM was used to teach and learn earth science.

By means of the review of the old physics books, we can have a look into the world of the physics education of a century ago. We know that the teachers preferred to use qualitative explanations and experimental exhibitions rather than theoretical demonstrations as nowadays. In fact, the physics texts show a lot of illustrations of physics instruments that could be found in the physics laboratory of every educational centre. This way of physics teaching was widely used in educational centres of different levels (primary school, secondary school, university) up to the end of the second World War. Nevertheless, in the course of the last years, a number of different authors have proposed using copies or redesigns of these old scientific instruments in order to make the process of learning physics more attractive. In Chapter 10, the authors carry out a review of some of the more interesting researches and experiences published in the last years. In some cases, new experimental devices have been designed in order to carry out qualitative exhibits (for example, colour disks based on Maxwell disks are used to show the colour vision mechanism). On the other hand, some old scientific instruments commonly used in qualitative exhibits have been redesigned in order to carry out quantitative evaluations (for example, an experimental device which permits the analysis of the transition from laminar to turbulent flow regime has been designed by means of the use of a Mariotte bottle). Also, the authors carry out a novel and introductory analysis of a quantitative intermittent fountain that is based on the well-known qualitative intermittent fountain. As a conclusion, an intermittent fountain can be used as an attractive and useful laboratory experiment in which the students can control the number and speed of the intermittences. In this chapter, it can be seen that some fields of experimental physics education (as mechanics, electromagnetism and optics) have been extended by means of the use of copies and devices based on old scientific instruments.

Research in science learning and teaching continues to show the difficulties in learning science and the challenges in changing/restructuring students' conceptions about scientific phenomena at various age levels. For the past decades, we have moved from behaviorism to constructivism, from Piaget and Bruner to cognitive scientists, from unitary representation to multiple representations, from a single explanation framework to multiple frameworks. Such changes make us wonder what other researchers have achieved and what challenges face us in science learning. Chapter 11 reviews the large body of literature on students' conceptions in science education over the past 30 years and summarizes what teachers and schools work effectively on developing their students' cognitive structures. The authors cover the following perspectives: (1) the trajectory of the development of students' conceptions in science, (2) the sources of students' scientific conceptions, (3) the theories of conceptual change in science learning, (4) ways of assessing students' learning, and (5) the learning and teaching strategies for promoting science learning. A particular model proposed by Chiu (2007) was also introduced. Finally, the implications for science education are discussed.

EXPERT COMMENTARY

In: Science Education in the 21st Century
Editor: Ingrid V. Eriksson, pp. 3-5
ISBN: 1-60021-951-9

OLD PHYSICS BOOKS AND SCIENCE EDUCATION

José Alberto Maroto Centeno[1] and Manuel Quesada-Pérez[2]
Group of Physics and Chemistry of Linares, Departament of Physics, EPS of Linares
(University of Jaén)
C/ Alfonso X el Sabio, 28, 23700 Linares, Jaén, Spain

Expert Commentary

A chapter in this book analyzes the use of old scientific instruments for physics teaching. A review of that chapter shows that a number of different authors have carried out very interesting works in this area in the course of the last years. Nevertheless, the use of old physics books in science education is an unknown area that needs to be explored. An example that shows the possibilities in this area is the work of R.C. Henry [1]. An extraordinarily simple and transparent derivation of the formula for the acceleration that occurs in uniform circular motion is presented in his work, and is advocated for use in high school and college freshman physics textbooks. The main aim of this article is to present Newton's method so that it can be used in effectively teaching the origin of the formula for the acceleration that occurs in uniform circular motion to the modern young student of physics. The essential problem that all the early physicists faced was the need to work a vectorial problem, not having a real theory of vectors available. What was available to them was the parallelogram law for combining vectors in two-dimensional problems; that result is as old as Aristotle. If one does have vector methods, there are modern methods for obtaining this formula that are delightful [2]. But the formula for the centripetal acceleration is so crucial that one does not want the beginning physics student to think that it can not be obtained except by sophisticated methods, with which the student is only beginning to become familiar. That is one reason why a clear exposition of Newton's wonderful method is pedagogically valuable. It must be underlined that the cinematic derivation showed in this work is based on the Principia of Newton. The method used by Newton, about 1665, is described by Westfall [3], and has been analyzed in detail, from an historical perspective, by Erlichson [4]. The aim of this work,

[1] E-mail address: jamaroto@ujaen.es
[2] E-mail address: mquesada@ujaen.es

however, is not historical; it is to present the method as it can and should be presented to the modern student. In a later work [5] Maroto and de las Nieves carried out a review of the Newtonian polygonal proof for centrifugal conatus in order to think up a new and original proof for centripetal force law, which permits one to extend the applicability of the pulse force concept, traditionally limited to some collisions analysis (golf club hitting a ball, hammer hitting a tack, ball bouncing on a rigid surface, and so on). On the other hand the authors think that it could be interesting to acquaint physics students with the polygonal proof of Newton, which shows the important role played by the centrifugal force law in the development of the inverse square law of gravitation.

In short, old physics books can be useful tools for the development of new proofs and physics problems. Also, the review of old physics books permits a suggestive understanding of the evolution of physics. In fact, we show a novel example of it below.

As we now know, X-rays are a type of electromagnetic radiation with wavelengths of around 10^{-10} meters. On November 8, 1895, Wilhelm Conrad Röntgen, a German physics professor, began observing and further documenting X-rays while experimenting with vacuum tubes. Röntgen, on December 28, 1895, wrote a preliminary report "*On a new kind of ray: A preliminary communication*". He submitted it to the Würzburg's Physical-Medical Society journal. This was the first formal and public recognition of the categorization of X-rays. Röntgen referred to the radiation as "X", to indicate that it was an unknown type of radiation. The name stuck, although (over Röntgen's great objections), many of his colleagues suggested calling them Röntgen rays. They are still referred to as such in many languages. Röntgen received the first Nobel Prize in Physics for his discovery. Röntgen was working on a primitive cathode ray generator that was projected through a glass vacuum tube. Suddenly he noticed a faint green light against the wall. The odd thing he had noticed, was that the light from the cathode ray generator was traveling through a bunch of the materials on its way (paper, wood, and books). He then started to put various objects in front of the generator, and as he was doing this, he noticed that the outline of the bones from his hand were displayed on the wall. Röntgen said he did not know what to think and kept experimenting. Two months after his initial discovery, he published his paper translated "On a New Kind of Radiation" and gave a demonstration in 1896. Rontgen discovered its medical use when he a saw picture of his wife's hand on a photographic plate formed due to X-rays. His wife's hand's photograph was the first ever photograph of a human body part using X-rays. Since Röntgen's discovery that X-rays can identify bony structures, X-rays have been developed for their use in medical imaging. Radiology is a specialized field of medicine. Radiographers employ radiography and other techniques for diagnostic imaging. Indeed, this is probably the most common use of X-ray technology.

Nevertheless, during the early years after the discovery of the X-ray, scientists thought that this radiation could be used not only as a diagnostic method but also as a therapeutic technique due to its antimicrobial power. The figures 1.a and 1.b show two images scanned from a physics book dated 1900 [6] in which Mr. Lortet, a doctor of the Lyon Faculty, relates a satisfactory medical treatment with X-rays. Mr. Lortet describes an intensive use of X-rays (from July 13 up to August 28) that permitted a hopeless patient to avoid death. So, this interesting book shows that science used to fail during its evolution. Science it is not foolproof: as now it is well known that a large exposure of X-rays causes cancer! As a conclusion, the review of old physics books shows suggestive historical data that can be used in science education.

References

[1] Newton, I.; Henry, R.C.. *J. Phys*. 2000, 68 (7), 637-639.
[2] Ninio, F. *Am. J. Phys.* 1993, 61, 1052.
[3] Westfall, R.S. *Never at Rest: A Biography of Isaac Newton*; Cambridge U.P.: Cambridge, 1980.
[4] Erlichson, H. *Am. J. Phys.* 1991, 59, 842-849.
[5] Maroto, J.A.; de las Nieves, F.J. *Rev. Esp. Fis*. 2002, 16 (4), 33-36.
[6] Marcolain San Juan, P. *Curso Elemental de Física Moderna*; Tipografía de Emilio Casañal: Zaragoza, 1900.

RESEARCH AND REVIEW STUDIES

In: Science Education in the 21st Century
Editor: Ingrid V. Eriksson, pp. 9-68
ISBN 1-60021-951-9

Chapter 1

BUILDING ON OLD FOUNDATIONS WITH NEW TECHNOLOGIES

Randall D. Peters
Mercer University

Abstract

Modern science has departed radically from the foundations laid in the renaissance period. As is well known through documented history, religious dogma hindered the advance of physics early in that era. Although some of its characterists are antithetical to the dogma of the 17th century, modern scientific hegemony has formed its own sytle of 'religion'. The scientific dogma that grew to maturity by the end of the 20th century has now with its prejudices also become a source of unhealthy resistance. Through oversimplifications of theoretical type, often due to overly specialized education, too few scientists are open to the new ways of thinking required of the 21st century. The present article attempts to validate this claim, based on things learned in a case-study. This ongoing research has involved a multitude of differing mechanical oscillator types, studied intensely over a period of approximately two decades. Some of the discoveries from this work are far-reaching, impacting fundamental physics on the one hand and applied physics and engineering on the other. The study became possible for reason of a triad of modern technological tools. These comprise (i) new forms of an old instrument, the pendulum, (ii) modern electronics allowing remarkably better measurements than were possible just two decades ago, and (iii) personal computers outfitted with powerful, yet user-friendly acquisition and analysis tools. The pendulum was a key instrument in the early development of physics. Unfortunately, its continued positive influence is seriously hindered by those who view it only from a historical perspective. They fail to recognize the continued central role that it deserves for the study of material complexities. Those who like the author have been trying to improve the present state of science education, recognize a great need. The textbook descriptions of mathematically elegant models, that often miss the mark when it comes to predicting the behavior of real systems, need to be labeled as such and not viewed as sacrosanct. Material properties regulated by defect structures at the mesoscale are ubiquitous. Our understanding of internal friction from first principles is at best severely limited and at worst non-existent. One of the last frontiers of physics may prove to be one that was highly accessible and yet which lay fallow before us for the greater part of a century.

1. Introduction

The pendulum has a rich history of influence on the development of science, having been the focus of attention by famous individuals such as Galileo (1564-1642) and Newton (1643-1727). Its role has been central to the discovery of key foundation elements of physics during the last three and a half centuries. For example, in a simple form it enabled Newton in the first of these three centuries to demonstrate the equivalence of inertial mass and gravitational mass to about one part per thousand. About a century later a different type of pendulum, of torsional type, was used by Coulomb (1736-1806) to understand the inverse square law of the force between point charges. About a decade later, Cavendish (1731-1810), with help from John Michell used a similar torsion pendulum (sometimes called a balance) to measure the average density of the earth–the first experiment from which one is able to estimate the fundamental constant, 'big' G, in Newton's universal theory of gravity. Then a half-century later, the pendulum enabled Stokes (1819-1903) to describe the drag forces of air acting to rob it of oscillatory energy. Using experimental data generated by others, Stokes developed the following friction-force-law, the use of which was necessary for Millikan (1868-1953) to estimate the charge of the electron.

$$f = 6\pi \, \eta \, R \, v \tag{1}$$

Here R is the radius of a small spherical particle (oil drop in Millikan's case) moving at speed v in a gas whose viscosity coefficient is η. Interesting in relationship to present research is the name that Stokes coined for the parameter η; i.e., what he called the "coefficient of internal friction". [Stokes, 1850]. One has to wonder why the inventor's preferance was not followed, so that it instead came to be called viscosity. Therein lies one of the shortcomings of physics education; i.e., too little attention given to the history of our discipline. Many of the great advances of physics involved expertise of type that is better labeled "art" than it is "science", according to the modern view of the latter. The tendency to equate physical understanding with mathematics is widespread; discussions of such erroneous thinking is postponed to a later section of this article.

Various forms of the pendulum have become the mainstay of seismology; since the most common means for detecting earthquakes has always involved some type of a pendulum, even though the instruments look nothing like those studied by Galileo and Newton. In the material that follows, a number of experimental cases are described in which the data was generated with an instrument that was close in characteristics to, if not in actuality, a bonafide seismometer.

In the last century the pendulum has played a central role in the development of the 'new science of chaos'. Unlike the motion-regime appropriate to the previous discoveries (nearly isochronous), in which the displacement θ away from equilibrium is always less than roughly a milliradian; chaos requires large-enough drive forces for the 'bob' to 'move past vertical'; i.e., $|\theta| \; > \; \pi$. For such motions a bob on a string will not work, since the string can only support tension and not the compression that must also come into play. Although the 'elastic' anharmonicity that thus results, constitutes a condition of nonlinearity necessary for chaos, it is not sufficient of itself to result in chaos. Other conditions involving the drive amplitude and frequency, relative to the pendulum's natural eigenfrequency and damping–must be met to exhibit the 'period doubling route to chaos'. [Peters, 2007], [Baker & Gollub, 1996].

Careful studies with pendulums are still very much needed because anelasticity has been recently recognized to have impacted two significant 'modern' physics efforts: (i) attempts to improve the accuracy of the gravitational constant $G \approx 6.67 \times 10^{-11} N\ m^2/kg^2$ and (ii) attempts at LIGO to detect gravitational waves. A Physical Review Letters article [Kuroda, 1995] explains how significant errors in the estimate of G have resulted for some experiments because of anelasticity in the torsion wire of the Cavendish balance (pendulum) that was used.

In the LIGO work a critical requirement is to minimize the influence of ground noises, while looking for gravitational waves that are so weak as to require detector sensitivity in the neighborhood of $10^{-18}m$, yet still recognizable above the noise floor. To achieve this level of sensitivity requires isolation of the mirrors of the interferometer from seismic vibrations. Various mechanical oscillator types have been studied for this purpose, such as (i) an inverted pendulum, and (ii) a Monolithic Geometric Anti Spring (MGAS) [Desalvo, 2005].

An empirical formula for the quality factor of the inverted pendulum was determined by the LIGO group to be

$$Q \propto \frac{L}{g}\omega^2 \tag{2}$$

where L is the length of the pendulum, $g = 9.8\, m/s^2$, and ω is the angular eigenfrequency of oscillation. Being proportional to the square of the frequency means that the damping is of internal friction type and of an entirely different character than what is taught to physics students. This universal (hysteretic) damping will be discussed in follow-on sections of the present article. The author's theory of the origin of the damping (anelastic flexure) provides an explanation for Eq.(0.2). [Peters, 2005-1]

Studies by the present author, some of which will be described in the sections that follow–point to his belief that the pendulum is unlikely to ever become a relic of the past, destined for major consideration only by historians [Peters, 2005-2]. New technologies have allowed him to finally perform experiments that could only be contemplated by Stokes and other great figures of the 19th century, such as Maxwell. Clerk Maxwell was interested in the anelastic nature of the mechanical parts from which real-world pendulums are fabricated. In the simplest models of pendulum motion, these structures are always assumed to be rigid. Of course, real structures are not completely rigid; and when they flex, their stress-strain cycles are not in accord with Hooke's law to describe 'springiness' [Erber, 2001]. Special attention is here paid to the prefix 'an' operating on 'elastic' (meaning 'other than') and not the prefix 'in' (meaning 'not'). Whereas 'in' is appropriate to the description of particle scattering in which energy is not conserved, it is inappropriate to the description of the stress-strain hysteresis of real materials. A well-known reference to such systems, in which the strains are treated as being in accord with the fundamental theorem of calculus–is the classic reference book by Clarence Zener [Zener, 1948]. In actuality, real systems can exhibit discontinuous strains, so that the fundamental theorem is not obeyed. The 'jerky' behavior first observed in magnetic materials [Barkhausen, 1919], and in the strain of alloys [Portevin/LeChatelier 1923], result from the organization of defect structures that determine the internal friction that regulates pendulum decay at low energies.

These defect structures operate at the mesoscale, and no 'first-principles' theory to describe them has yet been developed [Peters, 2004]. To this point in time, their properties have only been discovered by experimental empiricism; and their complex nature related to creep– is so diverse that their meaningful study had to await the development of modern technologies. For the author's use, an essential triad of technological elements converged in the last two decades. These are (i) a non-invasive linear sensitive detector of pendulum motion, (ii) the advent of powerful personal computers for acquisition of the sensor's output data, and (iii) user friendly, powerful software capable of analyzing the acquired data in both the time-domain and the frequency-domain. The absence of any single element of this triad would make the studies impossible; which are now finally positioned to do sophisticated experiments that would no doubt have pleased Richard Feynman, whose celebrated three-volume series [Feynman, 1964] include references to physics discoveries that relate to various of the phenomena being discussed. For example, Feynman was intrigued with the 'jerky' behaviour of magnetic materials (Barkhausen effect), and also provided in its entirety a reprint of the Bragg brothers article dealing with 'bubble rafts'. The bubble rafts described are a two-dimensional manifestation of 'dislocation' ('crystalline'-defect) structures. The present author claims that such defect structures are responsible in the three-dimensional world for internal friction that is part of an unfinished solid-state physics frontier. He has postulated that a quantum of internal friction energy, operating at the mesoscale and having the value of $1.1 \times 10^{-11}\ J = 11\, pJ$, is responsible for the myriad mysteries of anelasticity that regulate creep and pendulum damping.

Resulting from damping anharmonicity, as opposed to the well-known elastic anharmonicity responsible for pendulum chaos; most physicists have been directed into education paths that left them ignorant of these mysteries. Who knows why, for example, the Poretevin LeChatelier effect is almost unknown to the physics community; even though it is the topic of many discussions among materials scientists? The paucity of information interchange between scientific disciplines has in some cases been so extreme– that it has caused the author to modify the popular secular statement about 'the right hand not knowing what the left hand is doing' to read 'the right hand doesn't even know there is a left hand.' Compartmentalization of specialized science (such as the practice of 'delta function' physics) wars against a healthier broader perspective, and can result in remarkable ignorance. An example in the author's experience nicely illustrates the point. An individual who is highly capable in his specialized field, once placed on a PhD qualifying exam a problem concerned with an air bubble placed at the bottom of Earth's deepest ocean. Students of the exam were asked to calculate the increase in size of the bubble as it floated to the surface. The method of solution intended by the one who posed this problem–involved the ideal gas law. Only those whose education was broad enough to expose them to real-world gas equations of state (involving the virial coefficients) recognized that the assumptions supplied with this problem were highly erroneous. Moreover, the supplier of the problem, whose level of arrogance exceeds that of most, had no idea that his offering was anything less than 'perfect'.

Defects are 'everything' when it comes to most systems capable of exhibiting complexity. A distinction is here made between 'chaotic' systems and 'complex' systems. A pendulum can exhibit chaos even though it is adequately described by a relatively simple mathematical formula. Its non-complex mathematical equation of motion is basis for describing the system by the words, 'deterministic chaos'–a combination of words that seem incompatible to those unfamiliar with the methods employed.

Consider the importance of defects to the semiconductor world. Without the use of trivalent and pentavalent substitutional impurities in silicon, the information age in which we live would be impossible; since integrated circuitry would be incapable of functioning. A case is made in the present article for the similar importance of a different kind of crystalline defects. Instead of being substitutional in character, they involve atom vacancies. Such vacancies are well known to the material science profession, even though their earliest consideration was by physicists. The simplest possible defect is known by the name Schottky, a vacancy that is always present in crystals at room temperature; it can be explained easily by means of statistical mechanics. The simplest of the many-body vacancy structures are dislocations of 'edge' and 'screw' types. They are known to be responsible for real materials being much weaker than theoretical predictions based on an assumed perfect lattice. Edge dislocations propagate from surfaces inward when metals are strained. An analogy with "ease of moving a large rug against floor friction (by propagating a cusp)" is commonly used as a conceptual means to explain why the strength of materials is greatly reduced when dislocations are allowed to freely move. Because defect structures involving dislocations are complex and poorly understood, they have often been simply ignored by physics theorists.

In the world of physics education, our desire for simplification has at times gone to extreme. Any topic involving complexity is rarely if at all mentioned. For the material that is treated, descriptions tend to be encyclopedia-like in nature, and the student is tempted to believe that surely the writer knows exactly what he/she is talking about; and that the material provided 'tells the whole story'. Nothing could be farther from the truth when it comes to conventional wisdom concerning the damping of a pendulum. In the material that follows, several examples are provided to document this claim.

Hopefully, in years to come, the healthier attitudes of some pioneering figures among us, such as Stephen Hawking, can assist a greatly needed change. He has gone on record [Hawking, 2000] stating that he believes the 21st century will be the century of complexity. Perhaps he believes, like this author, that physics education is in need of a major overhaul. It is not so much the 'methods' independent of content (as epitomized by U.S. academic schools of education), but rather attitudes related to philosophy that are at the heart of this challenge. The hegemony of reductionism has been at work in the generation of elementary physics textbooks. Overly-simplistic equations of motion are presented with the appearance of being 'canonical' (liturgical term suggesting 'God given'). Those who 'cook the books' are usually theorists who have little if any first-hand knowledge concerning the actual limitations of the mathematical formulas they present. We've inherited a legacy of unbalance between theory and experiment. At the graduate level, every student who qualifies for the PhD is required to demonstrate a nontrivial mastery of the major theories of our profession. One doing an experimental dissertation must therefore become a 'jack of all trades'. Contrariwise, equality of training breadth is almost never required of one doing a purely theoretical dissertation. Problems presented on PhD qualifying exams are routinely unbalanced, as though excellence of experimental technique is something anybody can learn on his/her own; whereas theory must be taught 'by masters to the elite'. The author heard with his own ears (in a public lecture) the following paraphrased comment by a well-known high-energy Nobel Laureate: "if you theorists can't do your work the way I do it, then you ought to go learn how to use a soldering iron". As stated in another large public lecture following the demise of the Waxahatchie super collider, this same prominent figure (name purposely witheld) observed "that complexity deserves very little attention as compared to experiments needed to complete the standard model". As the epitome of reductionist thinking–if his position should indeed prove valid, one could infer that the only thing necessary to understand even the most complex living system (ourselves?) is to apply more and bigger computers to the task.

Of course there are other, equally gifted and famous individuals, whose philosphy of science is of antipodal type. [Anderson, 1972]. Condensed matter Nobel Laureate physicist Paul Anderson has stated that particle physics and indeed reductionism has only a limited ability to explain the world. He notes that reality has a hierarchical structure, with each level independent, to some degree, of the levels above and below: "At each state, entirely new laws, concepts and generalizations are necessary, requiring inspiration and creativity to just as great a degree as in the previous one." "Psychology is not aplied biology, nor is biology applied chemistry." Anderson's comments speak to the matter of synergy, where the collective behavior is greater than the sum of the parts taken individually. The synergetic properties of defect structures are unknown from first principles. In the encyclopedia article on the anharmonic oscillator [Peters 2007] it is stated that that one of the last frontiers of physics could actually be one of the most accessible parts of our world.

The philosophy driving the generation of this article is one which aligns itself with the viewpoints expressed by Anderson. Some things learned through the practice of that philosophy are now presented.

2. Math-Madness

Too many individuals of the present age see little if any distinction between physics and mathematics. The folly of this view derives from a failure of teachers to provide a healthy, broad education for their students. Lest this claim be doubted, consider the following quotes from three stalwarts of our profession:

Einstein:

"Fundamental ideas play the most essential role in forming a physical theory. Books on physics are full of complicated mathematical formulae. But thought and ideas, not formula, are the beginning of every physical theory. The ideas must later take the mathematical form of a quantitative theory, to make possible the comparison with experiment. The mathematics dealing with the problem of waves of matter is extremely simple and elementary, but the fundamental ideas are deep and far reaching."

Dirac:

"Mathematics is the tool specially suited for dealing with abstract concepts of any kind and there is no limit to its power in this field. All the same the mathematics is only a tool and one should learn to hold the physical ideas in one's mind without reference to the mathematical form."

Feynman:

"Look, these differential equations–the Maxwell equations–are all there is to electrodynamics; it is admitted by the physicists that there is nothing which is not contained in the equations. The equations are complicated, but after all they are only mathematical equations and if I understand them mathematically inside out, I will understand the physics inside out. Only it does not work that way. A physical understanding is a completely unmathematical, imprecise, and inexact thing, but absolutely necessary for a physicist."

3. Overly-idealized Simple Pendulum Equation of Motion

In its simplest form of a concentrated mass swinging at one end of a lightweight, inextensible string; Newton's classical laws of motion are readily used to describe not only the pendulum's near- isochronous low-amplitude motion–but also the period lengthening that occurs when it is made to swing at larger amplitudes. Newton's second law is adequate even to the description of chaotic motion for which the pendulum is an archetype, when the string is replaced with a rigid rod so that motion past vertical is possible. Only with the allowance of 'winding modes' in this nonlinear regime of the motion can chaos be observed [Peters, 2007].

Students of physics are regularly exposed to at least the simplest of various solutions which assume the pendulum to behave as a simple harmonic oscillator damped by a linear external friction force. No textbooks are known to the author, in which the serious limitations of the damping model are even mentioned. Physics education is itself so limited in this arena, that the authors of these textbooks "don't know what they don't know"! Inherent to the "math-madness" that is being propagated, is an ignorance of the fact that great figures of physics were not so naive. Unlike modern practitioners of our discipline, individuals such as Maxwell gave deep thought to the internal friction of solids. In contrast between his world and the present age, for which technology advances finally allow meaningful study; Maxwell could only see solutions from afar-off. For whatever reasons, the "modern physics" perspective has viewed solid-state defect physics as either too (i) "dirty" to be attractive, or (ii) old and thus surely a 'closed' field to warrant consideration.

A common belief is that the only parameter needed to predict the quality factor Q of the pendulum's decay is the viscosity of the air surrounding the instrument. It is true that air is usually the dominant source of damping for many configurations of the pendulum labeled 'simple'; however, even these simple cases are not adequately described by the following famous differential equation:

$$\ddot{\theta} + 2\beta\dot{\theta} + \omega_0^2\,\theta = 0 \tag{3}$$

where θ is the displacement from equilibrium, ω_0 is 2π times the natural undamped frequency, and the dots imply time differentiation (first used by Newton). The factor of two is usually included with β because it describes the exponential decay of the turning point amplitudes of the motion via the expression $e^{-\beta t}$. Even with viscous damping, the decay is rarely completely exponential, because the drag force that robs oscillatory energy from the pendulum is not simply proportional to the first power of the velocity. Although one might envision a hypothetical oscillator for which the Reynolds number is low enough for Stokes' law to hold precisely, and thus allow Eq.0.1 to be reasonable; real-world, practical-sized instruments do not obey this equation as well as is generally believed. George Gabriel Stokes, who was responsible for laying key parts of the theoretical foundation of fluid mechanics (along with Navier) never trusted this equation the way it has been viewed by modern users [Stokes, 1850].

Even if the pendulum bob is a sphere, for which the air drag acting on it may be thought to be linear, as in Eq.(0.1), Eq. (0.3) still does not account for other important contributors to the damping. For example, the paper by Nelson and Olsen [1986] illustrates some of the 'rich physics' deriving from a variety of drag forces, some of which are nonlinear as well as linear. In sections of the present paper which follows, evidence is provided to show that long-period physical pendulums come nowhere close to being adequately described by Eq.(0.3).

Eq.(0.3) is mathematically elegant, and if valid requires the following to be true. If the damping parameter β were constant, then the Q of the pendulum must be proportional to the frequency. The Q is defined by

$$Q = 2\pi \frac{E}{|\Delta E|} \tag{4}$$

where E is the total energy of oscillation and ΔE is the energy lost per cycle due to friction. In terms of Q, the equation of motion can be rewritten in canonical form as

$$\ddot{\theta} + \frac{\omega_0}{Q}\dot{\theta} + \omega_0^2\theta = 0 \tag{5}$$

from which it is seen that $Q = \omega_0/(2\beta) = \pi f_0/\beta$. Thus for constant β, $Q \propto \omega_0$. In terms of the log-decrement defined by $\delta = \beta T_0 = \beta/f_0$ and which is also sometimes used, we see that $Q = \pi/\delta$.

3.1. Damping Redshift

The mathematical form of the linear equation of motion just described requires a damping redshift, in which the frequency of oscillation is altered because of the damping according to the expression

$$\omega = [\omega_0^2 - \beta^2]^{1/2} \tag{6}$$

Textbooks routinely make a big deal of this expression without describing the great challenge required to actually measure the shift. As β becomes large enough for such effect to be observed, the short duration of that part of the signal that is above noise becomes important. The uncertainty principle figures into the difficulties encountered.

For the case of hysteretic damping due to internal friction, which is next described, there is no such shift to begin with.

4. Hysteretic Damping

It is postulated that the processes of creep described in section 0.11 are intimately connected with a free-decay small-amplitude equation of motion given by

$$\ddot{\theta} + \frac{\pi\omega}{4Q}[\omega^2\theta^2 + \dot{\theta}^2]^{1/2} sgn(\dot{\theta}) + \omega^2\theta = 0 \tag{7}$$

where the damping is seen to be nonlinear, involving only the algebraic sign of the velocity.

Although the friction described by this equation is cause for an exponential decay of the pendulum's turning point amplitudes, the shape of the friction force is nearly that of an attenuated square wave, in which the amplitude of the nonlinear friction force also declines exponentially. An example set of waveforms is provided in Fig. 1.

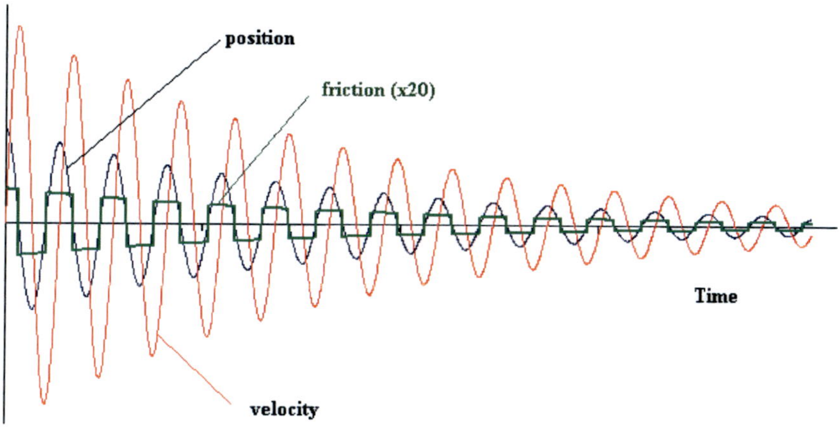

Figure 1. Waveforms generated by integration of Eq. (0.7) with Q = 20. The friction force has been, for purpose of visibility in the graphic display, multiplied by a factor of twenty. It illustrates the (nonlinear) square wave nature of the damping force.

Observe that no zero-subscript has been placed on ω, since the mathematical form of the nonlinear damping is incapable of producing a redshift. It can be shown [Peters 2004] that the only factor of importance to the determination of Q is the fundamental (lowest frequency) component of the Fourier series of the friction force. This is a consequence of the orthogonality properties of the harmonic functions (sine and cosine) which permit Fourier series to work. In Eq.(0.7) the fundamental of a unit-amplitude square wave ($\pi/4$) has been provided as part of the damping term. In a set of graphs restricted to the display of only position and velocity, very little difference can be observed between Fig. 1 and the result of integrating linear Eq.(0.3), subject to the same initial conditions. Moreover, direct measurement of the friction forces is not possible. We thus recognize a major reason for the temptation to believe only in linear damping; since a variety of different friction forms can all yield highly similar position/velocity waveforms.

The author has selected the dampened square wave described by Eq.(0.7) because of its similarity to classical Coulomb damping. By changing to zero the 1/2 exponent on the term in brackets; half-cycle decrements cause the turning points to fall along a straight line, rather than along an exponential. This case corresponds to Coulomb damping, which involves sliding with a constant coefficient of kinetic friction. Though usually referred to as 'hysteretic', the damping of Eq.(0.7) has also been referred to as 'modified Coulomb' damping. It is thought that internal friction of metals involves grain boundary 'sliding', but unlike classical Coulomb damping, the effective 'kinetic coefficient' is energy dependent. Consequently the decay is exponential and easily masquerades as linear viscous. There is a major difference, however, between hysteretic damping and linear viscous damping, which is now described.

5. Frequency Dependence of Pendulum Q

As noted previously, the linear viscous damping model predicts that the Q of a pendulum should vary in direct proportion with its eigenfrequency. The hysteretic damping model of Eq.(0.7) predicts that the Q should be proportional to the square of the eigenfrequency. Thus a means exists to test which damping type is more appropriate for a given oscillator.

The author has done such comparisons on a wide variety of mechanical oscillators, some of which looked very 'pendulum' in appearance, and others of seismometer type that looked quite different. In every case, it was found that a quality factor graph of the type shown in Fig. 2 was a much better fit to the data than one having a linear dependence on frequency as required by the linear viscous model.

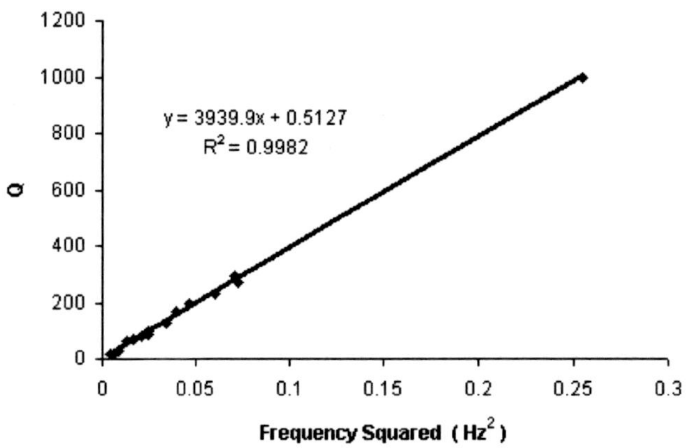

Figure 2. Quality factor plot showing the quadratic dependence on eigenfrequency that is the hallmark of hysteretic damping.

For a pendulum to oscillate at the longer periods necessary to generate a graph like Fig. 2 (the hallmark of "universal internal friction damping"), it must have a significant fraction of its total mass located (in a quasi-rigid structure) above the axis of rotation. Altering the mass distribution causes the period to lengthen as the center of mass draws closer to the axis. An example of such a pendulum, which was used to generate the Fig. 2 data is shown in Figure 3.

The sensor used is one version of the first fully-differential-capacitive type [Peters, 1995]. Its architecture allows great sensitivity while maintaining great linearity; an example of the latter, corresponding to the pendulum shown in Fig. 3, is the calibration data of Fig. 4.

6. Frontier at Your Fingertips

Approaching two decades ago, the term 'mesodynamics' was coined to describe the complex behavior of a physical pendulum oscillating at low energies. The prefix 'meso' was chosen by the author to emphasize the fact that the oscillatory energy of these instruments was "in the middle" between the atomic scale on the one hand (electron-volt level) and the classical scale on the other hand (joule level). A favorite of the author's statements to

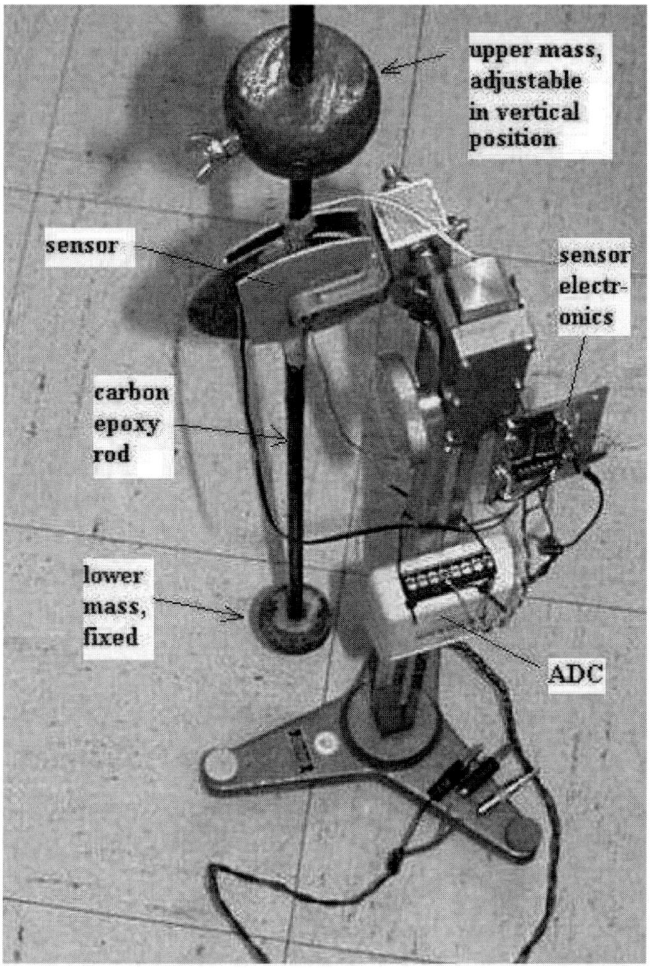

Figure 3. Physical pendulum capable of long-period motion that was used to generate the data plotted in Fig. 2.

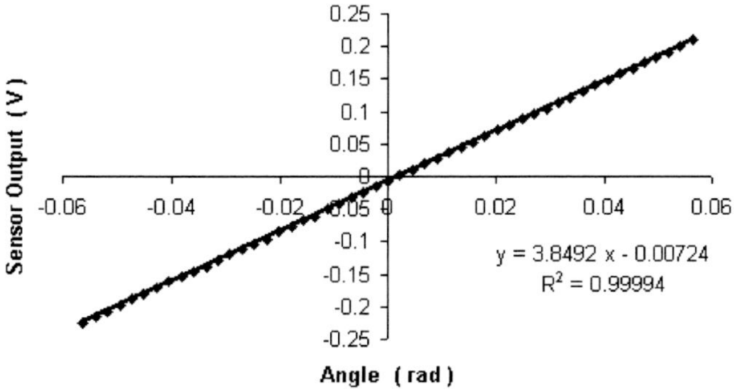

Figure 4. Calibration data corresponding to the setup of Fig. 3, used in the generation of the data of Fig. 2.

describe this energy range has been: "do you suppose in 19 orders of magnitude we might have missed something important?"

Recent claims by others is in strong support of this belief. For example, we read in an essay by Coleman [07]: "Between the nano- and micrometer scales, the collective behaviour of matter can give rise to startling emergent properties that hint at the nexus between biology and physics." He notes that Newton's laws account for single-particle dynamics, but it took Maxwell's development of the statistical description of atoms for an understanding of pressure. More startling, as Coleman observes, is "the potential for quantum matter to develop emergent properties." He notes that condensed matter physicists have taken a cue from evolution, and "believe that a key to understanding more complex forms of collective behavior in matter lies in competition not between species, but between different forms of order." The 'related' idea he mentions is 'criticality'–"the concept that the root of new order lies at the point of instability between one phase and another. So at a critical point, the noisy fluctuations of the emergent order engulf a material, transforming it into a state of matter, that like a Jackson Pollack painting, is correlated and self-similar on all scales..... today we are particularly interested in 'quantum phase transitions' involving quantum noise: jigglings that result from Heisenberg's uncertainty principle....critical quantum fluctuations (that) have a profound effect at finite temperatures." Coleman ends this essay by stating that "...the new reductionism that we need to make the leap into the realm between nano and micro will surely demand a new set of principles linking these two extremes.
(reductionism and collective behavior)

7. Compton Energy Scale of Friction Quantization

The author's studies of pendulum damping occasioned him (with influence from Tom Erber) to postulate the existence of a 'quantum' of internal friction that operates at the mesoscale, having an energy of 11 pJ. This value of the 'hysteron' energy is derived from dimensionality considerations involving the fine structure constant, and no first principles theory for its existence has yet been formulated. Consequently no mathematical formula is yet known to describe its properties, assuming that the hysteron actually exists. Its postulate is based strictly on experimental observations, in which a variety of functionally different mechanical oscillators have all pointed along the same path. The problem with that path is the problem of complexity. Reproducibility is challenged by the huge number of degrees of freedom accessible to atoms ordering themselves around defect structures.

The first observation of dissipation, in which values of energy loss or gain were found to be near-multiples of $E_h = 1.1 \times 10^{-11}\ J$, involved the free-decay of a long-period compound pendulum [Peters, 1990]. Similar tantalizing evidence for a 'hysteron' of internal friction quantization at the Compton-energy scale E_h was noted in the hysteresis that was observed in length fluctuations of metal wires as a function of temperature [Peters, 1993].

For the two experiments just mentioned, the friction responsible for the phenomena observed is thought to be of 'internal' type rather than the result of sliding friction (tribology). A recent experiment suggesting the possibility that even sliding friction might be quantized, is one that involved 'rolling friction' [Peters, 2005-4].

7.1. Basis for Associating E_h with Compton's Name

A summary of some of the author's mesodynamics research is found in a review article [Peters,2004]. Section 5.1 of this article tells how the hysteron energy was observed to relate to fundamental constants as follows:

$$E_h = \frac{m\,c^2}{\alpha} = 1.1 \times 10^{-11}\,J \tag{8}$$

where $\alpha = 1/137$ is the fine structure constant, m is the mass of the electron, and c is the speed of light.

Recently the author came to recognize an intimate direct connection of E_h with the Compton wavelength for the electron [$\lambda = h/(mc)$ where h is Planck's constant]; i.e.,

$$E_h = \frac{hc}{\lambda\alpha} \tag{9}$$

It is thus natural to refer to E_h as the Compton-energy scale of friction quantization.

7.2. Physical Interpretation

From Eq.(0.9) it is seen that E_h corresponds to the energy of a photon whose wavelength is smaller than that of the Compton wavelength by a factor of 137; i.e.,

$$E_h = hc/\lambda_{eff}\ , \qquad where \qquad \lambda_{eff} = \alpha \cdot \lambda = 2\pi\, r_e \tag{10}$$

where r_e is the classical electron radius.

That the value of $\lambda_{eff} = 1.8 \times 10^{-14}\,m$ is closer to nuclear than to atomic dimensions, could be important as now explained.

When a metal is strained anelastically, there is an alteration in the configuration of defect structures; which must involve significant displacement of groups of atomic nuclei. This process is very unlike atomic theory cases in which the Born-Oppenheimer approximation is reasonable; i.e., adiabatic approximation in which a nucleus remains at rest. A meaningful description of the 'jerky' (discontinuous) strains that are characteristic of many materials [Portevin, 1923], is hard to realize on the basis of quantization at the electron-volt level. The eV is roughly seventy million times too small to be a natural energy-decrement for the description of such jerky alterations.

Fine structure constant 'mystery'

The fine structure constant α is the fundamental physical constant characterizing the strength of the electromagnetic interaction. First considered by Arnold Sommerfeld in 1916, it can be thought of as the ratio of two energies: (i) the energy needed to bring two electrons from infinity to an arbitrary distance s against their electrostatic repulsion, and (ii) the energy of a single photon of wavenumber $k = 2\pi/\lambda = 1/s$ where λ is the photon's wavelength. Richard Feynman said of this constant, "It has been a mystery ever since it was discovered more than fifty years ago, and all good theoretical physicists put this number up on their wall and worry about it. ... It's one of the *greatest* damn mysteries of physics: a *magic number* that comes to us with no understanding by man. You might say the "hand of God" wrote that number, and "we don't know how He pushed His pencil". We know what kind of a dance to do experimentally to measure this number very accurately, but we don't know what kind of a dance to do on a computer to make this number come out–without putting it in secretly! " [Feynman, 1985].

A well-known trio of related fundamental-units of length 'revolve around' the fine structure constant:

$$r_e = \frac{\alpha\lambda}{2\pi} = \alpha^2 a_0 \tag{11}$$

where a_0 is the Bohr radius.

Since $2\pi\ a_0$ is a (deBroglie) wavelength that characterizes 'individual' electron interactions with stationary nuclei, it is postulated that $\lambda_{eff} = 2\pi\ r_e = 2\pi\alpha^2\ a_0 = 1.8 \times 10^{-14}\ m$ is a (photon) wavelength that characterizes 'many-electron' interactions with non-stationary nuclei. No first principles description of the mechanisms for such interactions presently exists.

8. Example Results Using the Technology Triad

Mention was made above of the critical importance of the three elements of (i) an outstanding sensor, (ii) high-resolution data acquisition, and (iii) effective analysis software. A recent example of a synergetic combination of (i) and (ii) is the integration of the author's patented SDC sensor and Analog Devices' new award winning AD7745 capacitance to digital converter chip. This combination has been used in two instruments that were brought to commercial market in the last year: (1) the first fully-digital seismograph (the VolksMeter) (online information available at http://rllinstruments.com), and (2) the computerized Cavendish balance (http://telatomic.com). Having 24-bit resolution, the chip represents a breakthrough in inexpensive technology for monitoring capacitive detectors. Data collection/analysis uses Cavendish-specific algorithms developed in Poland for that instrument, and for the VolksMeter, the software developed in RedWood City, CA by Webtronics, called WinSDR and WinQuake. Both software packages have stand-alone display capabilities. With the Webtronics code one can display records in both the time domain and the frequency domain. Both also are routinely used to export text files for analysis by means of Microsoft's Excel. Illustrated in Fig. 5 are examples from the Cavendish balance. The tungsten fiber (25 μm diameter) normally used with the instrument to measure $G = 6.67 \times 10^{-11}\ N\ m^2/kg^2$ was for this case replaced with a much larger diameter metal-alloy wire. Additionally, the small spheres that normally rest on the ends of the torsional boom were not in place for this experiment. Thus the eigenfrequency of oscillation was greatly increased, to simplify the measurement (both in time required and difficulty of setup) as compared to the design-configuration.

The theoretical fit of the upper graph required six parameters, all of which were estimated by trial and error. Two parameters were used to remove offset and drift (secular trend). Thereafter, an additional four parameters were employed; these are amplitude, frequency, phase, and quality factor Q. Captured in the screen-print of Fig. 6 is a portion of the Excel program that was used.

The theoretical fit can be accomplished quickly by trial and error. One parameter at a time is altered by typing in a new value (such as the frequency at position I2). The graph is

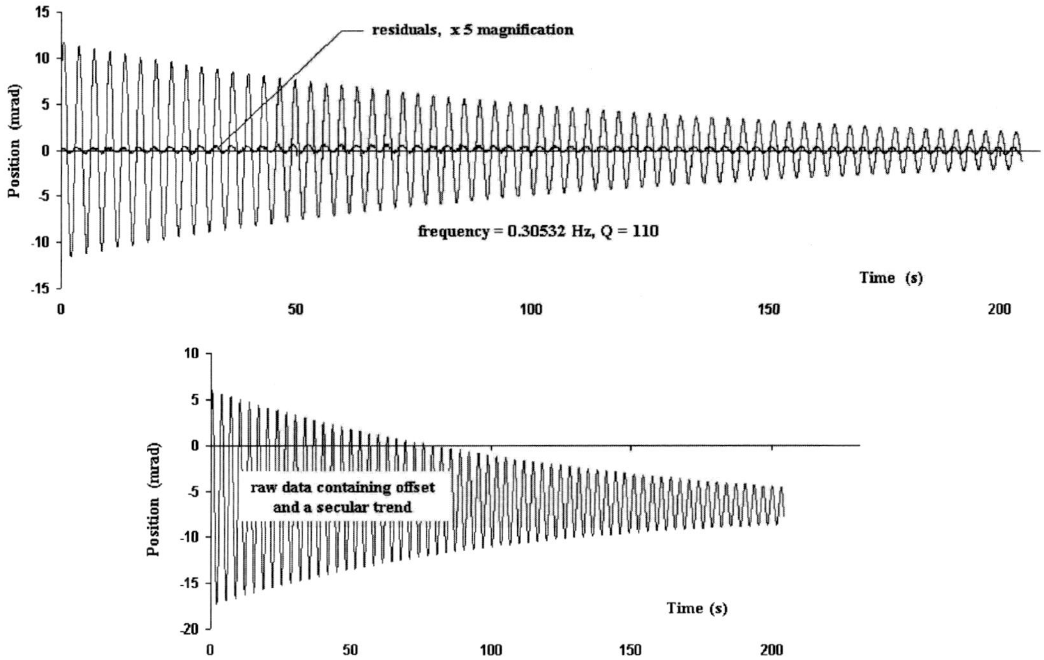

Figure 5. Free-decay torsional pendulum records. Offset and drift in the raw-data have been removed in the upper graph. A theoretical fit to this data was also performed using Excel, and the difference between theory and experiment (residuals) has been displayed with a five-fold magnification.

	A	B	C	D	E	F	G	H	I	J	K
1	time	corrected	theory	residuals	raw (mrad)	offset	secular	ampl	freq	phase	Q
2	0	6.999826	6.964747	0.175392	1.319825649	5.68	0.88	11.7	0.30532	5.35	110
3	0.1	8.668315	8.621929	0.231933	2.987885475						
4	0.2	9.999622	9.960159	0.197313	4.318761826						
5	0.3	10.95768	10.93094	0.133735	5.27639389						
6	0.4	11.53215	11.49932	0.164194	5.850434303						
7	0.5	11.68257	11.64516	0.187066	6.000423431						
8	0.6	11.39365	11.36386	0.148927	5.711066723						
9	0.7	10.67589	10.66647	0.04707	4.992878437						
10	0.8	9.580117	9.579289	0.004137	3.896677494						
11	0.9	8.147365	8.14284	0.022623	2.463495731						
12	1	6.406922	6.410393	-0.01736	0.722622871						
13	1.1	4.428025	4.445971	-0.08973	-1.25670433						
14	1.2	2.294084	2.321992	-0.13954	-3.391075134						
15	1.3	0.092444	0.116593	-0.12075	-5.593144894						
16	1.4	-2.12661	-2.08924	-0.18688	-7.81263113						
17	1.5	-4.27418	-4.21464	-0.29772	-9.960627556						
18	1.6	-6.24137	-6.18183	-0.2977	-11.92824841						
19	1.7	-7.96855	-7.91898	-0.24781	-13.65585327						
20	1.8	-9.40956	-9.36281	-0.23373	-15.09729624						
21	1.9	-10.5299	-10.4609	-0.34479	-16.21803045						
22	2	-11.245	-11.1736	-0.35673	-16.93354845						
23	2.1	-11.5201	-11.4755	-0.22318	-17.20912457						

entry at B2 "= E2+$F$2+$G$2*A2/204.7"

entry at C2 "=$H$2*COS(6.2832*$I$2*A2+$J$2)*EXP(-3.1416*$I$2*A2/$K$2)"

entry at D2 "=5*(B2-C2)

Figure 6. Excel architecture used to generate the theoretical fit of Fig. 5.

observed during the time of its rapid regeneration after striking the ‘enter’ key. One can tell immediately whether the change to the parameter was in the right direction to reduce the residuals. The ability of the brain to process features of the graph presented visually to it make the technique very effective. Other methods may also be used (such as Excel’s ‘goal seek’), and one could even employ multi-parameter math-methods using minimization of the residuals. Of the various methods he has tried through the years, the author’s personal preference is the simplicity/speed of the method shown in Fig. 6.

Material subtleties

The small residuals of Fig. 5 indicate excellence of the exponential theoretical fit that was assumed. One might thus conclude that a linear viscous damping model is an entirely appropriate theoretical description of the energy loss. This conclusion is not justified without additional experimental tests. If the damping were due only to air viscosity, then the conclusion is valid. Studies by the author in which mechanical oscillators were placed in vacuum demonstrated that there is always an internal friction (hysteretic) contribution to the damping; i.e., loss because of anelasticity of the torsion wire. There are no real macroscopic springs that obey Hooke’s law [Erber 2001][Peters, 2003-1].

The quality factor indicated in Fig. 5 (Q = 110) comprises two parts through the following relationship

$$\frac{1}{Q} = \frac{1}{Q_{air}} + \frac{1}{Q_h} \tag{12}$$

The subscript ‘h’ designates the hysteretic contribution. As with capacitors in series, the most influential component is the one with the smaller value. Although Q_h is expected to have been significantly larger than Q_{air} for the case shown, one must not naively assume this is always true. Moreover, as illustrated in Fig. 1, this hysteretic part of the damping is thought never to be properly modelled by means of a linear equation of motion.

It is possible that the drift in this example could be a tell-tale indicator of damping that is due to more than just air influence. If the result of creep which often accompanies hysteretic damping, then this would almost certainly be the case. A definitive conclusion is impossible without further tests, because changes in relative humidity of the air can also cause significant variations in the output of a capacitive sensor. The relatively short duration of the record (less than four minutes) suggests that creep is more likely the reason for the drift.

Not only do careful measurements of oscillator decay usually show creep influence, there is frequently a marked asymmetry in the turning points of the motion. In other words, the positive amplitude during one half-cycle of the motion does not match the negative amplitude of the associated half-cycle, after correction for attenuation has been made. The author’s studies have even provided evidence in some cases for positive- followed by negative-damping in each single-cycle of the motion, with the positive part being enough larger than the negative part to result in overall decay [Peters, 2004-2]. It is even possible to laser-condition crystals with radiation defects in such a manner as to yield short term net negative damping; i.e., the motion of a pendulum supported by the crystals able to increase through tens of cycles without an external drive [Coy, 1997].

8.1. Frequency Domain

As compared to the time-domain data shown in Fig. 5, frequency-domain data is oftentimes more valuable for purpose of judging between the relative importance of linear and nonlinear damping in a given oscillator. The Fast Fourier Transform (FFT) invented by Cooley and Tukey is the premier tool for generating spectral data, because of its incredibly greater speed than its discrete predecessor (DFT) [Peters, 2003-2]. Shown in Fig. 7 are the spectrum of the corrected waveform of Fig. 5, and also the spectrum of the residuals. An FFT of the uncorrected waveform would be worthless because of the offset and drift. An alternative to doing the correction before calculating the spectrum is to 'condition' the record with an 'apodizing function' such as the Hanning window. This 'cosine taper' significantly reduces the influence of the offset/drift by forcing the starting and ending values to zero. In the case of records viewed by WinQuake, near-equivalent conditioning is accomplished by doing a high-pass filter (infinite-impulse response type) with a low corner-frequency before doing the FFT.

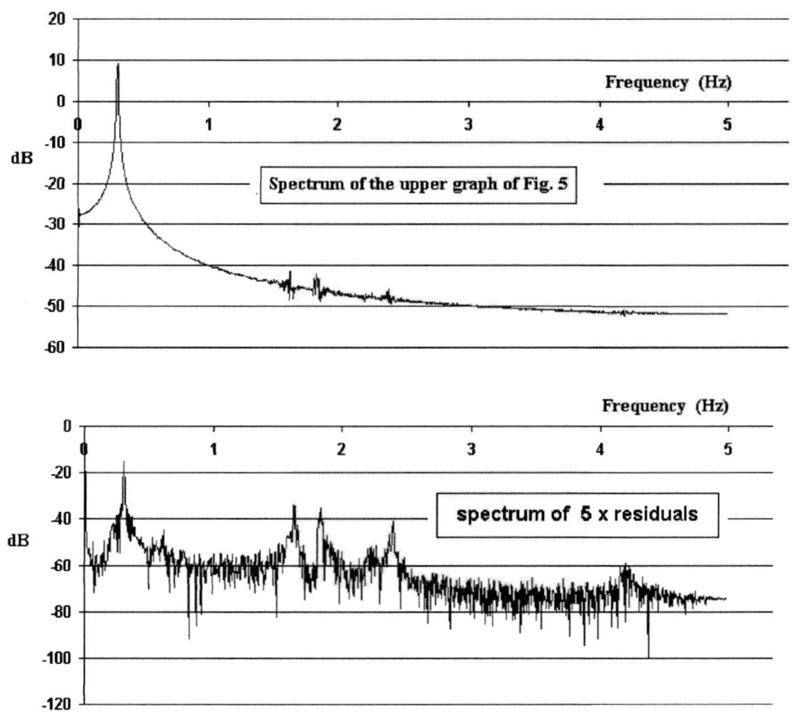

Figure 7. FFT spectra generated from 2048 data points of the Fig. 5 record shown in the upper graph. The ordinate values correspond to dB = 10 $log_{10}[2\ (FFT^2_{real}\ +\ FFT^2_{imag})]$. The maximum (Nyquist) frequency of 5 Hz is determined by the sample acquisition rate of 10 per s.

From the spectrum of the residuals (5 x magnification) three other modes are seen to be very clearly present, in addition to the primary (torsional) mode at 0.31 Hz. These are at frequencies 1.6, 1.8, and 2.4 Hz. The ones at 1.6 and 1.8 Hz are probably pendulous (boom-swinging) oscillations, and the highest frequency mode at 2.4 Hz is probably a 'rocking' mode of the boom. The 'splitting' of pendulous modes is expected because of differing moments of inertia (motion in the direction of the primary axis of the boom in one case and perpendicular to it in the other case). The instrument is designed (using two sensors on opposite ends of the boom, connected in electrical anti-parallel) to provide a "mechanical-common-mode-rejection-ratio" that significantly reduces the pendulous response as compared to the torsional response.

FFT normalization

The output from several popular FFT algorithms is not normalized, including the code embedded in Excel. Such codes are acceptable if one is interested only in relative amplitudes of spectral lines in a single record, but their use is a cause for much confusion. A fixed amplitude sinusoidal signal, for example, appears to grow in power with time as the duration of the analyzed signal is allowed to increase. Spectral calculations designed to yield proper absolute levels should be carefully evaluated using numerically generated test cases. When using Excel, each of the transform components must be divided by N, the total number of points ($N = 2^n$ where n = integer). The factor of 2 multiplying the sum of the squares of the real and imaginary components in the description of Fig. 7 is required because only the lower half (N/2) of the total set is considered. The upper half corresponding to negative frequencies and essential to computation, is not included when graphing. If the inverse transform is done with Excel, then no normalizing constant is employed if division by N was done on the forward transform.

It should be noted that the continuous Fourier transform is not as well understood by most as it should be. There are six different forms, all of which are permissible, depending on (i) where the negative sign multiplier on $i\omega t$ is placed, and (ii) how the factor of 2π is "distributed". Sometimes the $1/2\pi$ term is placed as a multiplier on the integral of the forward transform. Sometimes it is placed with the inverse transform. Sometimes it is split by way of the square root as a symmetric multiplier on both integrals.

9. Power Spectral Density

The FFT spectrum of output from an instrument excited by external forcing (usually with broad spectral content) is basis for generation of the power spectral density (PSD) function. In computing the PSD it is necessary to correct for the instrument response. This is a straightforward correction in those cases where the instrument is adequately approximated by a linear transfer function. In such cases, since superposition is possible, the total output is the convolution of the input signal and the instrument response. Our discussion of this subject is in the context of seismometers; the VolksMeter mentioned earlier is an 'ordinary' pendulum (physical) type. As such it responds at higher frequencies primarily to horizontal ground acceleration and at very low frequencies to tilt. Operating at near-critical damping, the pendulum of the VolksMeter can be simply treated, for purpose of transfer function correction, by considering its steady state response. If there were also a significant transient contribution to the response, then much more complicated Laplace transform mathematics would be necessary to address the matter.

9.1. 'Bad-Science' Nomenclature

Perhaps the earliest consideration of PSD's was in the field of electrical engineering. After measuring the voltage across a 1-ohm load as a function of time, if one computes the square of the FFT components generated from that temporal record, and then plots those squared values versus linear frequency, then the graph that results is a true PSD, meaningfully represented by units of W/Hz.

Unfortunately, the word 'power' has come to be routinely used to describe any spectral density function derived from FFT components that can be calculated, squared and then graphed. In this author's opinion, such practice is inexcusable, for reason of the widespread confusion that it has generated. In the case of seismology, the prevailing standard to describe instrument-generated (self) noise is one labeled PSD but having units of $m^2/s^4/Hz$. It will be shown in the material that follows that this is an illegitimate set insofar as proportionality to true power is concerned. Physics students are routinely taught that "if the units are not right, then something is wrong". For a seismometer, the specific power in W/kg/Hz is the most meaningful function and it is easily calculated. Isaac Newton would have been the first person equipped with the mathematics to do so, had such instruments existed in his time. Because specific power in a generic sense involves the product of acceleration and velocity, a legitimate seismic PSD can only have units of $m^2/s^3/Hz$ Seismologists actually graph this PSD when they use a log-scale in frequency plot. It is their label for the units that is incorrect.

It is convenient to work also with the cumulative spectral power (CSP), that is obtained from the PSD by means of integration. Because integration is a noise-reducing operation, the CSP is much 'cleaner'. To better understand the features of both functions, let us consider the following.

Math Background

The PSD and the CSP are related in the same manner as a pair of well known math functions used to quantify probabilities. The probability density function (pdf) has similarities to the PSD. In the case of the normal distribution, the pdf is the 'bell curve' familiar to many students; i.e., the normalized form of Gauss' function. For the area under the curve equal to unity, the function is given by

$$f(x) = \frac{1}{(2\pi)^{1/2}\sigma} exp(-[(x-\mu)/\sigma]^2/2) \tag{13}$$

Mathematicians also use the cumulative probability function (cpf), designated by F(x), and for which

$$\frac{d}{dx} F(x) = f(x) \tag{14}$$

These functions are illustrated in Fig. 8, where the curves were generated by Excel using a mean value $\mu = 1$ and a standard deviation $\sigma = 0.3$.

So that the PSD and CSP may be better understood (especially in terms of the density feature of the PSD, for which there is much confusion), the functions graphed in Fig. 8 are now described according to their meaning. The function f(x) specifies the probability that the magnitude of a randomly drawn variate x, governed by the normal distribution, will be located within the interval x to x + dx. Since the variate must have a value that is contained

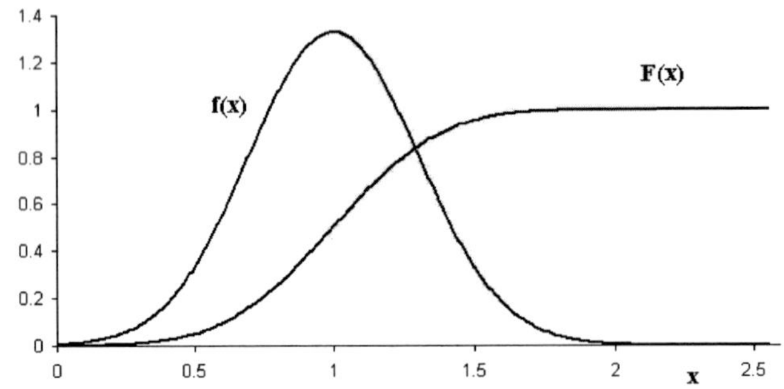

Figure 8. Illustration of the probability density function f(x) and the cumulative probability function F(x).

by the interval spanning plus and minus infinity, we see that the integral over f(x) must be equal to unity.

From Eq.(0.14) we obtain the following result for F(x):

$$F(x) = \int_{-\infty}^{x} f(x)dx \tag{15}$$

For a given value of x, the value of F(x) is seen to be the probability that a randomly drawn variate of the distribution will be less than x.

Although probability functions are inherently smooth, noise can be nevertheless arbitrarily introduced into f(x) to show that F(x) is relatively immune to the noise, as shown in Fig. 9.

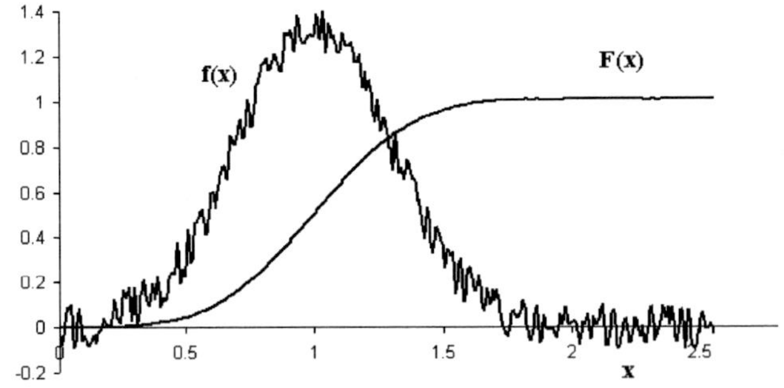

Figure 9. Illustration of the improvement in signal to noise ratio of a cumulative function as compared to the density function to which it is related.

We now generalize this discussion of f(x) and F(x) to the case of the PSD and the CSP, both of which can be plotted either versus frequency ν or versus period T. Because of the well-known study of earth noise in which 'acceleration power' is graphed versus period [Berger, 2004], we choose to express both functions in terms of T. Whereas the function F(x) has the lower bound 0 and the upper bound 1, the CSP is unbounded, although it is always greater than 0. At the shortest periods of earth motion to which a seismometer responds, much if not most of the power is due to localized, man-made activities. We thus restrict our attention to periods longer than about 2 s. There is also a restriction placed on the maximum allowed value of T. The longest periods allowed in graphing the CSP are determined by (i) the duration of the data-record, and (ii) the corner frequency of a digital filter used in the data processing. For the software presently employed, there is a minimum frequency limit of 0.5 mHz (T = 2000 s) placed on the high-pass digital filter through which the VolksMeter [http://rllinstruments.com] data passes before doing the FFT. This high-pass operation is used in lieu of other means for removing the adverse effects of secular trends in the data (end point differences causing problems, related to the Gibbs effect). The alternative to high-pass filtering is to 'condition' the data with a 'cosine taper' as noted in the earlier discussion of the Hanning window. It appears to have become a seismology standard [McNamara, 2003]. For those familiar with optical systems, the Hanning window is to the 1-dimensional FFT-world what an 'apodizing' function is to the 2-dimensional FFT-world. Apodization removes the 'sharp edges' of diffraction-type from images.

By analogy with the pdf, the function PSD(T) gives the amount of earth power in watts per kg operating in the period interval between T and T+dT. In the frequency-domain of the electrical engineering world, where functions of this type originated, the power spectral density is described by units of watts in the frequency interval from ν to $\nu + d\nu$. For example, it specifies the amount of Joules per second in the frequency-bin of width $d\nu$, starting at ν, that is converted to Joule heat in a resistive load (usually stated as W/Hz). An alternative engineering form is one in which the noise voltage is specified as Volts per root Hz. For a 1-ohm resistive load, the power $V^2/R \rightarrow V^2$ and so the connection between the voltage and power specifications is for this case obvious.

Graphing subtleties

In addition to the confusion that results from calling 'power' everything whose FFT components can be squared and then plotted, there is the additional challenge of properly accounting for how the density specification of the PSD depends on the nature of the plotting methodology.

Unlike the probability functions, which usually are adequately specified by means of a linear-scale for the abscissa, the PSD and the CSP generally require a scale that is logarithmic. This choice is necessary because the graphs normally encompass a total range of three or more decades of period (or frequency). Without 'compression' that results from the log-scale, the utility of such graphs would be severely limited. For reason of the large range of powers encompassed as a function of period, the ordinate values are also plotted with a log-scale in the form of the decibel.

To understand the influence of log-scale compression on a density function such as the PSD, it is necessary to understand what is being plotted. The PSD is derived from FFT output involving a sequence of N components, where $N = 2^n$, with n being an integer [Peters, 2003-2]. The N components are distributed as a 'comb' of frequencies, with equi-spaced 'teeth' separated from each other by $\Delta f = 2 f_N/N$, which involves the Nyquist (maximum) frequency $f_N = f_s/2 - f_s/N$, where f_s is the rate at which data is collected; i.e. the number of samples per s. Each of the FFT components has the same weighting factor in its contribution to the power, and the integral over frequency of the PSD (discrete approximation given by the sum of the components times Δf) yields the average power contained in the interval over which the integration was done. Parseval's theorem states that this must be the same power as obtained from an integral of the inverse transform in the time domain.

When the PSD is represented using a log-scale for the frequency, compression causes the spacing between points to no longer be constant as is true for the linear scale. When the discrete data points of the set are pictured as in the Fig. 10 illustration; it is immediately clear that the density specification (weighting factor) must be altered.

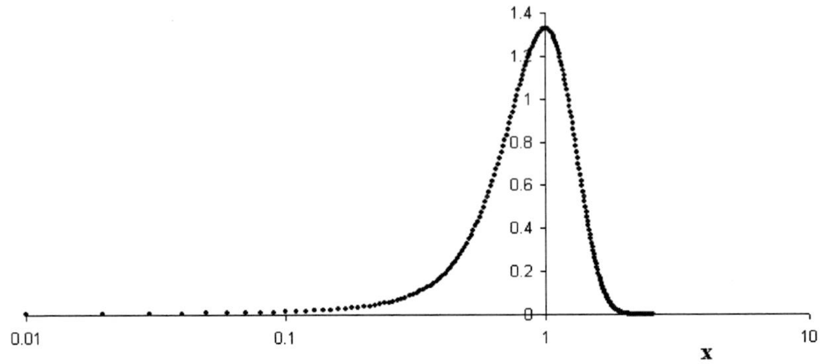

Figure 10. Illustration of how the probability density function of normal distribution type is altered when a log-scale is used for the abscissa.

The shape distortion of Fig. 10 is dramatic, and the number of points per decade increases as x increases. For example, the number of points in the interval from 0.01 to 0.1 is an order of magnitude smaller than the number of points in the interval from 0.1 to 1.

The function graphed in Fig. 10 is identical to the f(x) of Fig. 8, except that it has been plotted on a log-scale rather than a linear scale. If this were a PSD function, as opposed to the probability function shown, then it is clear that the density specification (the weighting factor per decade, or per octave, or per fraction of decade or of octave) must be frequency dependent for a PSD function graphed with a log-scale. For the power to come out right, the density specification must be increased as the frequency increases (proportional to f).

Another way to understand this frequency dependence of the density specification is to look formally at the mathematics. When using the log-scale, the following equality must be satisfied as a condition required for energy conservation:

$$\int f\ (logx)d(logx) \quad = \quad \int f(x)dx \tag{16}$$

Since $d(log\ x) \ = \ dx\ /\ x$, we see that as x increases, dx must be made smaller if the density specification is to remain constant. Since dx cannot be altered there must be a change to the density specification when using the log-scale. (Note: graphs we will discuss are generated with base 10 logarithms rather than the natural log as used in stating d(log x) = dx /x; the two differ only by the constant 2.30258 or its reciprocal.)

To plot versus linear T also gives a different density specification than linear ν. This is easily seen from the mathematics, since $T \ = \ 1/\nu$ and therefore $dT \ = \ -\ d\nu/\nu^2$. Figure 11 is a plot of the probability density function versus 1/x; it illustrates 'compression' even though the scale of 1/x is itself linear.

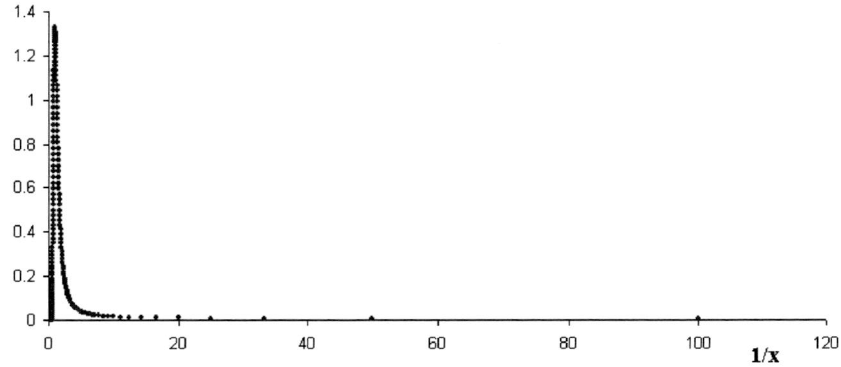

Figure 11. Plot of the probability density function versus 1/x using a linear scale. Compression at small values of 1/x makes this graph virtually worthless.

If we re-plot Fig. 11 using log(1/x), the result is seen from Fig. 12 to be the same as Fig. 10 except for the 'reversal' of direction. In other words, the density specification is the same for T and ν when using the log-scale. All one need do to convert from T to ν or vice-versa is to simply compute the reciprocal of the abscissa value.

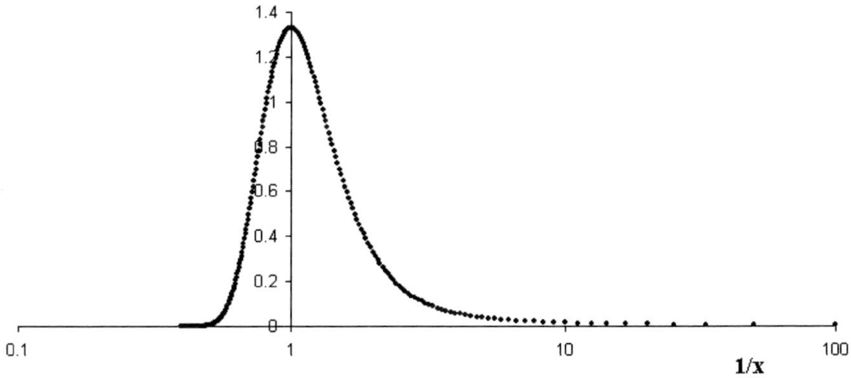

Figure 12. Same as Fig. 11 but using log-scale for 1/x.

10. Mechanical (Harmonic) Oscillator Theory

The response to acceleration of a pendulum is governed (linear approximation) by the following equation of motion:

$$\ddot{\theta} + \frac{\omega_0}{Q}\dot{\theta} + \omega_0^2\,\theta = -\frac{\omega_0^2}{g}a(t) \tag{17}$$

where θ is the angular displacement and a(t) is ground acceleration, related simply to ground displacement $A_G(t)$ for steady state motion at angular frequency $\omega = 2\pi\,\nu$ by $a(t) = -\,\omega^2\,A_G(t)$. This equation approximates actual damping (largely internal [Peters, 2005-1]) by means of the linear viscous term (involving the quality factor Q). For seismic purposes, Q is usually not much different from the value of 1/2 corresponding to critical damping. The natural free-decay (without imposed damping) period of the pendulum is given by $T = 2\pi/\omega_0$.

To specify displacement x in nanometers of a point on the physical pendulum, one multiplies the sine of the radian angular displacement θ by the distance from the axis to the center of percussion L_P. Then for small θ and acceleration at the angular frequency ω, the equation of motion becomes

$$\ddot{x} + \frac{\omega_0}{Q}\dot{x} + \omega_0^2\, x = -\frac{L_P\,\omega_0^2}{g}a(t) = \frac{L_P\,\omega_0^2\,\omega^2}{g}\,A_G(t) \tag{18}$$

Not just the pendulum, but virtually all conventional seismometers are governed by an equation of motion that is very similar to Eq.(0.18). By means of force feedback (including force 'balance') many conventional instruments tailor the response as a means for setting both the values of Q and ω_0. The actual frequency dependence of the instrument response is usually specified in terms of pole/zero values, for which Eq.(0.18) is overly-simplified. The difference between Eq.(0.18) and those using pole/zero specifications is thought by this author to be less important than the difference introduced by nonlinear (complex) properties of real world instruments that cannot be properly modelled by any of the linear equations employed [Peters, 2007-2].

The steady state solution to Eq.(0.18) is

$$\frac{x_0}{\omega^2\,A_{G0}} = \frac{L_P}{g}\,\frac{\omega_0^2}{[\omega_0^2 - \omega^2)^2 + \omega_0^2\,\omega^2\,/\,Q^2]^{1/2}} = \frac{L_P}{g}\,T_F \tag{19}$$

where x_0 is the amplitude of the pendulum displacement at angular frequency $\omega = 2\pi\,\nu$, and T_F is the transfer function of the instrument, an example shown in Fig. 13.

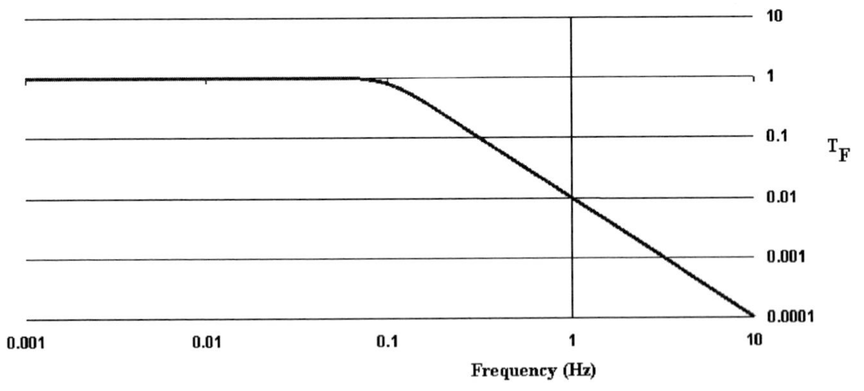

Figure 13. Transfer function (T_F, acceleration response) for an instrument with a natural period of 10 s.

It is instructive to look at Eq.(0.18) in the case of a simple pendulum; i.e., one in which $L_P \rightarrow \ell$ is simply the length of the string supporting a point mass. For this simple pendulum $\omega_0^2 = g/\ell$. In the limit of $\omega << \omega_0$ the pendulum displacement is equal to

$$x_0 = \frac{\omega^2}{\omega_0^2}A_{G0} = \frac{\ell}{g}\omega^2\,A_{G0} = \frac{\ell}{g} \times\ acceleration\ , \qquad \omega << \omega_0 \tag{20}$$

In other words, for earth acceleration at frequency lower than the natural frequency of the instrument, the angular deflection is just equal to the acceleration divided by the earth's field; i.e., $g = 9.8\ m/s^2$. It is important to understand that the pendulum also responds to tilt, which corresponds simply to the component of the earth's field that is perpendicular to the vertical axis of the case of the instrument. The difference between tilt response and response due to earth harmonic motion on the power spectral density is not treated here. The interested reader is referred to [Peters, 2007-3].

Power

Every sensor, no matter the type, responds to the stimulus exciting it according to one thing only–energy transfer (the time rate of energy being the power). Consider a mass element of the earth that oscillates in steady state. As shown above in Eq.(0.20), the amplitude of the acceleration measured by a seismometer observing the motion, is given by $a_0 = g\,\theta_0$, for frequencies of oscillation below the natural frequency of the instrument. Since the velocity of the mass is simply its acceleration divided by the angular frequency, one obtains the following expression for the total oscillatory energy of this piece of earth mass

$$E = \frac{1}{2}\,m\,v_0^2 = \frac{1}{2}\,m\,\frac{g^2\theta_0^2}{\omega^2} \tag{21}$$

Dividing this energy by the period $T = 2\pi/\omega$ gives a value of oscillatory power. A correction to this value is provided for the following reason. Power balance involving the generation of heat is central to our discussion. If the earth could oscillate without damping, then no external energy input would be needed to keep it oscillating. Similarly, if a seismometer were to operate without damping, no input of energy from the earth would be necessary to keep it moving. Of course damping is part of the dynamics of both systems. The damping of the instrument that monitors the earth is easy to describe in terms of the balance between power in and energy out per second in the form of heat. The quality factor is defined as $Q = 2\pi\ E/|\Delta E|$ where E is the total energy and $|\Delta E|$ is the loss of energy per cycle. When operating with crtical damping $Q = 1/2$, we see that the power that is converted to heat is a factor of 4π greater than the value that would be estimated on the basis of Eq.(0.21). Thus the following expression is used to define the specific power (power per unit mass):

$$P = \frac{g^2\ \theta_0^2}{\omega} \quad , \quad \omega < \omega_0 \tag{22}$$

A completely proper accounting for the power would have to consider every term in the frequency dependence of the Q of the earth, which is viewed as a daunting task. For persistent oscillations (fairly high Q eigenmodes) Eq. (0.22) is insufficient, but for most of the broadband noise of the earth it is a reasonable starting point for understanding how the oscillatory power of the earth is distributed among the frequencies allowed by its density of states.

Power Spectral Density PSD

The expression of Eq.(0.22) is valid only if the indicated condition is met. The general result is obtained by correcting θ_0 for the influence of instrument response; i.e., factor in the transfer function corresponding to Fig. 13. The frequency dependence of this specific power (units of watts per kg) is expressed using the decibel (involving log to the base 10). In the following equation, account has been given to both the transfer function T_F and the pendulum's calibration constant C_C (units of counts per radian).

$$P = 10\,log(\frac{g^2\,FFT^2_{mag}}{\omega\,C_c^2\,T_F^2}) + 10\,log(\frac{f_T}{2\,F_N/N}) \tag{23}$$

where the units associated with the argument of the first log term are inconsequential because the reference for the dB specification is $1\,m^2/s^3$. All the parameters are dimensionless with the exception of g with units m/s^2 and ω with 1/s.

Here $FFT^2_{mag} = 2\,(FFT^2_{real} + FFT^2_{imaginary})$ where real and imaginary refer to the complex components of the FFT of the measured values of pendulum angular deflection. The factor of two is necessary to accommodate negative frequencies which are normally not explicitly considered. Each FFT component corresponds to a specific frequency f_T and the term $2\,F_N/N$ is the minimum (lowest) frequency of the FFT set, seen to be equal to the Nyquist frequency divided by one half the total number of terms used in calculating the FFT. The second log term involving this ratio of frequencies is necessary when using the log-scale of either frequency or period, because of the compression described above in detail. In the example PSD graphs that follow, based on data from a fully-digital VolksMeter seismograph, the corner frequency of the transfer function is $0.918Hz = \omega_0/2\pi$, and $C_c = 2.5 \times 10^9\ counts/rad.$

Proper units of the PSD

The only acceptable units of power spectral density involving actual power rather than the illegitimate 'power' of abstract signals mentioned earlier — are m^2/s^3 per 'density-unit', where the 'density-unit' used in the material that follows is one-seventh decade. This density-unit is also the 'bin-width' that was employed by Berger et al [Berger, 2004].

Misleading labels of the PSD

It is now shown that seismologists work with bonafide (legitimate specific) PSD's, even though the units they employ are misleading. It appears that serendipity figured into the standard that developed years ago. As frequently stated to students of physics, if the units are not right, then something must be wrong.

The most common set of misleading units is $m^2/s^4/Hz$; which should instead be $m^2/s^3/Hz$. Evidently, the label evolved from the illegitimate practice of simply squaring FFT components and graphing them. Bear in mind that the (specific) power is really watts per kg per FFT component, which equals m^2/s^3 per FFT component. A frequency bin-width of 1 Hz is not the same as $\Delta f = f_{min} = 2\,F_N/N$ that separates adjacent spectral components of the FFT. When the FFT-squared values are plotted versus log-scale of either frequency or (more commonly) period; account must be taken for the compression discussed earlier. As noted, compression is included in Eq.(0.23). Moreover, by using properties of the logarithm, the equation can be rewritten as

$$P = 10\,log(\frac{g^2\,FFT^2_{mag}\,N}{4\pi\,C_c^2\,T_f^2\,F_N}) \quad , \qquad referenced\,to\,m^2/s^3 \tag{24}$$

In other words, following the historical practice and choosing to use a log-scale generates the true PSD by accident, but the units implied by that practice are misleading, because compression has been ignored.

Example PSD's

Shown in Fig. 14 are some example PSD's. The darker-shade graph was generated from a record that included the indicated earthquake. The duration of the record responsible for the lighter-shade graph ('background') was the same and collected at nearly the same time but shifted enough for none of the teleseism waves to be obvious.

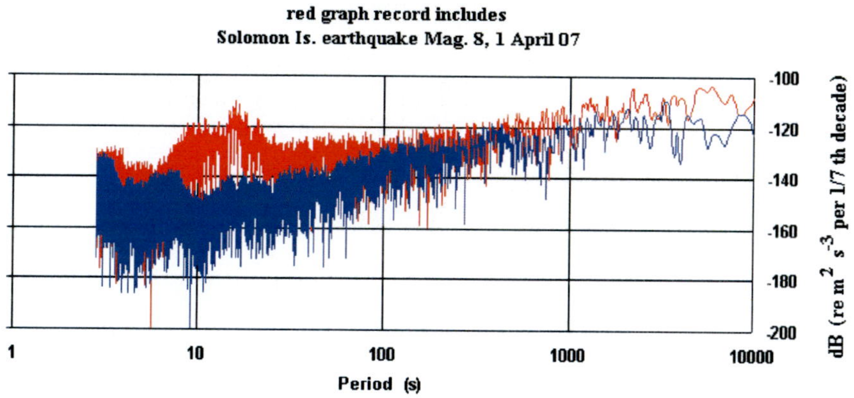

Figure 14. Example power spectral density graphs generated from data collected with a VolksMeter seismograph located in Macon, Georgia USA.

Cumulative Spectral Power

The noise that is typical of raw spectra and PSD's is quite obvious in the two cases shown in Fig. 14. As was stated earlier, such noise becomes much less significant when one works with the function that is based on the integral of the PSD; i.e., the Cumulative Spectral Power (CSP), some examples of which are shown in Fig. 15.

of Fig. 14.

Instead of using dB's as in the case of the PSD, the CSP's of Fig. 15 have all been plotted using a simple log-scale. These curves were generated with a small number of period-values (15) as compared to the number of points of the FFT (32,768) used to generate their associated CSP. The sum was performed using Excel after exporting the txt (comma separated variable) form of the FFT.

The value of the function for a given period-value is obtained by summing all the FFT components from the starting period (2.8 s) up to and including the component that corresponds most closely to the given period-value.

The smoothing that results from the integration process that defines the CSP is dramatic, as can be seen by comparing the earthquake curve that is common to figures 15 and 14. Also stunning is the simplicity of that earthquake CSP as compared to its PSD. Data representations of physics that have become famous through the years have always been those that interject simplicity into otherwise complex problems. The simplicity of the black curve of Fig. 15 suggests that earth motions derived from large earthquakes might be

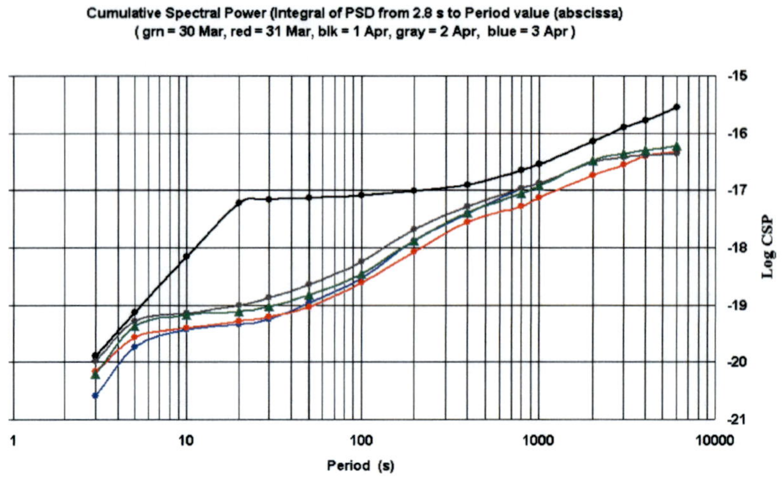

Figure 15. Examples of the Cumulative Spectral Power. The black curve was generated by numerical integration over the dark-shade curve of Fig. 14.

modelled more simply and thus effectively than most have wanted to believe.

Total Earth Power

The CSP is derived from specific power (W/kg) of the PSD. If the FFT used to generate the PSD is properly normalized, then the ordinate values of the Fig. 15 graph take on absolute meaning as opposed to the more typical view of spectra in terms of arbitrary units (for relative comparison). Moreover, multiplying a CSP value by the mass of the earth ($6 \times 10^{24}\ kg$) is not a meaningless exercise. The maximum value of the CSP has been found to be surprisingly constant for all cases other than those containing a large earthquake. Of the five cases (different days, all at the same starting time of a 12-h record) shown in Fig. 15, only the one curve containing the earthquake departs significantly from a maximum at 6000 s of about -11.2. A single seismometer at one site cannot adequately estimate the total vibrational power of the earth; however, an average over the maximum CSP times earth mass of a large number of instruments situated geographically such as to 'cover' the earth should yield a decent estimate of the average vibrational power in watts of the earth.

Secular trends in the maximum CSP

The maximum values of the CSP, based on records of the same type as Fig. 15, were used to generate Fig. 16. The data in all cases were from a VolksMeter located in Redwood City, CA. The remarkable feature of this graph is the secular increase that is evident for reason of the relatively small diurnal fluctuations in the maximum value.

11. Unconventional Creep Study

Shown in Fig. 17 is a pendulum that was used to study creep and also creep-recovery following low-level resonance excitation.

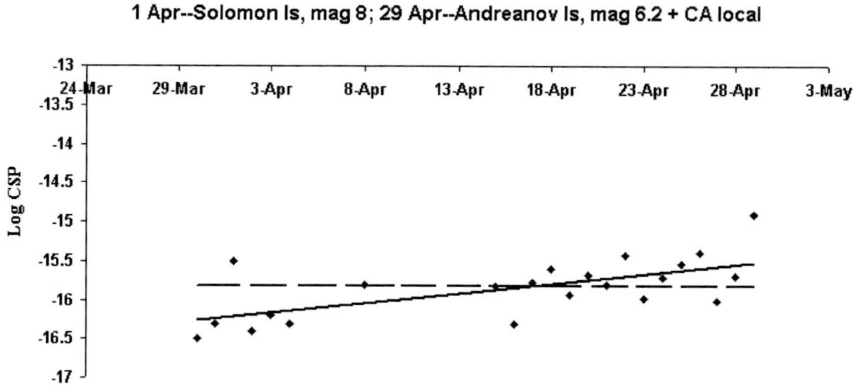

Figure 16. Graph of the maximum value of the CSP over a one-month inverval. Two teleseisms and one smaller, local earthquake occurred during this time. The solid line is a trendline fit to the data and the dashed line is the mean value of the data set.

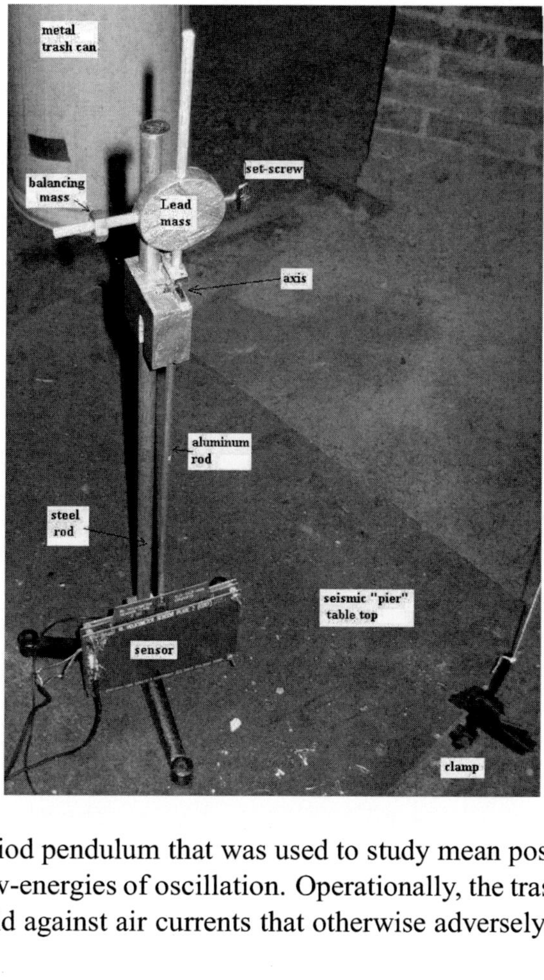

Figure 17. A long-period pendulum that was used to study mean position variations due to anelasticity at very-low-energies of oscillation. Operationally, the trash can was placed over the instrument to shield against air currents that otherwise adversely drive such a sensitive instrument.

In some of the experiments that will be described, the pendulum was driven by tilting the table at the instrument's resonance frequency (73 mHz; period = 13.7 s) using an actuator (above and out of view) communicating with the table via a string to the clamp shown. The actuator and oscillator used to provide the drive (Pasco PI9587C) are of the type used in physics general education laboratories to establish standing waves on a tensioned string. Of the possible highly-monochromatic wave forms available for the drive (sine, triangle, square), only the sine output was employed. Although the signal purity of the oscillator is thus great, the same was not true of the drive to the pendulum; since the tilt of the table was asymmetric. The asymmetry was due to the fact that the string connecting the clamp and the actuator could only support tension and not compression. We will see in some figures that follow that this caused considerable harmonic distortion in the output when the pendulum was being driven. By contrast, during free-decay (drive turned off) harmonic distortion was almost non-existent, consistent with the high-linearity of the sensor.

The pendulums pictured in figures 3 and 17 are mechanically similar, except for the following notable differences. The material of the major support structure of the former is carbon-epoxy whereas that of the latter is a strong aluminum alloy. The axis of the former is provided by a pair of low-friction ball-bearings attached to the plastic yoke that supports the static plates of the sensor. The axis of the latter is a pair of small tungsten carbide spheres resting on sapphire flats. The sensor of the former is positioned at the axis, whereas the sensor of the latter is positioned as far from the axis as allowed by mechanical constraints.

These design differences derive from the differing regimes of oscillation considered for study. The maximum amplitude of oscillation of the former pendulum is more than an order of magnitude larger than that of the latter pendulum. The former could thus be used to study viscous damping, as the lower sphere moved inside a container filled with liquid to surround it. [Peters, 2003-3]. By operating at larger amplitudes (and thus larger maximum speeds) it was insured that viscous damping would be significantly larger than anelastic damping due to strains of the solid materials of the pendulum. The pendulum was thus used to demonstrate that the harmonic oscillator damping term of Eq.(0.3) is overly idealized, even in those cases for which it is best suited. It was found that the coefficient β depends not only on the viscosity but also the density of the fluid; and the coefficient was even found to be frequency dependent! One should never therefore refer to β as the damping 'constant'. It is a coefficient that happens to describe the exponential character of a decay that may result from nonlinear internal friction processes, even though the equation of motion for which its inclusion has become famous–is itself linear.

The detectors of pendulum position are configured differently for the two pendulums being here discussed, even though they are both of the symmetric (fully) differential capacitive type. That of the former is a single-element SDC sensor, whereas that of the latter is a four-element array. Roughly speaking, if both sensors were positioned at the same place relative to the axis; the sensitivity of the array-type would be four times greater than that of the single-element type (assuming their overall electrode areas are the same). The four-fold increase in sensitivity is realized at the expense of a concomitant four-fold decrease in the largest displacement that can be measured.

The difference in sensitivity/dynamic range of the detectors used by these two pendulums can be understood as follows. If output reactance (based on the Thevenin equivalent circuit of a given single-element sensor) were inconsequential, then the sensitivity would be proportional inversely to the width of the electrode that moves with the pendulum between static electrodes positioned on opposite sides. In actuality, the output reactance increases as the width decreases (and thus degrades the sensitivity) unless one employs an array of identical units connected in electrical parallel. In the limit as the width goes toward zero as the number of individual units goes toward infinity (if large numbers were indeed allowed by the construction techniques employed), the sensitivity would thus be maximized relative to the allowable dynamic range.

By placing the sensor far from the axis, as with the pendulum of Fig. 17, there is additional sensitivity increase (independent of array-issues); since the smallest detectable displacement is proportional to the separation distance of sensor and axis. In the design of this pendulum, the goal was to maximize the sensitivity, subject to the constraint imposed by the total height of the instrument. The period can be lengthened substantially as compared to a simple pendulum, while keeping the height reasonably short, because of the upper lead (disc-shaped) mass which at 0.314 kg comprises 79.5 % of the total mass of the instrument.

For purpose of quantifying the energies of oscillation of this pendulum it was necessary to estimate the moment of inertia in its operational configuration. This was accomplished by measuring the period of the instrument at each of nine different positions of the lead-mass on the aluminum rod. For most of these period measurements (seven in number) the lead-mass was clamped by its set-screw at a place below, rather than the operational position that is above the axis. After first removing the lead-mass from the pendulum, the center of mass of the remaining structure was determined by balancing it on a knife edge in a horizontal orientation. This center of mass position served as the reference point for theoretical calculations.

The period of oscillation was measured with a stopwatch for each of the nine lead-mass positions mentioned above, and for which the graph of Fig. 18 was generated using Excel.

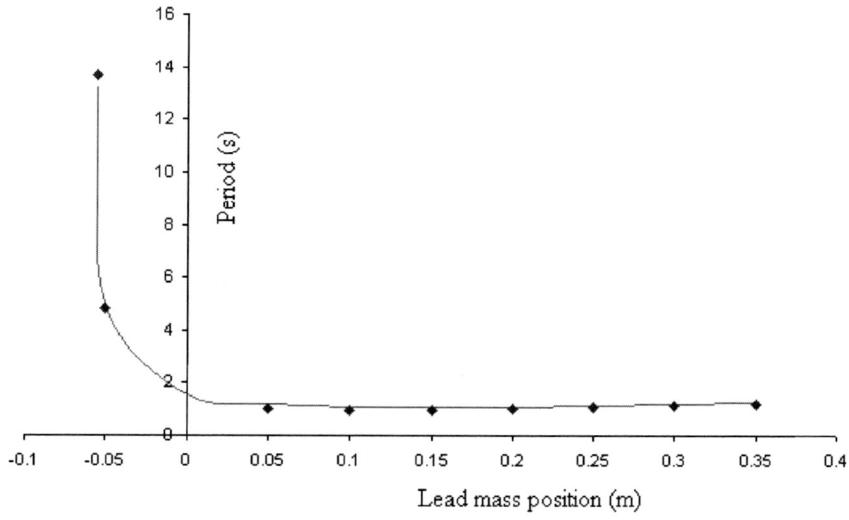

Figure 18. Theoretical fit (solid line) to measured period as a function of lead-mass position relative to a fixed reference point (center of mass of pendulum with lead-mass removed).

Using various component parameters such as: rod length = 56.8 cm, altitude of axis below top of rod = 17.3 cm, sensor moving electrode mass = 0.027 kg– a theoretical estimate for the period was generated for each of the nine points using the parallel-axis theorem.

Based on the reasonable theoretical fit observed in Fig. 18, the moment of inertia of the pendulum in its operational configuration was estimated to be $I = 9.20 \times 10^{-3}\ kg\ m^2$. The center-of-mass of the operational pendulum was in turn calculated to be 0.50 mm below the axis. These numbers are consistent with a pendulum having total mass of 0.3952 kg, oscillating with a period of $T = 13.7\ s$.

Using these numbers, oscillation energies are readily calculated by means of the expression

$$E = \frac{1}{2} I\, \omega^2\, \theta_0^2 \qquad , \qquad \omega = 2\pi/T \tag{25}$$

where the pendulum amplitude θ_0 is estimated from the sensor calibration constant that was measured to be $2.5 \times 10^7\ counts/radian$. The calibration was done by means of the classical optical lever technique using a He-Ne laser beam reflected from a small front-surface mirror (small piece of a non-processed, polished silicon wafer) attached to the balancing mass of the pendulum with a drop of super-glue.

11.1. Initial Creep

In the author's experience, the initial setup of every long-period mechanical oscillator is always followed by a time interval of usually tens of hours in which the instrument passes through a phase of primary creep. During this time a decrease in creep rate occurs while the process of work hardening is dominant. Following on the heels of the 'primary' creep, a balance between work hardening and recovery is established in the phase that is referred to as 'secondary' creep. Only engineering studies at enormously larger strains than the ones presently considered are likely to result in tertiary creep, where the formation and growth of cavities occur before material failure.

Except for strain levels so small that the associated displacements could only be measured with an atomic force microscope, and for which (as previously noted) Hooke's law is valid in a very limited (atomic) regime; there is no strain level below which creep does not occur. Stated differently, it is expected that defect organization, involving dislocation structures responsible for creep–should be observable with the highly-sensitive pendulum pictured in Fig. 17, operating near (or even below) the meso-scale energy of 11 pJ mentioned previously.

This scenario is one that must account for temperature changes during the course of measurements, since creep is intimately dependent on specimen temperature. The balance between processes mentioned above, concerning secondary creep, is one in which diffusion mechanisms having rates that increase with temperature, work in opposition to the mechanisms of work hardening. At some minimum of temperature, secondary creep must disappear for T below this threshold; since the product of Boltzmann's constant and the temperature ($k_B\, T$) is then less than the excitation level required to overcome the energies of dislocation entanglement.

In addition to the mean position change due to creep in Fig. 19, there is also a free-decay oscillation. Such oscillations are virtually unavoidable in the initial balancing of the highly-sensitive instrument, since it uses no externally imposed damping as is customary in the design of conventional seismometers.

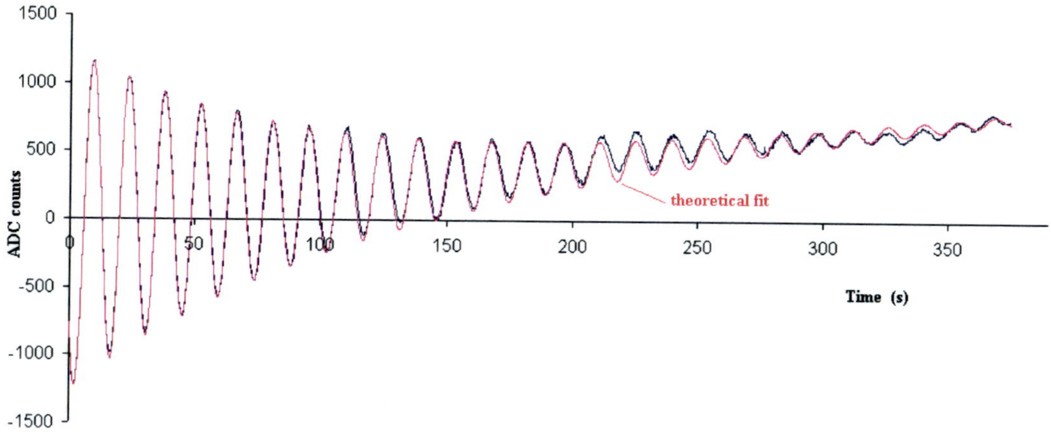

Figure 19. Illustration of primary creep (mean position change) following the setup of the pendulum shown in Fig. 17.

As can be seen from the 'best theoretical fit' to the experimental data, the oscillatory decay of this pendulum is nothing close to a simple process, when it is accompanied by creep. This pendulum was not designed to try and study the hysteretic damping quantization mentioned earlier in the section that discusses the Compton energy scale. Had that been the goal, then the 'high-Q' aluminum alloy main structural element of the pendulum would not have been the material of choice. It is this aluminum rod that is thought to be the primary source of both the creep that was observed and also the damping of pendulum motion.

It is interesting to consider the oscillatory energy of the pendulum during the approximately six-minute interval of the record displayed in Fig. 19. The amplitude of the motion at the start is about 1200 ADC counts, which for the calibration constant of the sensor corresponds to 48 μrad. Using this value for θ_0 in Eq. (0.25) one obtains an energy estimate of 2.2×10^{-12} J for the start of the record. At the end of the record, where the amplitude has decreased by more than an order of magnitude, the energy of oscillation is less than 10^{-14} J and yet still clearly visible to the sensor.

11.2. Creep Evolution

The change in character of the creep after the instrument had sat in its operational state for 20 hours was dramatic, as seen from Fig. 20.

The slope of the red curves in this figure are a measure of the creep rate, which on 2 March was 4.8 nrad/s; and which on 3 March had reduced to 1.1 nrad/s. The black temporal curves are what remains of the red curves after the secular trends were removed numerically using Excel. A statistical analysis of the data from which the black curves were generated showed (from the rms values) a decrease in fluctuation level during the 20 hour period starting on 2 March. The 14.4 rms counts in the starting case corresponds to amplitude fluctuations of 0.58 μrad. These have become slightly smaller at 0.52 μrad 20 hours later. One might reason that a change in seismic noise could be the cause for this difference in fluctuation level for the two days. It is thought that the spectral nature of the black curves

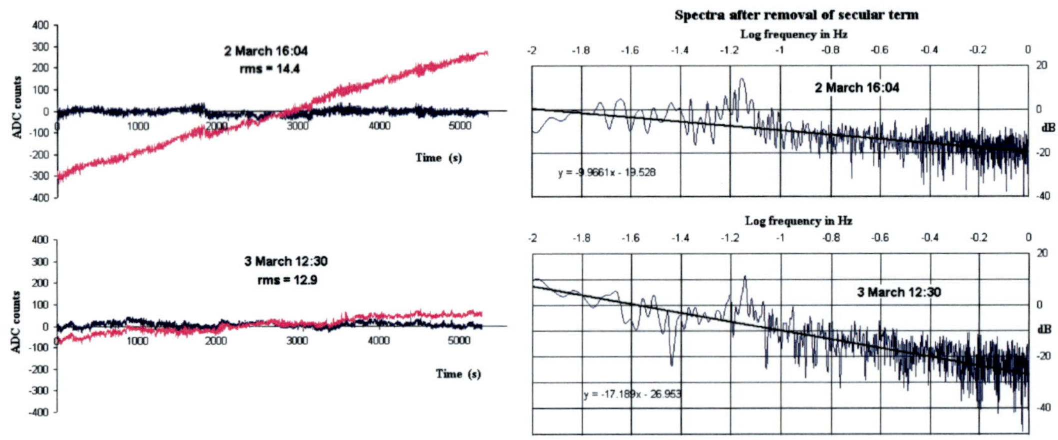

Figure 20. Illustration of the change in creep characteristics after the pendulum was allowed to stabilize for nearly one day.

tells us something about the matter. The change in slope of the trendline fit to the spectral data shown in the right pair of graphs of Fig. 20 speaks more probably of a difference that derives from the conversion of primary creep into secondary creep. Whereas there is about 7 dB lowering of the intensity of FFT components at 1 Hz in going from 2 March to 3 March, there is a raising by about the same amount of the 10 mHz components during the same period. The intensity of the fluctuations at 100 mHz is virtually the same for the two records. A comparison of stabilized-seismometer output data of similar duration, from one day to the next, does not usually show this much change in slope of a trendline fit to the spectra.

This example illustrates the considerable advantage of being able to look at both time-domain and frequency-domain results as opposed to years-past in which the Fourier transform could not be conveniently calculated. Of course the advent of analog to digital converters and the triad of technological tools that followed (focus of this article) have provided enormous advances in the means with which physics theory could be applied to the interpretation of data such as shown in Fig. 20–even if the time-domain data were the only thing available.

11.3. Example Tilt Drive at Resonance

Shown in the bottom graph of Fig. 21 is a sequence of three separate excitations for which the drive of the pendulum (set at the frequency of resonance) was turned on at a constant level for a time (69 s) equal to five cycles of oscillation, and then turned off to let the pendulum freely-decay to the level of noise. The amplitude at the end of the drive period, approximately 450 ADC counts, was for each case in the neighborhood of 18 μrad. In spite of this small amplitude, at least some portions of the motion show significant harmonic distortion involving the 2nd and 4th harmonics, as seen in the spectrum of the top graph.

In the Fig. 22 bottom graph is displayed a portion of the Fig. 21 record that contains only a single free decay. Inspection of the upper graph spectrum for this case shows that the harmonics of Fig. 21 derive from the asymmetry of the string drive, as was previously

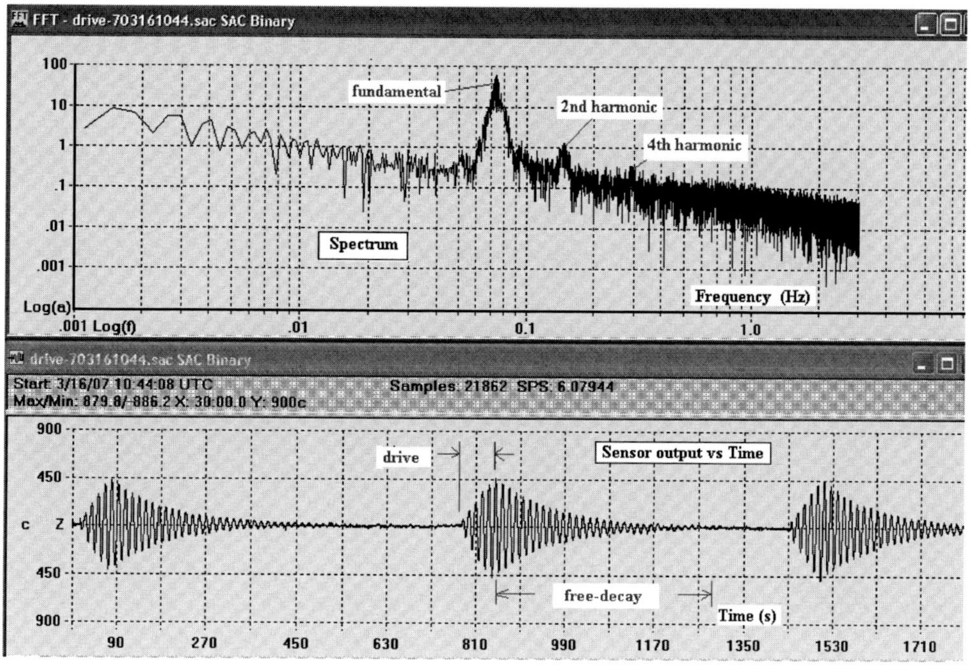

Figure 21. Resonance excitation of the pendulum, followed by free-decay, that illustrates the asymmetric nature of the tilt drive.

mentioned.

The Q quantifying the decay of Fig. 22 could be determined from a theoretical fit to the data as was done for figure 5 (with Excel details provided in Fig. 6). Another method uses the short-time- Fourier-transform (STFT). [Peters, 2005-1] The FFT is calculated for each of a sequence of sub-sections that are separated in time by equal amounts. The peak value of the oscillation's spectral line is graphed as a function of time. When the amplitudes are expressed in decibels, the graph is a straight line with negative slope, if the decay is exponential. From the slope of the trendline fit to these points the Q is determined. From a visual inspection of the plot one may also obtain a very rough estimate of Q as follows. Determine the number of cycles required for the amplitude to fall-off 63% and multiply this integer by π. For the pendulum configuration used to do the present creep study, the Q was measured from a set of a dozen free-decays in which the mean was calculated to be 22.1 with a standard deviation 1.5.

The set of decays from which the Q was estimated included some that were generated during the time of creep evolution as well as others after the pendulum had stabilized. For the set, no secular change as large as the indicated standard deviation was observed. On the other hand the frequency of the pendulum was not constant. From the starting time of primary creep until the end time (stabilized pendulum) where the mean frequency was 73.5 mHz, the frequency increased by 5.4%. This increase during the period of stabilization is more than an order of magnitude larger than the the uncertainty in frequency of the stabilized pendulum, which was estimated at less than 0.5% using theoretical fits to eight free-decays.

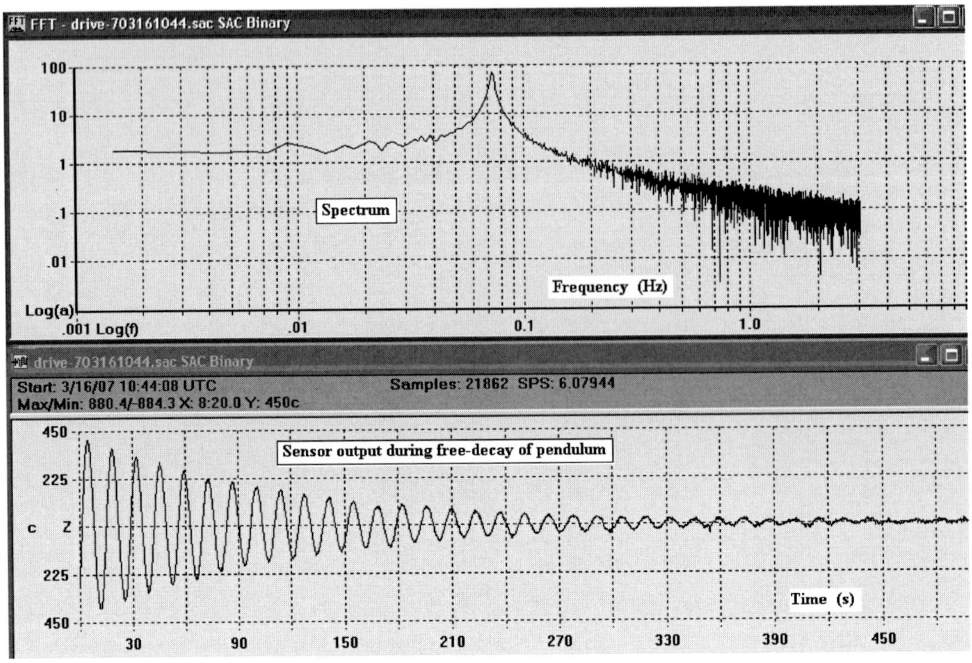

Figure 22. Illustration of the purity of the output (absence of harmonics in both pendulum and sensor) during free-decay oscillation.

11.4. Mechanism of the Damping

A significant contributor to the damping of a long-period pendulum involves the nonlinear (complex) coupling of energy between two different classes of oscillation. These are (i) the single low-frequency primary mode which has been the focus of discussion so far, and (ii) a set of high-frequency vibratory modes in which components of the pendulum above and below the axis move back and forth horizontally with the same phase relative to the stationary axis. As the pendulum moves in its primary mode, there is periodic flexure of the rod due to the downward weights of the upper and lower masses, acting in opposition to the upward normal force of the axis. If the flexural strains are devoid of Portevin Le Chatelier 'jumps', then the decay will be smooth. Sometimes a free-decay will be virtually textbook in character for this reason. On the other hand, resonance excitation of the pendulum is frequently followed by free-decay that is complex and anything like textbook. Examples are provided in Figures 24 and 25.

Whereas the Portevin Le Chatelier effect has been historically studied at large levels of strain, Fig. 25 suggests that the effect may be observed at the mesoscale following the cyclic introduction of almost vanishingly small stresses. Even as there is essentially no level below which creep doesn't occur, it appears that at the mesoscale every creep/damping process of internal friction type may involve discontinuous strains.

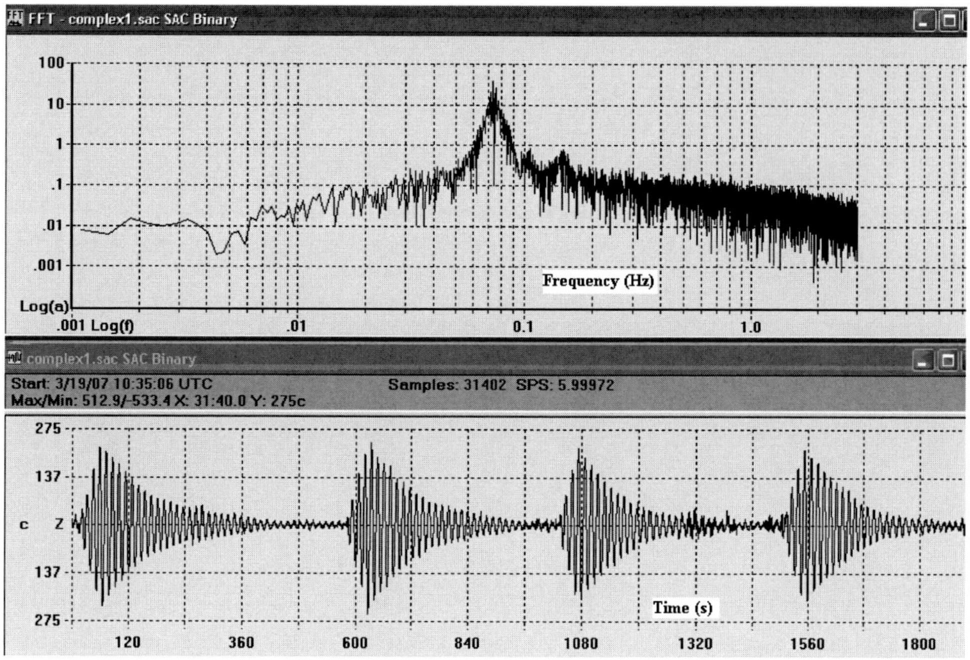

Figure 23. A sequence of four resonance-drive excitations that were followed by free-decay. Observe the differing features of the 'noise' (lower graph) once the oscillation has died mostly away. The upper graph of the spectrum again illustrates the asymmetry of the drive that was previously mentioned.

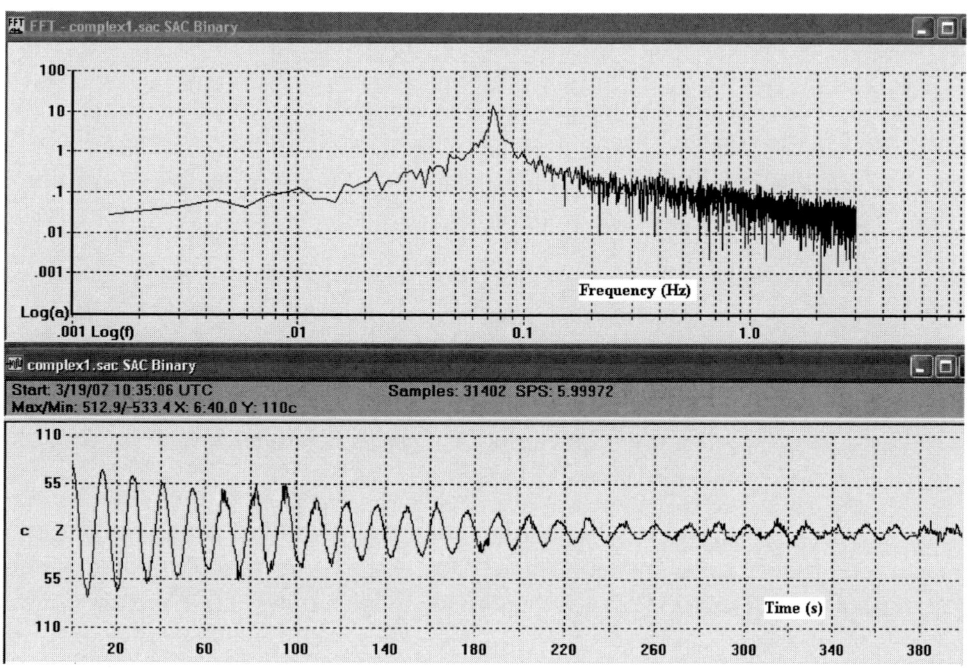

Figure 24. Free-decay-only following the first excitation of Fig. 23. Compare this case to Fig. 25.

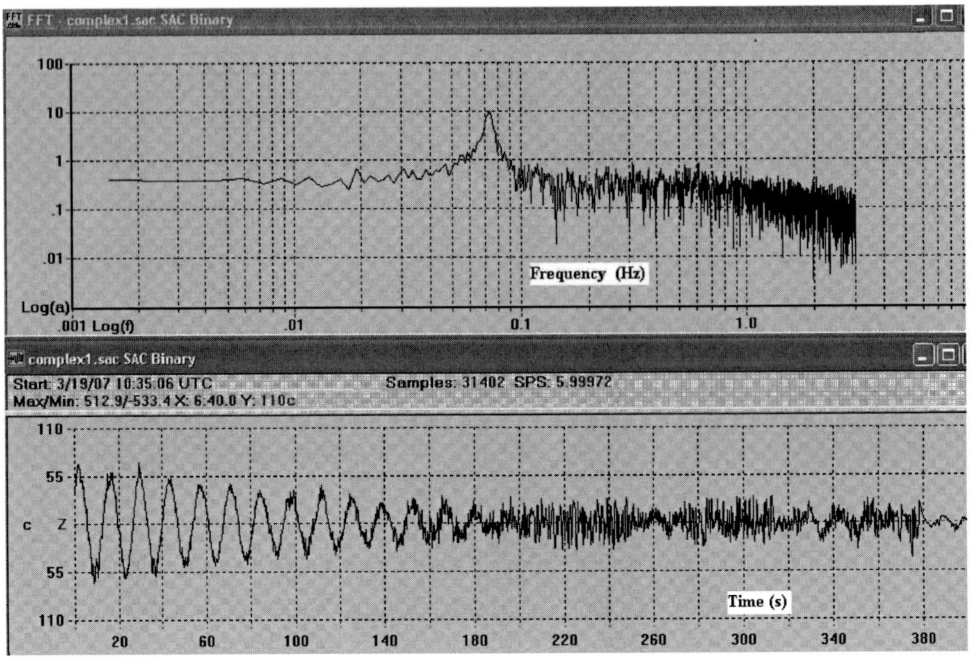

Figure 25. Free-decay-only following the third excitation of Fig. 23. Compare this case to Fig. 24.

11.5. Seismic Noise

The pendulum of Fig. 17 is sensitive to seismic disturbances, and for that matter, convective flow of air inside the trash-can cover when thermal gradients are such as to allow convection. (An incandescent light bulb was eventually placed at the top of the trash can to raise its temperature and help to suppress convection.)

The table supporting the pendulum sits on a home-made pier that goes about five feet into the basement soil of the author's house. As such, there is decent coupling of earth-waves to the instrument, which is not a 'bonafide' seismometer because of its excessively high Q.

The challenge of determining the cause for a given, observed low- level pendulum response is illustrated by the spectrum of Fig. 26.

Microseisms at 70 mHz are readily observed by the pendulum because of amplification that occurs at the resonance frequency as compared to the response at other frequencies. The amount of amplification is proportional to the quality factor; for this case Q = 22.

The microseisms thus visisble are referred to as "primary". The same pendulum, tuned to operate at a period closer to 4 s (with a higher Q) was found to be a sensitive detector of the "secondary" microseisms. It was postulated [Peters, 2006] that a pendulum so tuned could be used in hurricane forecasting.

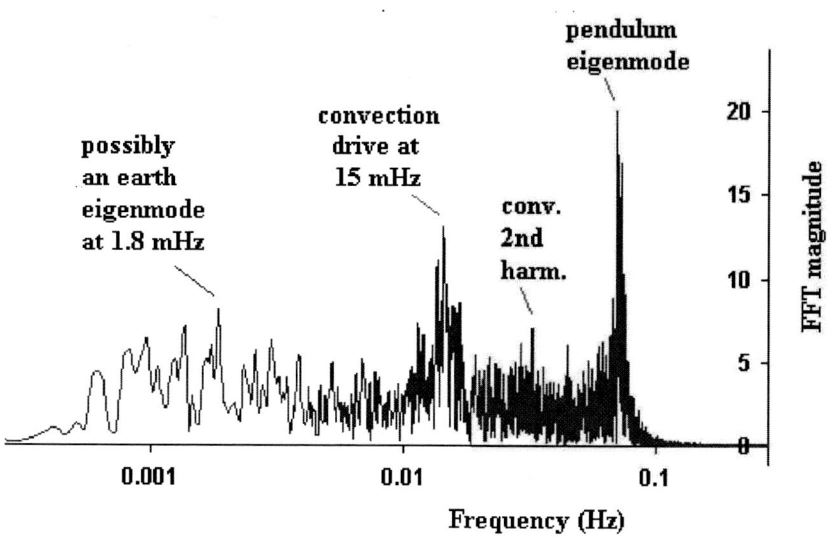

Figure 26. Spectral lines showing two clear sources of pendulum excitation: (i) air convection at 15 mHz (and associated 2nd harmonic at 30 mHz), and (ii) motion at the eigenfrequency of the pendulum due to microseisms.

12. Altering Secondary Creep by Low-Level Resonance Excitations

Temperature variations have a great influence on creep, and thermal control was not used for the experiments that are next described. Two cases are examined of the influence on creep due to disturbance of the pendulum. The first case is one in which the pendulum is driven in a sequence of resonance excitations near the time of daybreak, after the basement has cooled overnight. The other case is a similar sequence of resonance excitations near sunset, after the basement has warmed during the day. The diurnal change in temperature of the outside air during these experiments was approximately 15 Celcius degrees; however, because of the large heat capacity of the author's brick house, the swing in temperature experienced by the pendulum was only about 5 Celsius degrees.

Fig. 27 is representative of how the displacement of the pendulum would vary in the absence of imposed external forcing.

The ADC used in this study is the 24-bit instrument (single channel) sold by Symmetric Research. Kip Weiss provided the author with a a specialized algorithm that allows the PAR1CH to operate with the Amaseis software produced by Alan Jones. The graph of Fig. 27 was generated with Amaseis, which is common among amateur seismologists.

The description for the reason of the diurnal variation observed in this figure is based on the author's experiments with a similar but different pendulum, in which records were collected continuously for more than a year. Toward the end of those experiments, a nearly identical pendulum was placed well below ground in a California location in San Francisco (the Presidio), and the magnitude of the effect was found to be dramatically reduced. This suggests that thermal expansion of surface soils to a depth of a few meters is important insofar as tilting influence.

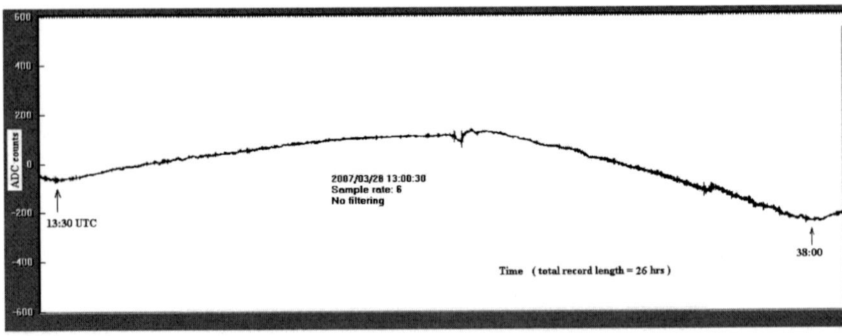

Figure 27. An example of diurnal motion exhibited by the 'undisturbed' pendulum. The 12 μrad (300 ADC counts) peak-to-peak diurnal change is thought to derive mainly from tilt associated with heating/cooling of the ground supporting the house.

These studies grew out of a problem long-considered by Jet Propulsion Laboratory scientist, James Shirley. The question with which he was concerned involves the offset from the center of the earth of the earth/moon barycenter. Could it be responsible for detectable diurnal accelerations whose magnitudes should vary seasonally according to sidereal rather than synodic time? The earliest experiments to address this hypothesis, and for which no definitive conclusions could be reached, were those conducted by George Darwin. The son of famous biologist Charles Darwin, George was influenced to consider his experiments by the famous physicist, Lord Kelvin. As in the time of Kelvin/Darwin, so likewise a century later in the time of Shirley/Peters, no definitive answer to this question could be reached–due to the incredible complexity of earth's influence on a pendulum.

12.1. Disturbance after Cooling

Shown in Fig. 28 is a sequence of four resonance drives, separated from each other by about ten minutes. The amplitude of pendulum motion at the time drive was turned off was for each of the four excitations approximately 1000 counts (40 μrad).

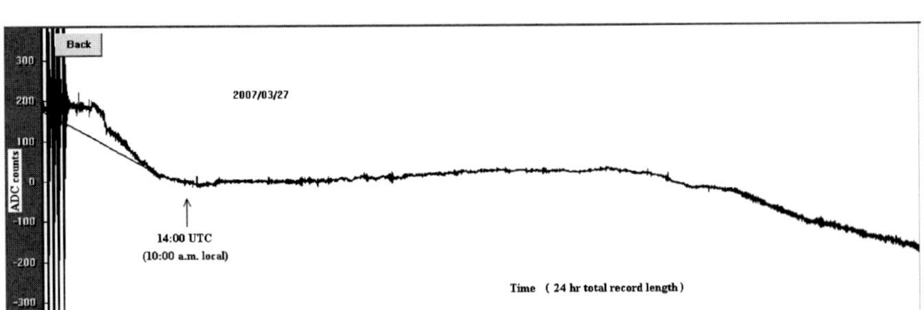

Figure 28. Influence of resonance excitations after the pendulum has cooled overnight.

The scale of the graph was selected to clearly show what happens after the last drive signal was turned off. For this scale-choice the largest displacements of the pendulum while being driven are not visible due to clipping.

It is seen from Fig. 28 that the driving excitations result for this case in an unmistakeable upward trend, followed by a downward recovery over a total time of about three hours. The change is 'highlighted' by means of the red line suggestive of the path that would have been followed in the absence of any pendulum disturbance. An interesting feature of this case involves the maximum energy with which the pendulum was oscillating during each of the four resonance drives; i.e., only $\approx 10^{-12}\ J$, based on Eq.(0.25).

The behavior illustrated in Fig. 28 was found to be essentially repeatable, as seen in Fig. 29. Nearly the same steps were performed (except six rather than four excitations) at about the same time of the day. The scale of both abscissa and ordinate have been altered to show the features of the effect more clearly.

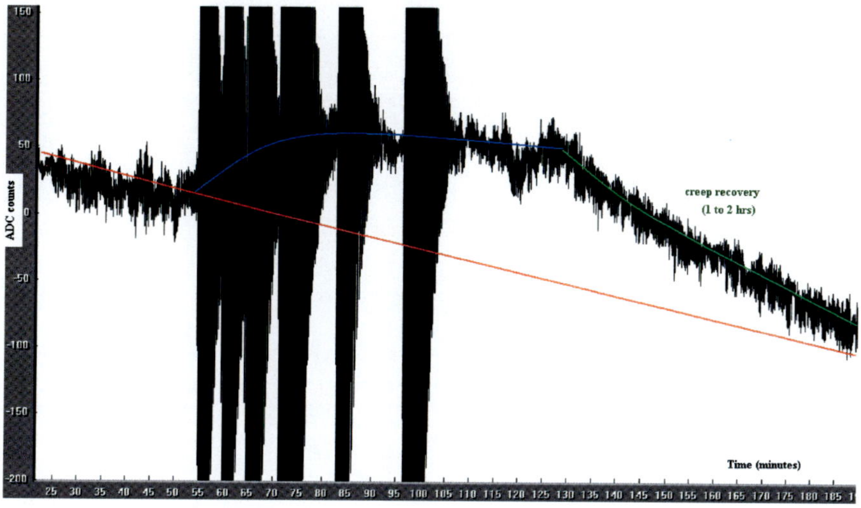

Figure 29. Demonstration of the repeatibility of the phenomenon observed in Fig. 27. The data of this figure were collected a day later than that of Fig. 27.

12.2. Disturbance after Warming

A totally different pendulum response was observed when the resonance excitations were performed around sunset as opposed to sunrise. As with the cooling case, the influence of drive after pendulum warming was found also to be repeatible; Fig. 30 is an example.

It is seen that the effect is significantly smaller, and also oppositely directed, as compared to the drive after cooling.

13. Influence of Creep on Seismometers

The frequency response of a seismometer is typically assumed to be the textbook case of a viscous damped, simple harmonic oscillator. As we have documented above, real mechanical oscillators are not ideal, and the damping at low frequencies, due to internal friction, is presently too poorly understood to describe from first principles. Even if the low-level motions were smooth (which they are not), the mean position of a seismic mass changes because of creep and creep recovery.

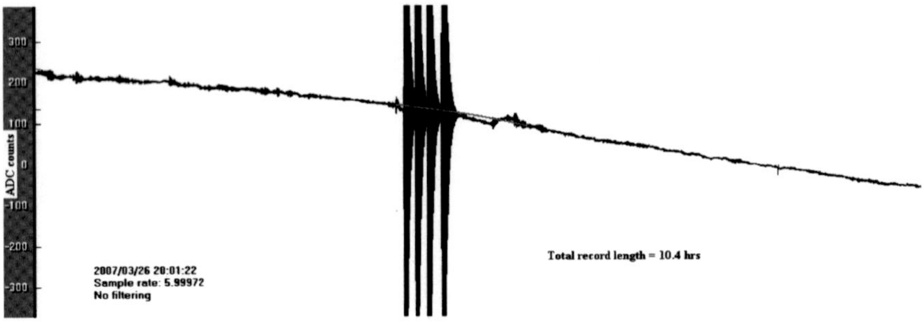

Figure 30. Influence of resonance excitations after the pendulum has been warming during the day.

Consider Eq.(0.18) re-expressed in generic form with the right-hand-side acceleration $a_d(t)$. We assume a harmonic drive $a_d(t) = -\omega^2 A_d e^{j\omega t}$ having amplitude A_d and $j = (-1)^{1/2}$. Using the Steinmetz phasor method, one obtains the steady state solution to this well known equation. The result is frequently provided in terms of the Bode plot shown in fig. 31.

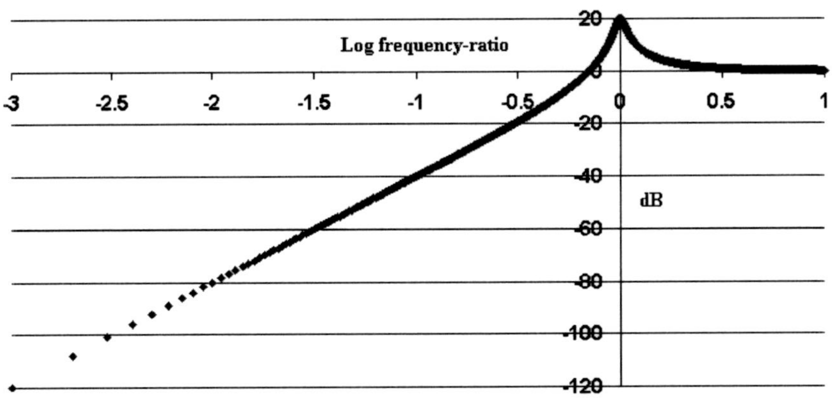

Figure 31. Graph of the ratio of steady state oscillator amplitude to drive amplitude as a function of drive frequency-ratio r, where $r = \omega/\omega_0$. The graph was generated with Q = 10.

The equation used to generate Fig. 31 is

$$|\frac{A}{A_d}| = \frac{\omega^2}{\sqrt{(\omega_0^2 - \omega^2)^2 + \frac{\omega_0^2}{Q^2}\omega^2}} \tag{26}$$

which the reader may wish to compare with Eq.(0.19).

13.1. Creep Influence on the Equation of Motion

Creep and creep recovery cause the length of the spring to vary; consequently the equation of motion is changed to

$$\ddot{x} + \frac{\omega_0}{Q}\dot{x} + \omega_0^2(x - x_0) = a_d(t) \tag{27}$$

Once a system has 'stabilized', primary creep (exponential) is no longer important except in response to major stress changes. Secondary creep, however, never disappears; since it is well known by material scientists that there is no level below which creep vanishes.

The secondary creep/creep recovery rate is assumed to be proportional to the acceleration; i.e.,

$$\dot{x_0} = c\, a_d(t) = -c\,\omega^2 A_d\, e^{j\omega t} \tag{28}$$

which yields for the steady state (variable) mean position

$$x_0 = j\, c\,\omega\, A_d\, e^{j\omega t} \tag{29}$$

Substituting Eq.(0.29) into Eq.(0.27) and solving for the steady state solution with creep, yields an expression involving the sum of two terms. The first term of the new solution yields a Bode plot identical to Fig. 31, and the second term of the new solution yields the Bode plot shown in Fig. 32.

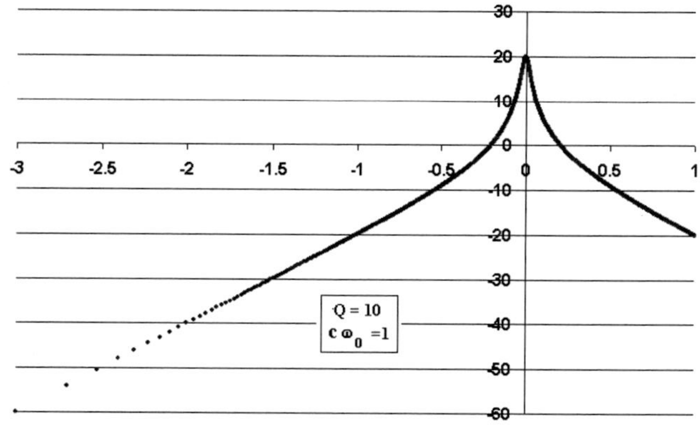

Figure 32. Normalized Bode plot for the creep component of the new solution. The complete response is obtained by adding $c\,\omega_0$ times Fig. 32 to Fig. 31.

The equation used to generate Fig. 32 is

$$|\frac{A}{A_d}|_{creep} = \frac{c\,\omega_0\, r}{\sqrt{(1-r^2)^2 + r^2/Q^2}} \quad , \quad r = \frac{\omega}{\omega_0} \tag{30}$$

13.2. Consequence of Creep

The difference in shape of figures 31 and 32 gives rise to practical consequences. When driving at frequencies above, or in the vicinity of resonance (r = 1), creep is inconsequential. For frequencies below resonance, creep can yield an increased amplitude response. Observe that in this regime, the fall-off of Fig. 31 is 40 dB/decade, whereas that of Fig. 32 is only 20 dB/decade. If the constant c is not vanishingly small, then the creep could allow signals to be detected at low frequencies, that would otherwise be below noise.

13.3. Practical Matters

This work is still too preliminary to quantify the practical importance of what has been observed. The following figure shows that creep should not be summarily disregarded.

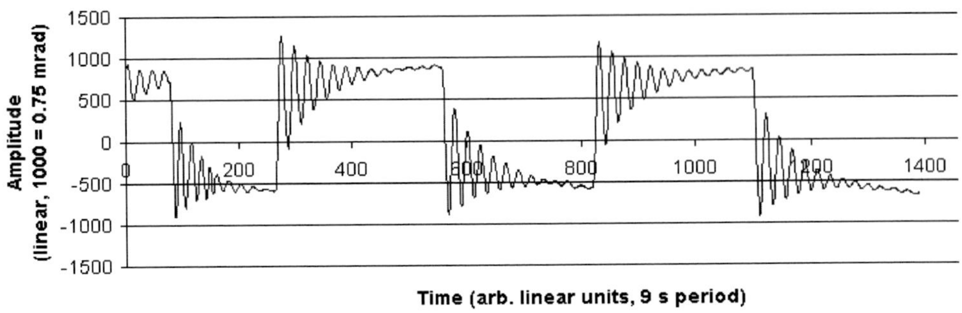

Figure 33. Illustration of creep in an unconventional torsion pendulum.

The instrument in this case was a torsion-gravity pendulum acting as a tiltmeter (remotely similar to a 'garden gate' horizontal seismometer). Because the tungsten torsion wire of the pendulum is tied to both the top and bottom of the case, the instrument is tilt sensitive. For that reason the mean position could be easily adjusted using a piezo-translator, since the restoration depends on the Earth's gravitational field as well as the shear modulus of the wire.

It is seen that changes of only a few hundred micro-radians are accompanied by a significant creep response–for the ratio of drive frequency to natural frequency in the vicinity of 0.04. Clearly the creep coefficient is for this case too large to be ignored for its influence on the instrument's response at low frequencies.

14. Legacy of 'Delta Function' Specialization

The tendency in physics to focus only on very narrow features of a problem has resulted in adverse consequences. For example, classical Hamiltonian mechanics is taught almost exclusively (if at all) as a 'spring-board' into quantum mechanics of the type encountered in the 'modern' era. This is tragic for reason of the fact that nonlinear classical mechanics problems, which generally can be solved only by numerical means, are best treated using Hamilton's canonical equations. These are summarized in Fig. 34.

An additional reason students need to know about Hamilton's method is because of the well-known-to-engineering concept of the Kalman filter. It is remarkable that such a powerful and elegant tool is almost unknown to the physics community. After all, the optimization scheme by which it functions is based in Hamiltonian mechanics.

Hamilton's Canonical Equations

$$\frac{\partial H}{\partial p} = \dot{q}$$

where $H(p,q) = \frac{p^2}{2m} + U(q) = K.E. + P.E. = E_{total}$

$$\frac{\partial H}{\partial q} = -\dot{p}$$

p is the momentum conjugate to the generalized coordinate $q \to x$.

It is obtained by taking the partial derivative of the Lagrangian $L(q,\dot{q}) = K.E.(\dot{q}) - P.E.(q)$ with respect to $\dot{q}$

EXAMPLE: The simple harmonic oscillator without damping:

$$L = \frac{1}{2} m \dot{q}^2 - \frac{1}{2} k q^2$$

$$p = \frac{\partial L}{\partial \dot{q}} = m\dot{q} \text{ gives } H(p,q) = \frac{p^2}{2m} + \frac{1}{2} k q^2$$

[for a simple pendulum of length l, the term $\frac{1}{2} k q^2$ would be replaced with $m g l\ (1 - \cos\frac{q}{l})$ where $q \to \theta$ (angular displacement)]

Operating with the canonical expressions on this Hamiltonian yields

$$\frac{\partial H}{\partial p} = \dot{q} = \frac{p}{m}$$

$$\frac{\partial H}{\partial q} = -\dot{p} = kq$$

If we combine this pair of equations, we get $m\ddot{q} + k q = 0$ which is usually re-written as $\ddot{q} + \omega_0^2\, q = 0$ where $\omega_0 = 2\pi f_0$ in terms of the natural (eigen) frequency of the oscillator.

[For the simple pendulum this becomes $\ddot{q} + g \sin\frac{q}{l} = 0$ which for small q becomes $\ddot{q} + \frac{g}{l} q = 0$ where $\frac{g}{l} = \omega_0^2$]

Figure 34. Basic equations of Hamilton's canonical type.

Whereas a regression analysis (trendline of Excel) is a fit to static data, the Kalman filter is a fit to data that is dynamic rather than static. Immediately after the filter is initialized, measurements dominate the output generated by the algorithm. As time progresses, however, variances decrease (in roughly exponential manner), causing the measurements to become less important as more stock is placed on how the system is supposed to evolve on the basis of theory (without noise corruption).

Shown in Fig. 35 are the equations to describe a viscous damped simple harmonic oscillator.

The single observable (measurement x_m) in this case is the oscillator's displacement. To illustrate the use of the equations, Fig. 36 is provided. The case treated was a decay of the same type shown in Fig. 5.

The effects of filtering can be readily seen by comparing the spectra of the data (with and without filtering) in the last half of the decay, as shown in Fig. 37.

Equation of motion, viscous damped simple harmonic oscillator
(x is displacement, p = $\dot{x}$):

$$\begin{bmatrix} \dot{x} \\ \dot{p} \end{bmatrix} = \begin{bmatrix} 0 & 1 \\ -\omega^2 & -2\beta \end{bmatrix} \begin{bmatrix} x \\ p \end{bmatrix}$$

Kalman Filter Equations, SHO with viscous damping (hat implies filter generated value, x_m is the measured value of x):

$$\dot{\hat{x}} = \hat{p} + \psi_v^{-1} V_{11}(x_m - \hat{x})$$

$$\dot{\hat{p}} = -\omega^2 \hat{x} - 2\beta\,\hat{p} + \psi_v^{-1} V_{12}(x_m - \hat{x})$$

Time rates of the Variance Equations
(ψ_v^{-1} is the reciprocal of the measurement variance) :

$$\dot{V}_{11} = 2V_{12} - \psi_v^{-1} V_{11}^2$$

$$\dot{V}_{12} = V_{22} - \omega^2 V_{11} - 2\beta V_{12} - \psi_v^{-1} V_{11} V_{12}$$

$$\dot{V}_{22} = -2\omega^2 V_{12} - 4\beta V_{22} - \psi_v^{-1} V_{12}^2$$

Figure 35. Equations of Kalman filter for the simple harmonic oscillator with viscous damping.

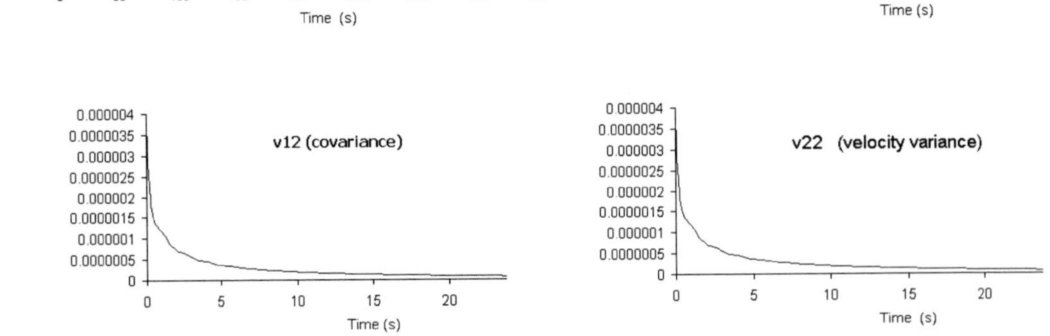

Figure 36. Results from an application of the equations of Fig. 35 to a torsional oscillator. The raw data is very similar to the free-decay shown in Fig. 5 except that the initial amplitude was about twice as large.

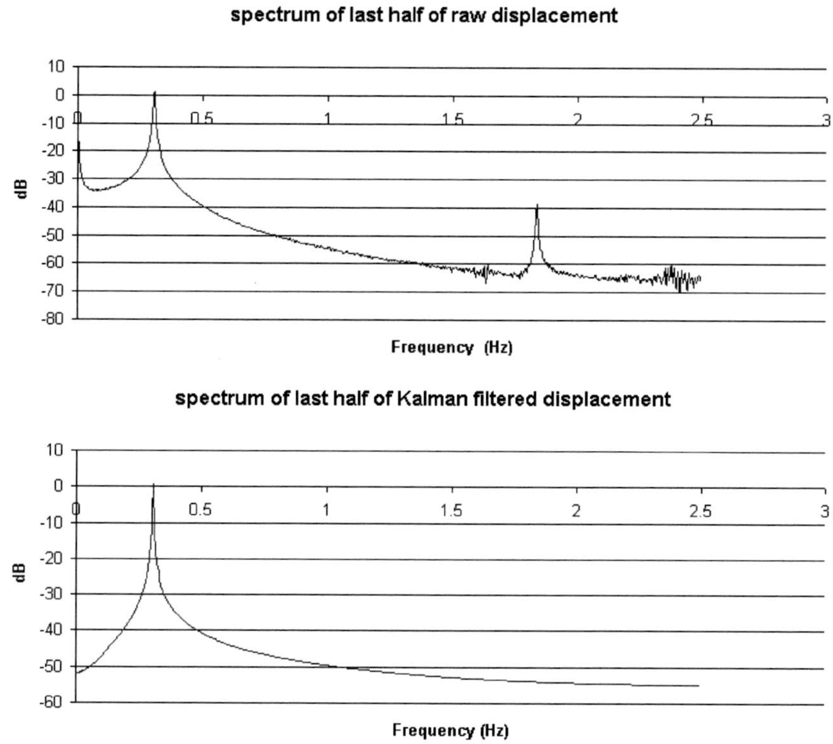

Figure 37. Comparison of the raw data spectrum (top) with with the filtered data spectrum (bottom). The high frequency line, which is not present in the filtered output, is due to pendulous swinging of the boom of the torsion balance.

From the many additional experiments conducted by the author, we consider now the last study to illustrate the power of the technology-triad that was first mentioned in the abstract of this article.

15. Correlation Measurements of Atmospheric Pressure Variations and Seismicity during Hurricane Dennis

During the passage of hurricane Dennis (July 2005), two functionally different instruments made continuous recordings from their location in Macon, Georgia. These were (i) a non-conventional vertical seismometer, and (ii) a low-level differential pressure sensor. Data were collected for 72 hours, beginning when the eye of the storm was located in the vicinity of Key West, Florida. Over the ensuing three-day period, fluctuations in the outputs from the two instruments were found to be highly correlated, with the exception of one eight hour interval.

15.1. Background

In a previous study it was shown that hurricane Charley not only generated microseisms as many would expect, but it also at times excited oscillations of some of the free-modes of our planet [Peters, 2005-3]. It was hypothesized that passage of the eye of a hurricane across a region of the earth where there is a sharp change in acoustic impedance, should result in a driving force for eigenmodes. The forcing function for this case involves friction and the sustained reduced pressure in the eye of the storm is the source of power.

Consider the situation of a large hurricane for which the eye is in the vicinity of the eastern seaboard, as for example in the vicinity of Charleston, SC. Cyclonic rotation about the eye results in easterly winds for geographic points north of the eye; i.e., in North Carolina. At the same time, westerly winds are experienced south of the eye; i.e., in Georgia. Storm-surge off the N.C. coast is responsible for large breaking waves. The buildup of wave amplitude is readily explained by the conservation of energy. As waves approach the coast and the water depth decreases; there is a decrease in their speed of propagation, since the phase velocity is proportional to the square root of water depth for shallow water. As speed decreases, energy conservation requires the amplitude to increase, since the transmitted power is proportional to the product of velocity and the square of wave amplitude. We thus see a mechanism by which the earth must experience large oscillatory driving forces from the oceanic waters. The mechanism is due to the large gradient of wave activity across the eye of a large hurricane when located near a coastline.

One expects the excitation of localized earth crustal motions to also generate pressure disturbances in the atmosphere. Resulting infrasonic waves should travel large distances from their source and, at a given observation point, cause significant pressure fluctuations. The present study offers support for this conclusion by demonstrating a strong correlation between the outputs from the two instruments that were monitored: (i) a modified Sprengnether vertical seismometer, and (ii) a highly sensitive differential pressure detector invented by the author.

15.2. Instruments

15.2.1. Seismometer

The seismometer is described in detail in an article [Peters, 2005-3], except that for the present study the (unconventional 'weak') force-feedback described there was not utilized. In other words, the instrument was operated open-loop—-a mode that is thought by professionals of geoscience to be completely unacceptable! The reason is that conventional professional seismometers use capacitive sensors that are severely restricted in mechanical dynamic range. Thus the standard modus operandus of these instruments is one involving force-balance. A disturbance is monitored by means of the error signal required to keep the mass of the instrument virtually stationary.

As noted in various of the author's publications concerned with internal friction, strong force-feedback of this type, when coupled with the PID (proptional integral, differential) network that is employed–severely restricts the low frequency performance of the instrument.

The area-varying (as opposed to gap-varying) architecture of the author's patented fully differential capacitive sensor [Peters, 1995] allows a much greater mechanical dynamic range. In turn, this permits the seismometer to operate open loop (without destruction of sensitivity) and allow the non-ideal spring of the instrument to 'find its own best' equilibrium position, as it is being continuously influenced by anelasticity responsible for creep.

In previous experiments the open-loop mode was not feasible because the ADC employed was limited to only 16-bits (Dataq DI-700). For adequate low-level sensitivity, the rail voltage was set at +/- 0.1 V. For large earthquakes this value was too low to prevent saturation of peak signals. The present study used a Symmetric Research single-channel ($\approx$ \$100) converter having 24-bits, so that the rails could be increased to +/- 10 V. As seen in later graphs, this permits open-loop diurnal voltage swings as large as 1 V to be accommodated while signals at the 1 mV level (such as by distant earthquakes) can still be resolved. Most large earthquakes should also be observable without waveform clipping.

Seismologists are accustomed to viewing 'velocity' output as opposed to the 'displacement' output of the present instrument. If one insists on working with the former, it is a simple means to take the numerical derivative of the filtered records. To collect the raw data as velocity rather than displacement causes the signal at low frequencies to be attenuated 20 dB per decade. Though commonly practiced, this severely restricts an instrument's ability to study the most interesting eigenmodes of the earth.

15.2.2. Pressure Sensor

The pressure sensor operates as a gap-varying fully differential capacitive detector that uses a diaphragm. The principle of its operation is described in the referenced patent literature. The diaphragm is fabricated from thin aluminized mylar held in tension. The static plates on either side of the diaphragm were fabricated from printed circuit board material. As manufactured and sold by Tel-Atomic Inc. (no longer available), the sensor was operated with their SDC electronics box. For the present work this box was replaced with the autozeroing electronics package designed by the author, shown in Fig. 38.

The 'return to zero' time constant is about 20 s. This limits the low frequency response of the instrument; in a Bode plot, the falloff occurs at a value close to the lower corner frequency of early-generation seismometers referred to as 'long-period' type.

The sensitivity of this differential pressure instrument was estimated $\approx 100mV/Pa$. For the noise level of the electronics this yields a resolution significantly less than one part per million of atmospheric pressure. This is much smaller (about two orders of magnitude) than the resolution of a typical barometer. Barometers are absolute pressure instruments, and their resolution is limited by friction effects, whether internal to their more massive diaphragm, or Coulomb friction (older mechanical instruments) in the mechanism used to communicate diaphragm strain to a recording indicator.

15.3. Method of Differential-Pressure Measurement

One side of the diaphragm is open to the atmosphere. The other side communicates through a tube to a bottle. The bottle is housed in a thermally insulated container, so that temperature fluctuations are minimized. With ideal electronics, the lower corner frequency is determined by the leak rate of the bottle. This rate determines the properties of this mechanical

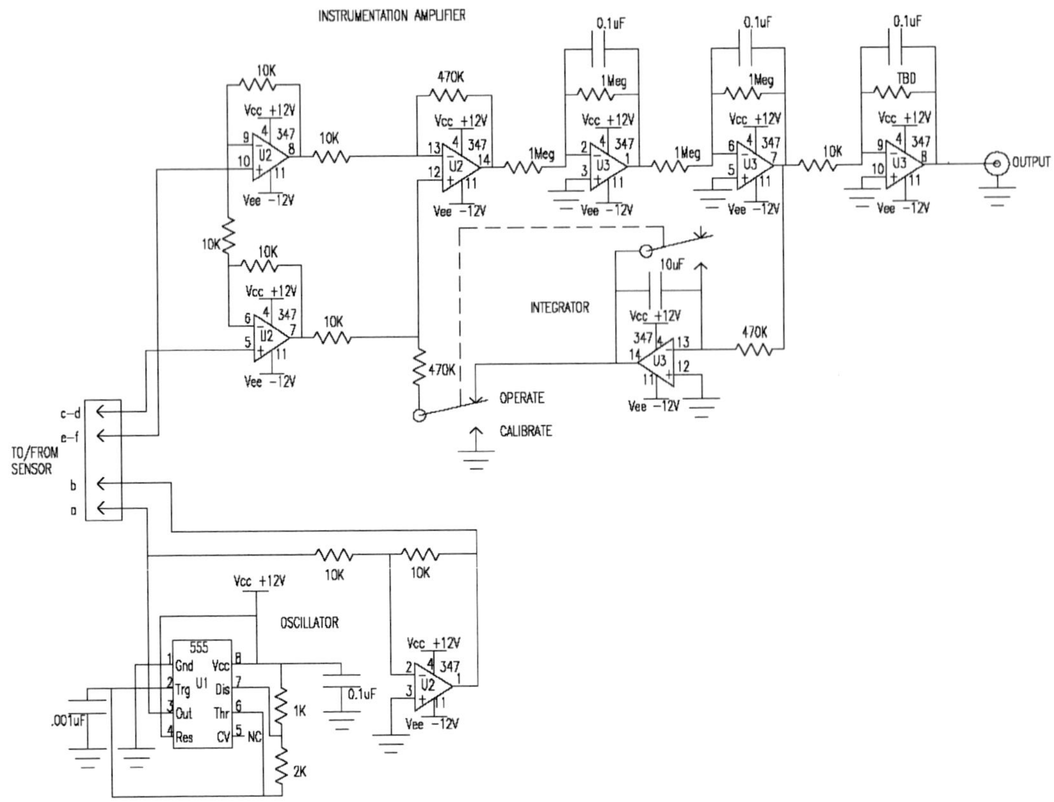

Figure 38. Autozeroing electronics used with the pressure sensor. This circuit requires that the sensor be configured with diodes and resistors (1 Meg) in accord with the equivalent circuit of Fig. 1 at http://telatomic.com/sdct2.html

high-pass filter. Even if the electronics of the instrument was not low frequency limited as mentioned earlier, this method of measurement would still place a limit on the maximum period of pressure variation that could be detected.

15.4. Results

15.5. Pressure Data

The two instruments were not co-located, but rather separated in distance by two miles. The pressure sensor was located in the Willett Science center of Mercer University and the seismometer was located in the quieter location of the author's home basement.

Shown in fig. 39, for the pressure sensor, are two days worth of temporal data from the three-day recording session.

Indicated with the red arrows are various places where coherent oscillations were found by using frequency-domain software created by Dataq, Inc (http://www.dataq.com). When present, these oscillations in the range from 5 to 13 mHz must be large, since electronics attenuation has reduced them by a factor of about 25 dB (rough estimate). An example oscillation is provided in fig. 40.

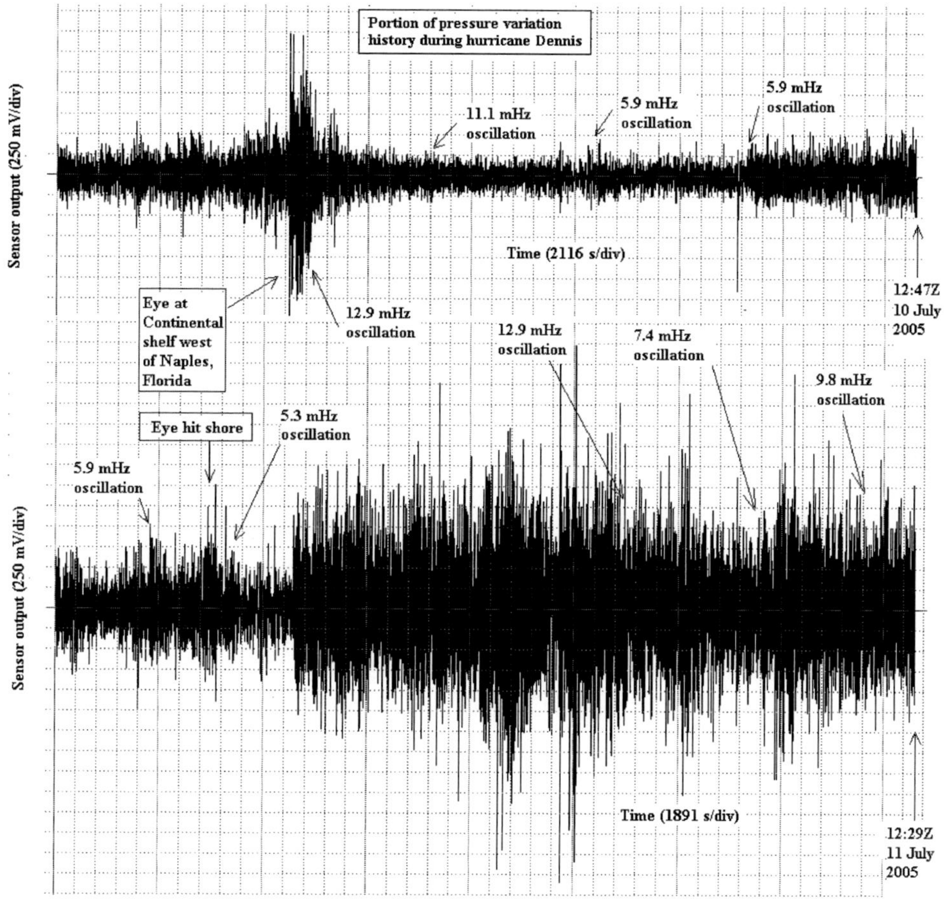

Figure 39. Pressure vs time for two of the three days of recording.

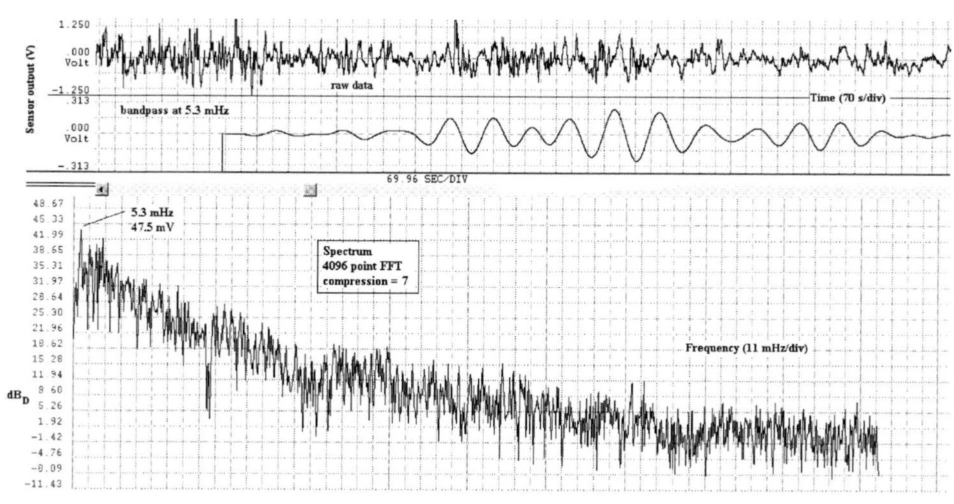

Figure 40. Oscillation thought to be triggered by passage of the eye onto shore.

The decibel scale for this graph differs (subscript 'D' for Dataq) from the other cases of this paper. It is referenced to the full scale (10 V) gain setting of the 16-bit ADC (DI-700). To convert a dB_D value to the conventional scale, one simply uses (for this gain setting) $dB = dB_D - 70.31$.

The convenience of the Dataq software makes easy the task of finding oscillations in the pressure data, by locating spectral lines after FFT calculation with a Hanning window. Different regimes of frequency require different 'compression', which Dataq refers to as 'input averaging'. Exploration is straightforward by means of rapid 'mouse click' transformations between the time and frequency domains. Once a line is identified, the raw data can be manipulated by a variety of filter possibilities, such as the bandpass capability employed for fig. 40. Uncorrected for electronics attenuation, the 5.3 mHz oscillation level, at 47.5 mV; corresponds to a pressure amplitude of approximately 0.5 Pa. In units common to meteorology this is 5 μbar, 5 parts per million of atmospheric pressure. For comparison, the pressure in the eye of Dennis at this time was estimated from NOAA advisories to be about 960 mbar; i.e., a deficit relative to std pressure of 53 mbar.

Just as oscillations were easily discovered in the pressure data, so oscillations could also be easily found in the data of hurricane Charley [Peters, 2005-3]; since a Dataq ADC was also used with the seismometer for that study. Unfortunately, such ease is not yet available for the present seismic data, which was collected with the ADC sold by Symmetric Research; the output of which was analyzed using QuickBasic generated code written by the author. Because of the user 'unfriendliness' of this software (severe as compared to the Dataq algorithms) no serious effort was mounted to try and find similar oscillations in the records generated by the seismometer. If such oscillations could be located and shown to correlate in both frequency and time with the pressure data, such a discovery would lend even greater support to the present hypotheses.

In addition to the oscillations noted in fig. 40, a number of prominent features are evident within the time-varying level of the signal. The discussion of these is postponed until later graphs, where the same features manifest themselves in data from the seismometer.

15.6. Seismic Data

Instead of plotting seismic temporal data in the manner of the pressure results of fig. 39, the seismic results are given in terms of frequency plots as a function of time. These display spectral evolution as hurricane Dennis interacted with various land masses. For example, a noteworthy graph is one of the set from fig. 41, labeled in time as 9.1 h.

From the Advisory Archive data provided by NOAA's National Hurricane Center, it was determined that the eye of the storm was at this time located near the edge of the continental shelf west of Naples, Florida. As with hurricane Charley, Dennis grazed a continental shelf. With Charley the grazing produced a pronounced disturbance in both the seismic time and frequency plots. In the present case with Dennis, only the spectrum appears to change significantly during this shelf interaction. The transient appearance at this time, of a lower frequency 'hump' in the spectrum, is thought to be consistent with the short-lived 12.9 mHz oscillation that occurred in the pressure data at this time (refer to fig. 39).

In the majority of the graphs there is an unmistakeable presence of microseisms–the sawtooth structure with a peak typically near 0.2 Hz. This structure is very nearly the same

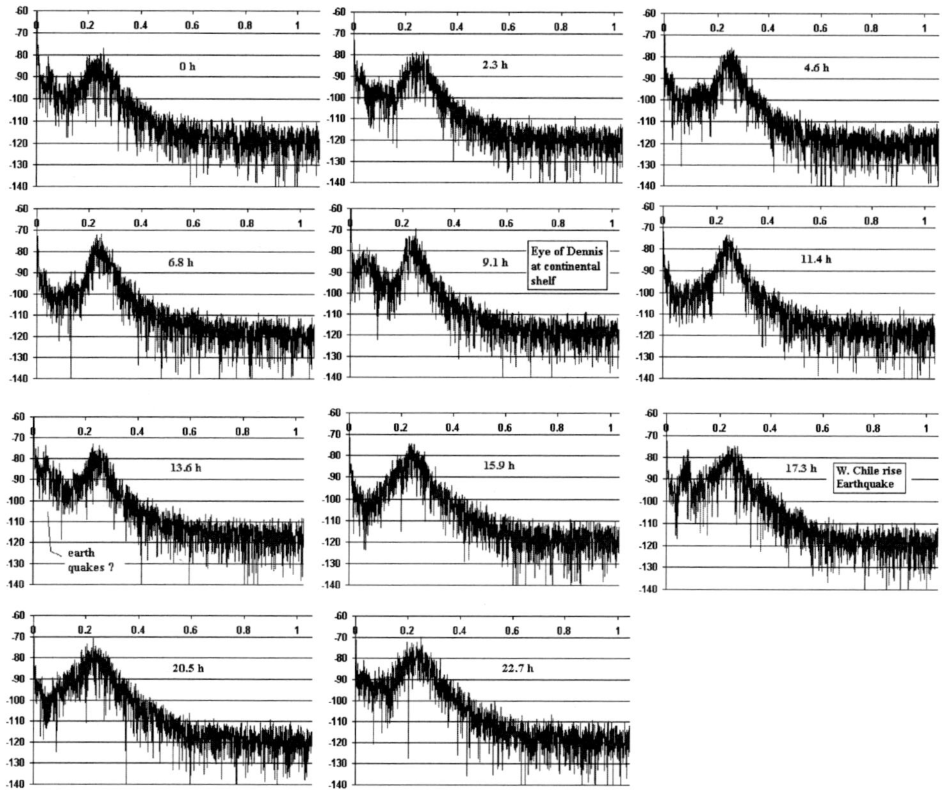

Figure 41. Seismogram spectral characteristics the first day.

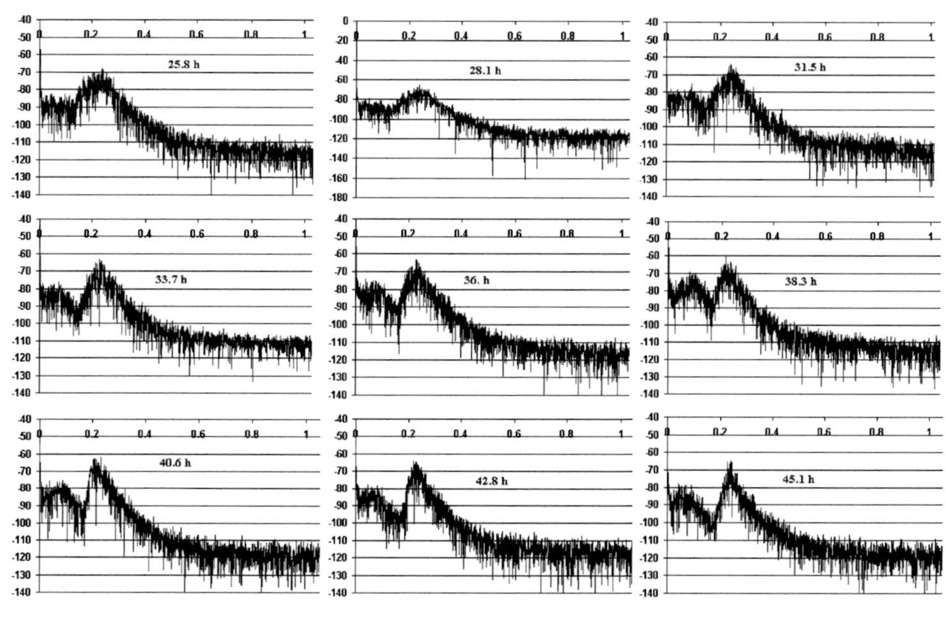

Figure 42. Seismogram spectral characteristics the second day.

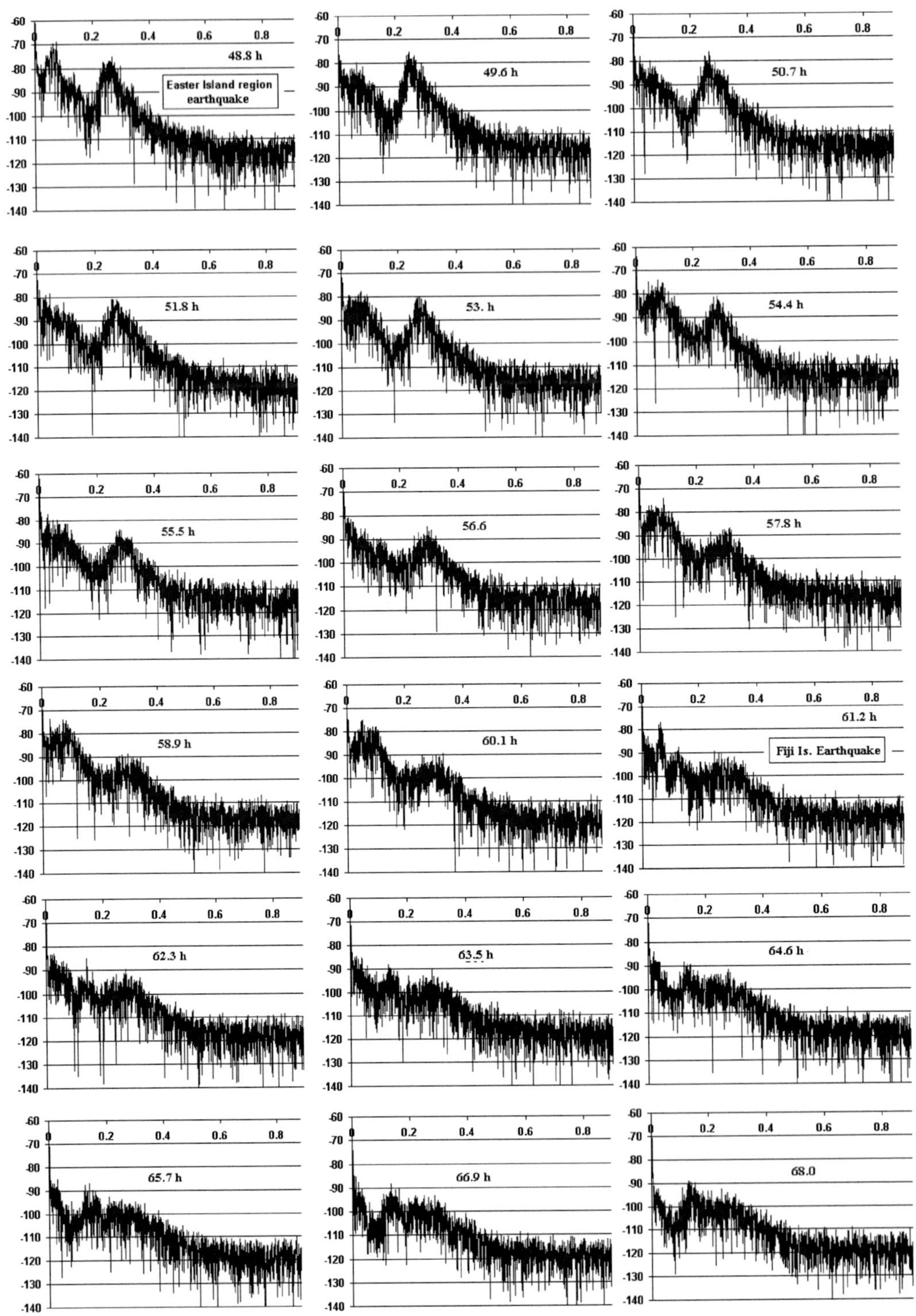

Figure 43. Seismogram spectral characteristics the third day.

as was observed with Charley when the eye of the storm was located in the Gulf as opposed to the Atlantic. As noted [Peters 2005-3], the difference in character is attributed to ocean depth differences.

15.7. Earthquakes

Three earthquakes were large enough to be obviously registered by the seismometer during the three days of data collection. In each case they show a spectral feature with frequencies lower than that of the microseisms, as seen in figures 41 and 43. In figure 41, time-case 13.6 h, the structure labeled with the question mark may or may not be associated with quakes in Japan and Indonesia. Records from USGS do not show a sizeable earthquake that could have been responsible for the hump that is attributed to interaction of hurricane Dennis with the continental shelf.

In Fig. 44 are shown both time domain and frequency domain results for the Easter Is. earthquake. The low frequency variations in the time domain graph would not be visible in the output from a conventional seismograph.

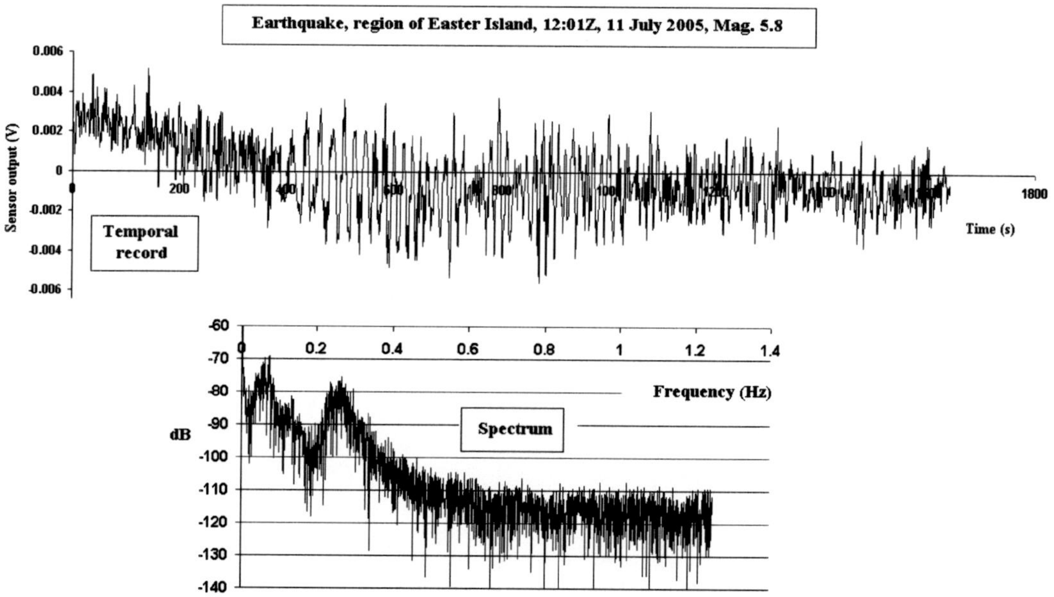

Figure 44. Records from the Easter Island earthquake that occurred at 12:01:35 UTC on 11 July 2005.

15.8. Correlation

A strong basis for the claims of this paper is found in the results shown in fig. 45.

Surprisingly strong correlation between seismometer output and pressure sensor output is seen to occur everywhere except for the short segment that was excised before doing the right-side correlation plot. With a correlation coefficient (R^2 from the linear fit) of nearly

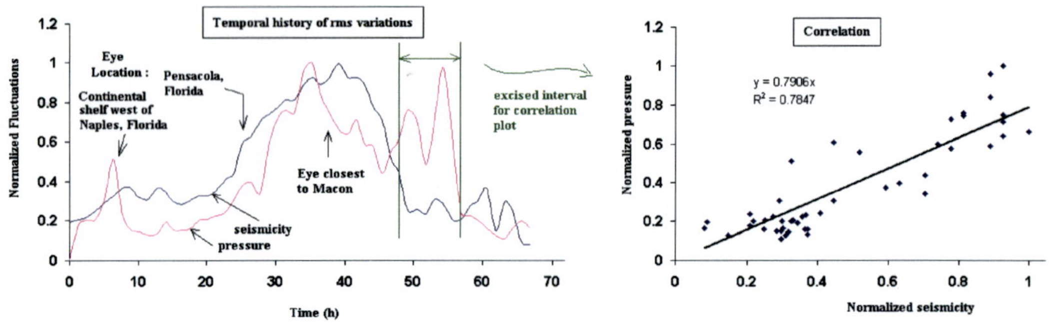

Figure 45. Correlation of fluctuations from the pressure sensor and the seismometer.

80%, there is evidence for significant coupling between ground motion and atmospheric dynamics.

For fig. 45 the seismic data was high-pass filtered. Motions with periods greater than about 200 s were suppressed as follows. After reading each individual record of length 1024 binary-data-words (sample rate 5 per s), both mean value and linear trend were numerically removed using Excel. By concatenating every fifth value of these adjusted records in a composite of length 4096 s , the frequency regime of the seismic data and the pressure data were more closely matched. The effect of this unconventional filtering operation can be appreciated by comparing the earthquake data of fig. 46 (Fiji Island) with that of fig. 44 which used no filtering. In particular, notice the difference in the spectra.

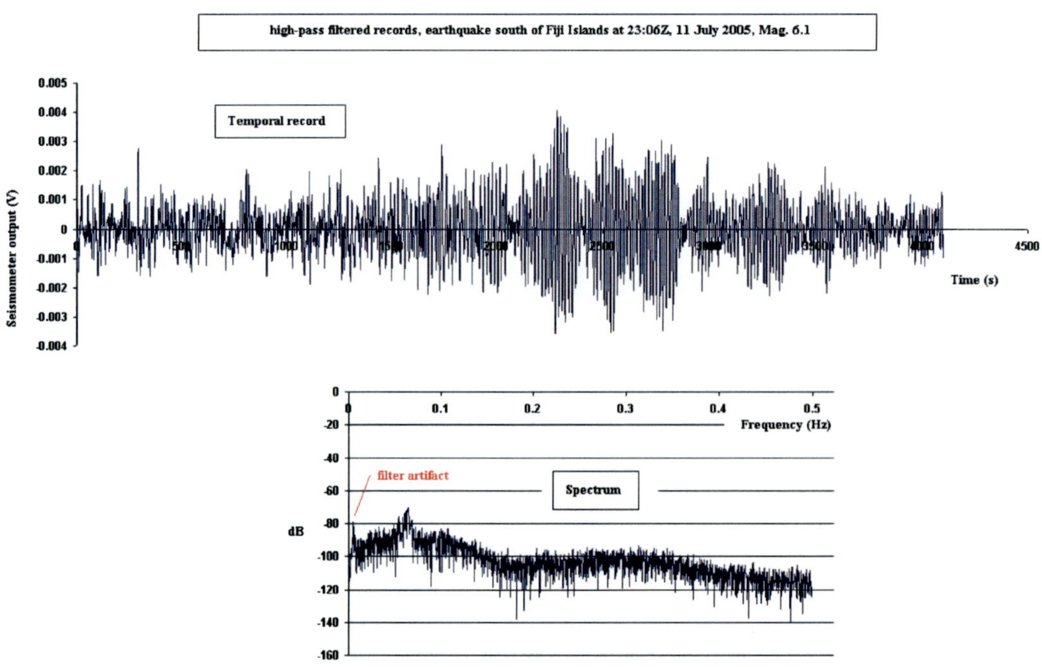

Figure 46. Fiji Islands earthquake; observe that the filtering technique has introduced a numerical artifact (line with frequency of 5.2 mHz).

All spectra shown in this paper were calculated using the FFT algorithm of Excel. The artifact shown in the spectrum of fig. 46 is not due to any anomalous behavior of the excel software. Rather it results from the concatenation of segmented records (filtering process mentioned earlier, in which mean and trend were removed from each of the individual segments).

If the low frequency information in the seismic data is not removed, then some of the 'event' features are even more pronounced, as seen from fig. 47.

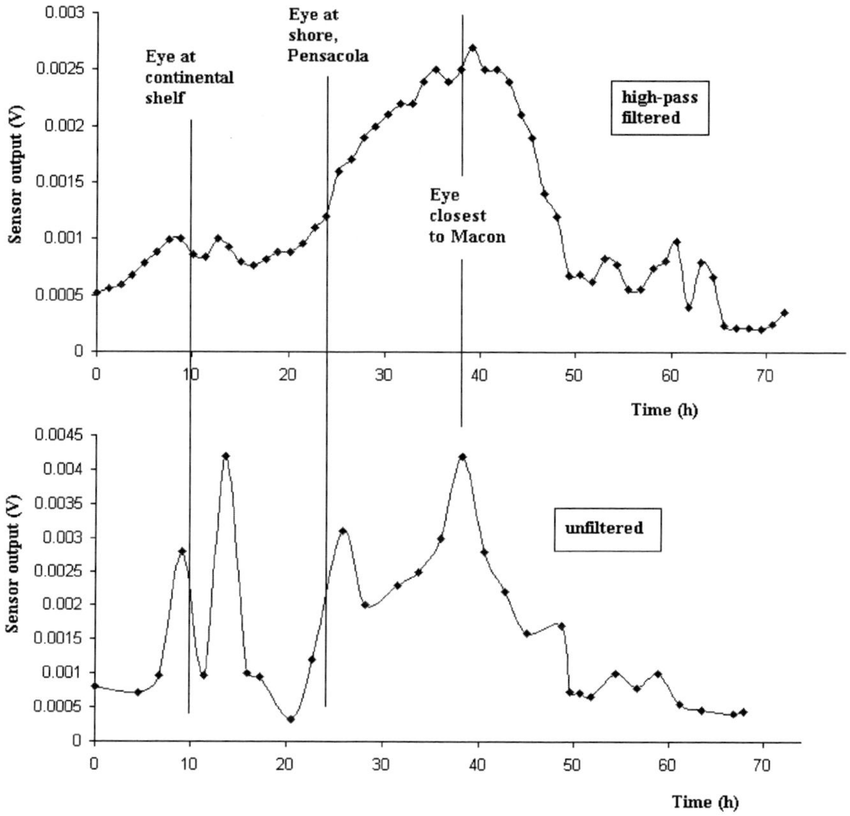

Figure 47. Comparison of seismic fluctuation data after filtering (top) and without filtering (bottom).

Here the double hump associated with grazing of the continental shelf is dramatic. The two other humps are also reasonably explained; i.e., when the eye (i) hit the gulf shore and when it was (ii) closest to the instruments located in Macon, Georgia.

15.9. Anomalous Segment

The reason for legitimately excising the segment (left-side) of fig. 45 from the correlation plot can be understood as follows. During this time the storm was responsible for considerable turbulence in the Macon area. For example, a close neighbor's house was the only one severely damaged by a huge sweetgum tree that was broken by strong winds at its half-height point. The influence of such localized turbulence is of much greater influence on a pressure sensor than it is on a seismometer. Its small spatial extent is incapable of greatly disturbing the ground.

15.10. Arrival Time Issues

In the correlation analysis the pressure and seismicity data were adjusted timewise to maximize the correlation coefficient. It was observed that peaks of the pressure data preceded those of the seismic data in all cases by a couple of hours. Explanation for this arrival time difference can probably be understood in detail only with additional experiments involving more than one pair of sensors, preferably an array of sensors. It is thought that the pressure variations are the result of buoyancy waves that travel at much lower speed than seismic waves responsible for the seismometer response. Thus the tardiness of the seismic data suggests that the origin of excitation is different for the two. In particular, it is deduced that the pressure data derives mainly from the outer bands of the storm (between eye and sensors), whereas the seismometer data derives from its eye. Because the outer bands encounter a land discontinuity before the eye gets there, the pressure should be first to announce the encounter.

16. Conclusion

It is hoped that this article has awakened readers to the realization that defect-structure complexities are an important, mostly unstudied field of physics, including the phenomenon of creep. Outstanding physicists of previous centuries would be disappointed that so little attention has been given to the subject. One might believe that this is an unfair claim, since creep has been studied extensively by materials scientists and engineers. However, their studies have been restricted to much larger levels and/or different conditions than the ones described above. It is the author's expectation, that even those familiar with engineering creep studies are likely to be surprised by the results of figures 27-29.

Moreover, it is hoped that this article will draw the attention of some readers to the terribly inadequate (textbook) descriptions of the pendulum. Although it is permissible to refer to certain classes of the pendulum as 'simple', real instruments need to be at least recognized as anything but simple, even if their complexities are only mentioned without descriptive detail.

References

Anderson, P. W., 1972: "More is different, *Science* **177**, 393-396.

Baker, G. L. & J. P. Gollub, 1996: *Chaotic Dynamics, An Introduction*, Cambridge Univ. Press.

Barkhausen, H., 1919: *Z. Phys*. **20**, p. 401.

Berger, J. and P. Davis, 2004: *J. Geophys. Res.,* Vol. 109, B11307.

Coleman, Piers, 2007: *Nature*, vol 446, 22 March.

Coy, D. and M. Molnar, 1997: "Optically driven pendulum", *Proc. 11th Nat'l Conf. Undergrad. Res.*, **1621**. (R. Peters was a faculty sponsor for this work that was conducted at the United States Military Academy.)

Desalvo, R. et. al., 2005: "Study of quality factor and hysteresis with the state-of-the-art passive seismic isolation system for gravitational wave interferometric detectors", *Nucl. Inst. Meth. Phys. Res. Sec. A: Accel., Spectr., Det. & Assoc, Equip.*, Vol 538, Issues 1-3, pp 526-537.

Erber, Thomas, 2001: "Hooke's law and fatigue limits in micromechanics", *Eur. J. Phys.* **22**, 491-499.

Feynman, R., R. Leighton, & M. Sands, 1964: *The Feynman Lectures on Physics*, Vol 2 (Addison-Wesley, Reading, Mass.) A recent article concerned with bubble rafts by Vandewalle, N, S. Trabelsi, & H. Caps, "Block-to-granular-like transition in dense bubble flows", *Europhys. Lett.* **65** (3), p.316 (2004) is online at http://www.iop.org/EJ/article/0295-5075/65/3/316/epl7991.html.

Feynman, R, 1985: *QED–The strange theory of light and matter*, Princeton University Press, Ch. 4, Loose Ends, p. 129; useful associated information in Wikipedia, online at http://en.wikipedia.org/wiki/Fine_structure_constant

Hawking, S., 2000: On January 23, he is quoted in Jan Jose Mercury News, "I think the next century will be the century of complexity". http://www.comdig.de/Hawking.htm

Kuroda, K, 1995: "Does the Time-of-Swing Method give a correct value of the Newtonian Gravitational Constant?" *Phys. Rev. Lett.* **75**, 2796-2798.

McNamara, D. and R. Buland, "ambient noise levels in the continental United States, manuscript in review, BSSA, online article at http://geohazards.cr.usgs.gov/staffweb/mcnamara/PDFweb/McNamaraBuland.pdf

Nelson, R. and M. Olssen, 1986: "The pendulum–rich physics from a simple system", *Am. J. Phys.*, Vol 54 (2) 112-121.

Peters, R. 1990: "Metastable states of a low-frequency mesodynamic pendulum", *Appl. Phys. Lett.* **57**, 1825.

Peters, R. 1993: "Fluctuations in the length of wires", *Phys. Lett. A*, Vol. 174, no. 3, 216.

Peters, R. 1995: "Symmetric differential capacitive (SDC) sensor, U.S. Pat. No. 5,461,319. Online information is available at http://physics.mercer.edu/petepag/tutorial.html.

Peters, R., 2003-1: "Study of Friction at the Mesoscale using Nitinol Shape Memory Alloy", online at http://arxiv.org/html/physics/0308077

Peters, R., 2003-2: "Graphical explanation for the speed of the Fast Fourier Transform (FFT), online at http://arxiv.org/html/math/0302212 . Other relevant papers are at http://web.mit.edu/redingtn/www/netadv/XfastFouTr.html

Peters, R., 2003-3: "Nonlinear damping of the linear pendulum", online at http://arxiv.org/pdf/physics/0306081

Peters, R., 2004: "Friction at the mesoscale", Contemp. Phys., vol. 45, no. 6, pp. 475-490. Related online material in the article "Study of friction at the mesoscale using Nitinol Shape Memory Alloy", http://arxiv.org/html/physics/0308077

Peters, R. 2004-2: "Beyond the linear damping model for mechanical harmonic oscillators", online at http://physics.mercer.edu/petepag/debunk.html

Peters, R., 2005-1: Eq. 20.31, p. 20-39, *Vibration and Shock Handbook*, ed. Clarence de Silva, CRC Press.

Peters, R., 2005-2: "The pendulum in the 21st century–relic or trendsetter?", pp. 19-37, in *The Pendulum: Scientific, Historical, Philosophical and Educational Perspectives*, ed. M. Matthews, C. Gauld, and A. Stinner, Springer, Dordrecht.

Peters, R., 2005-3: "Hurricane Excitation of Earth Eigenmodes", online at http://arxiv.org/html/physics/0506162 .

Peters, R., 2005-4: "Mesoscale quantization and self-organized stability", online at http://physics.mercer.edu/hpage/cube.pdf.

Peters, R., 2006: "Compound pendulum to monitor hurricanes and tropical storms", online at http://arxiv.org/html/physics/0610092

Peters, R., 2007: Two articles in the 10th Ed. of the McGraw Hill Encyclopedia of Science and Technology: (i) "Chaos (Some mathematical foundations in physics, Vol. 3, pp.744-748", and (ii) "Anharmonic Oscillator", Vo. 1, pp. 675-678. An online interactive chaotic pendulum with remote frequency adjustment is available for hardware experimentation at http://physics.mercer.edu/PENDULUM/

Peters, R., 2007-2: "A new tool for seismology–the cumulative spectral power", online at http://arxiv.org/pdf/0705.1100

Peters, R., 2007-3: "State-of-the-art Digital Seismograph", online at http://seismicnet.com/volksmeter/State-of-the-art_Digital_Seismograph.pdf

Portevin, A. & M. LeChetelier, 1923: "Tensile tests of alloys undergoing transformation", *Comptes Rendus Acad. Sci.* **176**, 507.

Stokes, Sir George Gabriel 1850: "On the effect of the internal friction of fluids on the motion of pendulums", Read December 9, 1850, *Trans. Cambridge Phil. Soc.*, Vol. IX, p. 8. Online at http://www.marcdatabase.com/~lemur/lemur.com/library-of-antiquarian-technology/horology/stokes1850.pdf

Zener, Clarence, 1948: *Elasticity and Anelasticity of Metals*, Chicago Press.

In: Science Education in the 21st Century
Editor: Ingrid V. Eriksson, pp. 69-106

ISBN: 1-60021-951-9

Chapter 2

OF MODELS, MERMAIDS AND METHODS: THE ROLE OF ANALYTICAL PLURALISM IN UNDERSTANDING STUDENT LEARNING IN SCIENCE

Keith S. Taber*
University of Cambridge Faculty of Education

Abstract

Research into student ideas and learning difficulties in science education is undertaken to inform developments in pedagogy. Complex phenomena may be best understood by exploring them from a range of viewpoints, and it is argued here that using a battery of 'analytical lenses' to illuminate research data may be an appropriate strategy in researching student learning. The approach is considered methodologically sound provided the 'lenses' derive from perspectives that are congruous with the theoretical framework underpinning the research. The present chapter illustrates such 'analytical pluralism' by applying three 'analytical lenses' ('modelling mentality', 'learning impediments' and 'student ontologies') to illuminate data from research into student understanding of the orbital model of atoms and molecules. The three 'lenses' focus on different features of the data in order to offer possible explanations for student learning difficulties and recommendations for improving teaching.

Key words: analysing qualitative data; analytical pluralism; analytical lenses; curriculum models; meaningful learning; learning impediments; cognitive structure; student ontologies; orbitals; quantum theory.

Introduction

One key area of research in science education seeks to understand how students learn science. It is recognised that this is a complex phenomenon, such that there are many factors that can

* E-mail address: kst24@cam.ac.uk; http://www.educ.cam.ac.uk/staff/taber.html. http://people.pwf.cam.ac.uk/kst24/index.html. Dr. Keith S. Taber, University of Cambridge Faculty of Education, 184 Hills Road, Cambridge CB2 8PQ

influence the course of a students' learning. Research into aspects of learning science has been a particularly active field for the last two decades or so, and so there is a now a considerable literature providing insights for teachers and researchers (Duit 2007). Indeed, it is argued here that there is sufficient maturity in the field (see Taber, 2006a) for research into the learning about any science topic to be informed by quite well developed theoretical perspectives – e.g. conceptual change, learners' ideas, problem-solving etc. Each perspective provides explanations for some of the reasons learning takes the course it does.

The argument explored in this chapter is that data concerning complex phenomena (such as learning) may be sometimes best understood by when examined in the light of a number of 'analytical lenses', i.e. analytical tools deriving from distinct theoretical perspectives. Such 'analytical pluralism' is justified where each theoretical perspective only provides a partial picture of the process being studied, and where the distinct perspectives may be considered to be congruous.

The notion of analytical pluralism will first be compared to two related and well-accepted notions: triangulation and the 'battery-of-tests'. The approach will then be briefly illustrated by examining data from an interview study that explored learners' developing understanding of a recognised 'problem' target concept area, i.e. the orbital model of atomic and molecular structure (Taber, 2004). Three different analytical lenses will be applied to the data to illuminate learners' difficulties. The case being made here is not that this approach necessarily provides *definitive* interpretations of the data (rather, like most such enquiry, it provides recommendations for changes in teaching approach and emphasis that may themselves provide the focus for a further cycle of research) but rather that analytical pluralism provides a more inclusive interpretation of the data than is possible from any single perspective.

It is recognised that not all theoretical perspectives can be considered to be coherent, and therefore it is important that researchers utilising analytical pluralism should be confident that the analytical lenses they select do not derive from perspectives based upon inconsistent assumptions. However, it is argued here that, providing researchers are confident that the analytical lenses applied derive from congruous perspectives, analytical pluralism can lead to a more thorough interpretation of research data.

Methodological Triangulation: Slices of Data and Layers of Interpretation

The term triangulation derives from the method of locating a beacon by finding its bearing from different locations: the several directions give limited information individually, but collectively establish the source of the signal. By analogy, the term is also used to describe a common approach to data collection in the social sciences (such as education). Triangulation is considered to be one of the criteria that marks out high quality studies in chemical education (Eybe & Schmidt, 2001).

In *methodological triangulation*, different sources of data are collected (Cohen, et al., 2000; Hitchcock & Hughes, 1989; McNiff, 1992; Schwandt, 2001). For example, a teacher may be interviewed about her teaching, but the interview data is compared to that from other interviewees (e.g. her students and colleagues), observations of her teaching, examination of the comments she makes on student work etc (cf. Elliot, 1991). The assumption is that an informant, even an honest informant, will have biases, and will have access to only some of the relevant perspectives, and so sound conclusions can only be drawn from evidence that is corroborated from several data sources.

Triangulation is primarily used, then, due to considerations of validity - for the purpose of raising the 'trustworthiness' of interpretations (Ely, et al. 1991). In 'qualitative' research 'trustworthiness' has been defined as the quality of a study that makes it noteworthy to an audience (Schwandt, 2001).

Similar approaches are used in educational evaluation (e.g. Hopkins et al., 1999). So for example, in the process of inspecting schools in the UK inspectors are required to collect such a range of evidence:

> "A school inspection is a process of evidence gathering in order to provide an assessment of how well a school is performing. This is achieved using data exchange, lesson observation, interviews with teachers, analysis of pupils' work, meetings with parents, pupils and governors" (OFSTED, 2003).

What is suggested in the present chapter is that, just as different 'slices of data' (Glaser & Strauss, 1967) may be useful in providing different perspectives on a research focus, so may *complementary analytical frameworks* be useful in 'interrogating' a specific data set. It is not suggested that this is an alternative to collecting slices of data from several sources: depending upon the study either or both may be appropriate.

It should also be pointed out that the purpose of the *analytical pluralism* proposed here is somewhat different from the use of multiple data sources. Whereas the rationale for triangulation is to see if the same interpretations can be supported from different 'slices' of data, the purpose of analytical pluralism is *to offer alternative insights from the same data set.*

The Battery-of-Tests Principle

The notion of using a battery of tests is certainly familiar from medicine and science. In general medical practice, a family doctor has available a well established set of standard diagnostic tests that can be used to help identify a medical condition (e.g. Pescar & Nelson, 1994)

The principle is also well-established in analytical chemistry, for example, where an unknown sample may be subject to a wide range of distinct tests, including determinations of melting and boiling temperatures, tests for reaction with a range of reagents, and a wide variety of chromatographic and spectroscopic investigations (Finar, 1973; Wheatley, 1968; Williams & Fleming, 1973). The assumption is that few of these individual tests will provide definitive evidence of the identity of the unknown (especially if it is a mixture), but each test may provide clues.

Some tests will exclude many possibilities; others may suggest the presence of specific features. Given a *sufficient range of tests* it is often possible to make confident identifications of the sample. This is a well-established principle where the different analytical techniques available are seen as "tools [which] give different types of data [and] which are most effectively used in conjunction with each other" (Allinger et al., 1976, p.76).

Although the subject matter of general medical practice or analytical chemistry may seem remote from the central concern of this present study, viz. the learning of science, the 'battery of tests' principle is well established in a field cognate to research into student learning. In psychology a number of tests used together (for example when selecting individuals for educational programmes or employment) is known as a 'test battery' (Anastasi, 1982). It is an accepted principle in psychological testing and evaluation that "important decisions are seldom made on the basis of one test only. Psychologists frequently rely on a test battery - a selected assortment of tests and assessment procedures" (Cohen, et al., 1996, p.153).

It is suggested here, that when interrogating data from studies into student learning, a similar approach may be useful. Rather than subjecting an individual to a series of tests (as in psychological evaluation), a data set is 'subjected' to inspection through a series of analytical 'lenses'. These analytical lenses will each be suitable for uncovering a particular potential feature of the data set, in an analogous way to how infrared, ultra-violet and mass spectroscopy have the potential for revealing distinct structural features of a chemical unknown.

Clarifying the Philosophical Basis

It is recognised that the notion of analytical pluralism could be *suspected* of being underpinned by a view that there is no objective basis for knowledge, and that therefore multiple (and potentially inconsistent) interpretations of the data could be presented as valid results of a study. A philosophical basis for such an approach *could* be found in those modern thinkers who reject the notion of the world as a single reality that can be known in an objective sense. As previous science education research (in the constructivist tradition) has incited similar criticisms (e.g. Matthews, 1994; Scerri, 2003 – see Taber, 2006b), it seems appropriate to explore the issue at this point. Whilst recognising the importance of such debates, the type of analytical pluralism espoused here does not rest on the acceptance of such a philosophical basis.

The idea that the world simply 'is' as we perceive and construe it, would be seen as a naïve and child-like notion. It is generally well accepted that our understanding of the world is mediated by the nature of our perceptual and cognitive apparatus (Ellis & Hunt, 1989) and our existing conceptual frameworks (Kelly, 1963). Thus our internal representations of the world are mental models (Johnson-Laird, 1983) that are contingent upon our biological heritage, and our culture and personal life experiences, *as well as* upon any 'external reality' that may be observed.

Some thinkers conceptualise this argument as relating to limitations in the extent to which we *can know* such reality: where others go beyond this to suggest that the notion of there being an objective external reality becomes untenable (cf. Gergen, 1999). Although this debate may seem somewhat esoteric in the context of the present study, it has become important in considering the future direction of research into learning in science (Taber, 2006b), and so is worthy of consideration here.

The exploration of learners' ideas about science topics has been a major research activity for the past two decades. Much of this work has centred on the identification of alternative conceptions and alternative frameworks (Pfundt & Duit, 1994). This research programme has drawn upon a constructivist thesis for much of its justification (Duit, 1991; Taber, 2006a), and the constructivist stance has been the subject of some major criticisms.

The constructivist position taken by many science educators (which I will refer to as 'contingent-constructivism') is based upon work exploring the psychology of learning (Taber,

2006b). This work suggests that human learning about a topic is necessarily a 'piece-meal' process, which takes place over extended periods of time, and where the meaning of new information (and so how readily and in what way such information is integrated into cognitive structure) is highly contingent on the interpretive frameworks (i.e. prior knowledge) available (e.g. Driver, 1991). In other words what can be learned depends upon what is already 'known' (i.e. thought to be known, whether consistent with accepted formal knowledge structures or not), as existing understanding provides the 'conceptual goggles' (Pope & Watts, 1988) for making sense of new information, and the substrate or bed-rock for anchoring new learning. Within this contingent-constructivist position, existing knowledge is a major determinant of the ease and direction of future learning, and so learners' ideas become a key focus for research.

However, there is a wider constructivist movement that goes beyond the contingent-constructivism of science education. This view, which may be referred to as 'radical-constructivism' after von Glasersfeld (1989; Nola, 1998), is at least agnostic about external reality (i.e. an objective reality can not be known, and so our constructions are all we can know, and perhaps all there is).

When related to science, and taken to an extreme, this radical-constructivist viewpoint would not see developments in scientific understanding as progress towards greater knowledge, or more accurate models of the universe, but rather as reflections of the prevailing cultural milieu. So Feyerabend argues that "Scientific entities...are projections and thus tied to the theory, the ideology, the culture that postulates and projects them", (1988, p.265). This view, that scientific knowledge is relative to the prevalent culture, is anathema to some commentators, and has led to criticisms of the constructivist movement *in science education* (Matthews, 1994). Yet such criticisms would seem to be misplaced for two reasons.

Firstly, the importance ceded to learners' ideas in science are due to their significance *for learning*, and not because science educators wish to raise the naïve or misconceived notions of pupils to be the equal of established science (Taber, 2006b). Secondly the raison d'être of science *education* is not directly concerned with developing our species' knowledge and understanding of the world (à la science), but rather concerns developing learners' knowledge and understanding *of curriculum models* (Gilbert, 1998) of science, i.e. of an existing body of knowledge that is formalised, defined and objectified in curricula and syllabuses. Even if some science educators or teachers do have a relativist/radical-constructivist epistemology of science, their pragmatic concern is primarily with helping learners construct versions of curriculum science that match the objectified versions as well as can be. *Science* is about developing reliable public knowledge (Ziman, 1991), where *science education* is about developing personal knowledge (Kind & Taber, 2005).

The present author's position is that of contingent-constructivism, i.e. that individual learners' must construct their own internal representations of scientific knowledge, and that their learning will be channelled and constrained by their existing knowledge and understanding, as well as by the limitations of the perceptual-cognitive apparatus available to them, and by features of the learning environment (e.g. teacher knowledge and pedagogic skills, nature and quality of learning activities including student-student interactions, etc.) Any analytical perspectives applied therefore need to be seen to be consistent with this underlying position.

Understanding Multifaceted Phenomena

So recommending analytical pluralism is not tied to a relativist view of science or a radical-constructivist epistemology. Even from a naive realist viewpoint (considering the world as objective and fundamentally knowable) analytical pluralism can be justified when researching complex phenomena.

A key area of concern to science education researchers is, naturally enough, student learning in science. By understanding the process by which learning takes place, we are better able to offer pedagogic advice to teachers and curriculum planners. Yet, science learning is a very complex phenomenon, and (as implied above) there are many factors that could be the focus of such studies.

If a student's ideas are elicited at the end of a course of study, it is not sensible to look for '*the* cause' of some particular feature of learning. The factors that contributed to a particular understanding may be legion: the learner's pre-established intuitive ideas about a topic; the degree of match between the learner's preferred learning styles and the teacher's teaching style - including the sequencing and pace of the teacher's presentation; the extent and nature of any peer discussions that accompanied the lesson; any associated informal learning from the media; the idiosyncrasies of the way the teacher and/or the student use certain relevant vocabulary; the way the student recognised connections with other lessons (in the same or other subjects); the frequency and nature of the revision activities undertaken by the student; the extent to which the teacher sought to reinforce learning in later classes etc. It would be possible to extend this list considerably (at the risk of boring the reader), but perhaps a few examples will suffice.

Factors totally beyond the teacher's control, such as the quality of the student's diet and sleep patterns (Sousa, 2001), will influence learning. The time of day of a class, the nature of previous lessons and the weather conditions (not to mention ill health or romantic infatuation) could all influence the student's level of concentration and attention.

Some of these factors are outside the normal focus of research in science education, but they may all be relevant variables in determining the learning that ultimately derived from a teaching episode. Learning (from the contingent-constructivist position established above) is a multivariate and multifaceted phenomenon, highly contingent on features of both the learner's internal 'conceptual ecology' (Hewson, 1985), and the external learning environment, and no single causal explanation of a specific feature of student learning is ever likely to be a complete or sufficient explanation. When dealing with such complex phenomena it is sometimes appropriate to look for different levels of interpretation. A number of parallels can be drawn with other areas of study.

In cognitive psychology, an information processing perspective on cognition often considers three different 'levels' (Dawson, 1998), 'computational' (considering the information processing 'problem' being 'solved' by the system), 'algorithmic' (concerning the method of processing the information) and 'implementational' (the physical properties of the system). Each of these levels of analysis is an accepted focus for research, with its own terminology, techniques of inquiry etc., but ultimately cognitive psychologists would wish to develop an over-arching theory incorporating and relating these levels.

In chemistry we often use several levels of description and explanation to explore reactions: considering the molecular and molar levels for example (Jensen, 1995; Johnstone, 1991) and in an analogous way - exploration of science learning should consider both conceptual factors (e.g. alternative frameworks) and cognitive factors (e.g. limitations due to the capacity of working memory) as complementary levels of interpretation (Taber, 2000c; in press). Similar approaches are seen in aspects of biology, where, for example, understanding the functional role of an organ within a system, understanding the structure of the organ, and appreciating the biochemistry of its component cells are complementary in providing an overall understanding of the organ (e.g. Beck, Liem, & Simpson, 1991).

Failure to appreciate such multi-level analyses can also occur. Lovelock's Gaia hypothesis (or 'geophysiology') posits the earth as a single system, *akin to* an organism, with a high level of interdependence among its constituent parts - geology, atmosphere, oceans and all the biota (Lovelock, 1987/1979). The system has evolved a wide range of feedback loops (that may be considered asanalogous to homeostasis within a biological organism) and this perspective can be a fecund way to think about topics such as biodiversity, climate change and pollution.

It is possible to talk about Gaia as being (to some extent) a self-correcting system, in the same way that many organisms are able to adapt to changes in their environment. For example, changes in the sun's radiation output since the formation of life on earth would have made the planet too hot for life, had the ecosystem not evolved in response.

However, some critics have seen this argument as teleological, and consider Gaia as a mystical idea that imbues the earth with its own consciousness and will, because it is able to adapt like an organism (Sattaur, 1987). This is an error of confusing different levels of analysis: the feedback loops that regulate Gaia can all be explained in terms of well-accepted physics and chemistry (e.g. changes in evaporation rates altering cloud formation), but sometimes it is more appropriate to focus on the system level, where the emergent properties may be effectively discussed in terms of the 'geophysiological' model. The Gaia hypothesis provides a perspective for studying the environment that may act as a useful way of thinking and can suggest fruitful research approaches (i.e. a 'positive heuristic' in the sense of Lakatos, 1970).

Modelling Complex Phenomena

Given that learning (like the biosphere) is a complex phenomenon where research can only be expected to provide partial understandings, it is appropriate that the results of such research should be considered as models. Models can be representations that are simplifications and generalisations. Many of the 'alternative frameworks' reported in the literature should be seen in this way: simplifying complex data to provide models that can be readily communicated, understood and applied; generalising similar features from the distinct ideas of different individual students to provide models of the types of thinking that teachers are likely to encounter (Gilbert & Watts, 1983; Pope & Denicolo, 1986; Taber, 1998a).

Models are recognised in science as often having a heuristic value, as being useful for guiding further research (and in this sense often being of worth even when they are clearly inadequate in some respects). Providing that the users of the models realise that this is their nature - often incomplete, imprecise but useful ways of organising or summarising complex data - then they are appropriate outcomes of scholarly research. As an example, Bohr's model of the atom was recognised as having problematic aspects, yet was still a significant development, having heuristic value to the research community. Bohr's model was known to lack coherence, being 'a kind of mermaid' as a synthesis of classical and quantum principles (Niaz, 1998), but its novel features, and its internal incoherence were both able to stimulate further research.

Theoretical Maturity and Research into the Learning of Science

Research into learning science has now been an active field for several decades, and may be considered to have developed some theoretical maturity (Solomon, 1993a; Erickson, 2000; Taber, 2006a). In other words, there are now a number of well-grounded theoretical perspectives that can usefully inform studies into the learning of science. It is suggested that (at least some of) these different perspectives are complementary, and can illuminate different aspects of the phenomenon by 'asking different questions' of the data. A comparison could be made with wave-particle duality: to fully understand the nature of an electron one would need to explore both its particle and wave aspects (Atkins, 1974).

For example, studies of learners' intuitive ideas about the world (e.g. Driver et al., 1994) provide information about one important determinant of later learning pathways; research considering the way learners process information (e.g. Johnstone, 1989) acknowledge another important constraint on the learning process, and ideas about how to judge the status of competing conceptions (e.g. Hewson & Hennessey, 1992) provide insights into how conceptual change occurs. Although each of these perspectives may be distinct, they are all consistent with a contingent-constructivist position.

An Example of Analytical Pluralism

The principle of analytical pluralism will be illustrated here by showing how several different (but compatible) perspectives provide suitable analytical tools to interpret data into student learning. The example discussed is based upon research into student learning in one area of the curriculum - learning about quantum models of atoms and molecules in college chemistry. The findings of this research (summarised below) have been published (Taber, 2002a, b, 2004), and this research will be referred to as the 'source' study.

Three analytical perspectives (modelling mentality; learning impediments; student ontologies) will be briefly considered to show how each illuminate different aspects of the reported data. The present chapter does not set out to provide a comprehensive analysis of the data set but rather to demonstrate the value of the approach. The purpose of *this* present study is to show that different perspectives, each drawing on distinct established literature, but each consistent with the author's underlying position ('contingent-constructivism', as described above), can provide complementary 'analytical lenses' to interrogate data and contribute to the construction of a more thorough (and so more authentic) understanding of the research focus. The topic area, and published research findings will first be briefly reviewed, before the different analytical lenses are introduced, and then employed to illuminate features of the data.

Learning about Quantum Models of Atoms and Molecules

It is recognised that students find the particle model of matter as problematic during school science (e.g. Johnson, 1998; Johnston & Driver, 1991; Lijnse et al. 1990; Taber, 2002c), even though at this level the quanta of matter - atoms, molecules etc. - are treated as if 'classical' particles. At college (senior high) level students are taught about an even more counterintuitive model of matter, where 'particles' - perhaps better labelled as 'quantum objects' (Ponomarev, 1993) or 'quanticles' (Taber, 2004) - such as electrons are considered to be wave-particle hybrids with non-classical properties. The structure of the atom, usually taught at the previous educational stage as a planetary-type system with electrons arranged in shells, is now explained in terms of electron orbitals which can technically stretch to infinity - although with infinitesimal probability of finding electrons there! There is little in students' experience of the perceptible world that can be used as a starting point for making sense of such ideas.

Studies into college and university students' understanding of the curriculum models of atoms and molecules presented at this level have revealed that learners commonly have difficulty adopting key features of the ideas presented to them (e.g. Cervellati & Perugini, 1981; Cros et al., 1986, 1988; Mashhadi, 1994; Petri & Niedderer, 1998; Tsaparlis and Papaphotis, 2002). In view of the abstract and unfamiliar nature of the subject matter (Feynman, 1985; Petruccioli, 1993) it is not surprising that attempts to explain and counter students' learning difficulties have met with limited success (Buddle et al. 2002a, 2002b; Fischler & Lichtfeldt, 1992; Jones, 1991; Niaz, 1998; Shiland, 1997; Tsaparlis, 1997). *This is a topic area where illumination from a range of perspectives can fruitfully contribute to directing further research and informing pedagogy.*

This present chapter draws upon evidence from the source study of UK college students' understanding of the quanta/orbital model of atoms and molecules. That study derived from a research project that explored the developing understanding of chemical bonding among students taking the 'A level' chemistry course in a Further Education college in England (Taber, 1998a). This course is at 'university-entrance' level, and is typically studied by 16-18 year olds. The principle informants in the research were 15 students who were interviewed in depth during their studies. The specific (source) study of learners' understanding of the quanta/orbital model drew upon the interview comments of 12 of the students, and detailed findings have been published (Taber, 2002a, b).

It was found that the learners had *particular* difficulties in making sense of aspects of the orbital concept and related ideas. These college level students (1) did not appreciate why quantization was introduced into the atomic model; (2) had difficulty forming an adequate concept of electron orbitals; (3) confused related concepts such as shells, sub-shells, orbitals, energy levels, etc.; (4) did not appreciate what was meant by electronic spin; (5) found the designations of orbitals confusing (Taber, 2002a); (6) did not clearly distinguish molecular orbitals from atomic orbitals; and (7) held alternative notions of what resonance structures were meant to represent (Taber, 2002b). Some of these features will be briefly considered in this present chapter, which draws upon the previously published results.

Introducing the Analytical Lenses

The three 'analytical lenses' that are applied in this study are:

- modelling mentality
- learning impediments
- student ontologies

These lenses draw on perspectives that are discussed in the science education literature:

1) modelling mentality: student understanding of the roles of models and representations;
2) learning impediments: the role of prerequisite knowledge in channelling and supporting new learning;
3) student ontologies: the arrangement of concepts within a learner's cognitive structure, and the nature of conceptual change.

Each of these distinct perspectives is consistent with the contingent-constructivist position, and so it is argued that each perspective, *inter alia*, can help illuminate aspects of the data.

Analytical Lens: Modelling Mentality

Understanding the Role of Scientific and Teaching Models

As alluded to above, modelling is a central aspect of the scientific process (Gilbert, 1998). Scientific models are constructed to explain or organise existing data, to allow prediction in areas where empirical data is not yet available, and so to suggest fruitful directions for further research and theory development. Scientific theories must be open to falsification (Popper, 1989), and so they should never be considered final or proven (cf. the position of radical constructivists, see above); and the status of our models of the world should reflect this. Although, *in practice*, well-established models do become accepted as part of the description of nature, *in principle* they are our constructions, our exploratory and explanatory tools, and should be considered as partial and fallible representations of the world.

Scientific models of the atom provide a case in point. The historical progression in proposed models demonstrates their heuristic value (Justi & Gilbert, 2000; Niaz, 1998). Ideas for modelling atomic systems were introduced because they seemed to work - they helped develop theory that fitted exiting data, and to suggest fruitful directions for future research (Petruccioli, 1993). However, had these ideas come to be viewed as conceptual cul-de-sacs (like phlogiston or caloric) then they could *still* have played a useful role in *the development* of science.

Models are mind-toys that help us explore: a model that does not match new data is still informative, even though it may fall into disuse. And some models which are found *not* to match new data may still retain their usefulness in limited contexts: so Newtonian mechanics is still very widely taught and used even though it leads to increasingly inaccurate predictions as velocities approach the speed of light. An orbital model of atoms is taught in college level chemistry even though it is only theoretically valid for hydrogenic systems (atoms and ions with a single electron), and the extrapolation to apply the 'orbital approximation' to more complex systems seems to be justified by the extent to which it 'works' (Scerri, 1989, 1991; cf. Sánchez Gómez & Martín, 2003).

The nature of curriculum models (Gilbert, Boutler and Rutherford, 1998) is similar to that of scientific models with an important distinction alluded to earlier in this chapter. Scientists develop models to help them come to know more about the world. Teachers use models to help learners come to know more *about the scientific models* that are accepted and represented in curricula. In teaching, the target is usually defined in curriculum documents and is already known to the teacher, who has the 'problem' of moving student understanding towards the known 'target' knowledge (Kind & Taber, 2005).

It is worth stressing these points because it is important for science teachers to appreciate the role of the models they use in teaching (cf. Justi & Gilbert, 2002), i.e. as tools for helping students come to understand aspects of scientific models and theories. The two key points here are that (a) as models are tools they are not intended to be represent absolute replicas of scientific knowledge; and (b) that the target knowledge *is itself a set of models*, which may sometimes only map onto the available data in a limited way. What the literature does suggest is that school *students* may often have quite limited appreciation for the provisional nature of theory and the role of models in science (Driver, Leach, Millar & Scott, 1996; Grosslight, Unger, Jay & Smith, 1991; Harrison & Treagust, 1996, cf. Treagust, Chittleborough & Mamialia, 2002).

Certainly if teachers rely upon the models of the atom found in many textbooks then they might well be presenting hybrid models that have dubious scientific or historical validity, and so limited educational value (Justi & Gilbert, 2000). For example, the atom is often modelled as a tiny solar system, but this analogy is inherently problematic both in terms of the need to explore the negative aspects of the analogy, and indeed, in terms of the 'classical' picture of the atom presented (Harrison & Treagust, 2000a; Taber, 2001c): students need to appreciate that they are being taught about one partial scientific model with limited application. This particular example is even more questionable as a teaching tool when it is realised that pupils often hold alternative conceptions of both the target concept and the supposedly familiar analogue (Taber, 2002c).

The Role of Multiple Models

A perspective on models that sees them as tools - which are partial representations with limited ranges of application - allows the modeller (or model user) to accept that several different, and apparently incoherent, models may collectively give greater insight than an approach that relies on a single model. Given, then, that research suggests that pupils often have naïve appreciation of the role of models in science, it is likely that they will not readily accept that they should use several clearly inconsistent models of the same phenomenon. Rather, this can be a source of confusion (Carr, 1984).

Research into students' ideas often apparently reveals the presence of 'multiple frameworks' for concepts such as energy or force (Pope and Denicolo, 1986), and there is some evidence that - at least at college level - students who hold manifold conceptions of a concept area *can* learn to see them as alternative explanatory 'stories' which may all potentially apply, but which have different strengths and limitations (Taber, 2000a cf. Harrison & Treagust, 2000b; Petri & Niedderer, 1998; Treagust et al., 2002). This would seem to be an attitude that science teachers would do well to encourage.

Interpreting Chemical Representations

This may be particularly important in chemistry where students typically meet a wide range of graphic representations of atoms, molecules and other structures. These representations vary along at least two separate dimensions. For one thing they may be based upon different underlying scientific models (Platts, 1968). So a figure of an atom may show electrons on shells as in a Bohr-type orbital model (ignoring the notion of orbitals), or may show the shapes of orbital lobes. In addition, the way in which the same type of model is *represented* varies considerably.

At a relatively trivial level, electrons may be shown as e, e^-, •, ×, O, etc. *Most* college level learners seem to 'see past' such differences, although this is not true of some academically weaker students (Taber, 2002c). More significantly, representations of atoms modelled with orbitals may show orbital lobes with sharp cut-offs, or alternatively may attempt to show the gradual change in electron density. Some figures show overall electron density for a species (molecule, atom, ion), others show the distinct orbitals, and it is quite common for students to meet hybrid figures (which may reflect most of the electron density in terms of lines for bonds, but explicitly show certain atomic or molecular orbitals). Such a range of representations for molecules and related species is quite common in textbooks used by students at the level in the UK system where the source study was undertaken (e.g. Andrew & Rispoli, 1999; Clugston & Flemming, 2000; Lewis & Berry, 2000).

There may often be good reason why particular types of representations are used to illustrate specific points in texts. However, authors and teachers may not recognise the extent to which this increases the 'cognitive load' on the students (cf. Johnstone, 1989). Whereas interpreting the multitude of representations used in college level chemistry is largely a subconscious process for the expert chemist or teacher it can provide a major challenge when seen 'at the resolution' of the novice student.

Modelling the Molecular Structure of Benzene

In the source study of students' understanding of orbital and related ideas, it was found that some of the college students had particular difficulties making sense of the structure of molecules such as benzene that are considered to have 'delocalised' electrical structures. Students may meet various types of textbook illustrations of the structure of the benzene molecule (including a common approach that represents *part* of the bonding in terms of lines, and *part* in terms of the overlap of atomic orbital lobes, e.g. Sykes, 1986).

Two types of structural formula are commonly used to represent the benzene structure, either showing a resonance between several canonical forms (figure 1) or using a circle (figure 2) to symbolise the delocalised electrons (the pi-system of molecular orbitals).

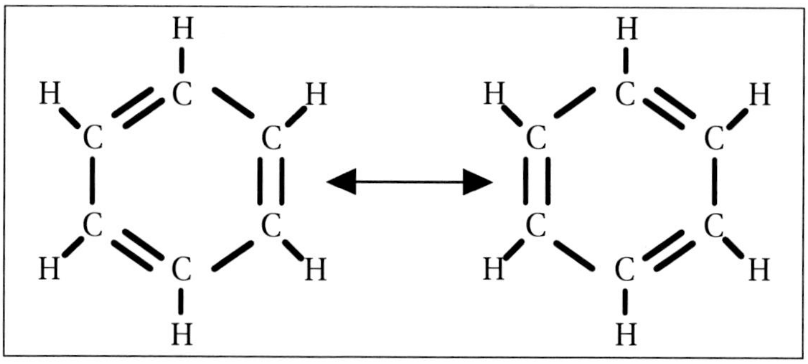

Figure 1. Representing the structure of the benzene molecule as a 'resonance'.

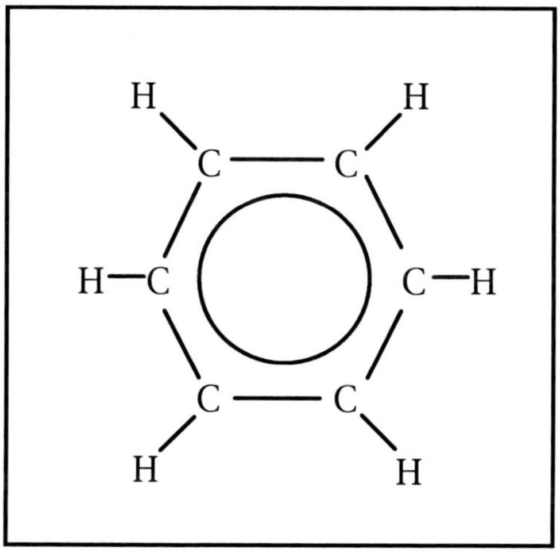

Figure 2. A representation of *the* benzene molecule.

Some students (Taber, 2002b) misinterpreted diagrams of resonance between canonical forms as showing an actual alternation between discrete alternative structures ("you can either have a double bond, or a single bond", "and where there was a double bond in one diagram there would be a single bond"). The students recognised the symbols for single and double bonds, and interpreted the figure as showing that the single and double forms swapped positions: "it can change from one to the other". This is clearly a reasonable inference in terms of the students' *existing* models of molecular structure and how such structures are usually represented. So one student (Carol) described how,

> "It will be double bond, single bond, double bond, single bond, double bond, single… and where there was a double bond in one diagram there would be a single bond in the other one...you can either have a double bond, or a single bond, … sometimes single, sometimes double"

Another student, Quorat, who was able to talk of "resonance structures" and "canonical forms" construed these notions as devices *for overcoming ignorance* about *which* bonds were double and which were single, "since the actual positions are not known, it is better shown as a delocalised system".

Several students interpreted figure 2 as implying that there were 'spare' electrons in the molecule. In terms of their existing understanding of bonding models and representations this figure showed each carbon centre to be involved in three bonds (rather than four) leaving a spare electron not used in the bonding. These spare electrons were considered to occupy a *reservoir* in the ring, the "six spare electrons in the middle", indicated by the circle on the diagram. For example, Carol (C) explained to me (I) how each carbon was only shown to have three bonds:

I: How many bonds are there in that diagram, altogether?
C: Twelve.
I: Twelve. What kinds of bond are they?
C: Covalent. And there's six spare electrons in the middle.
I: Ah - whereabouts are these spare electrons?
C: Well, they're represented by a circle, and they're dawn in a like diagram form. And they're just spinning around.
I: You say in the middle?
C: Well, I don't really know whether they are actually in the middle, in real life, but they're shown to be in the middle in the diagram.
I: Where do you think they might be, these spare electrons?
C: They're probably orbiting the carbon. Each, each electron, well the carbon, it's got a valency of four, hasn't it? And, because three bonds are only shown, it's got to have one still whizzing round itself. I think. But you can't, but to show that on a diagram it'd be messy, so you put that in the middle, by a circle.

It is suggested here that part of the difficulty that these students had of making sense of the figures was related to their rigid interpretations of the representations. These students had learnt models of single and double bonds, and of how to represent them, but they did not think of them as models - as *partial*, as *provisional*, as *tools* to aid thinking. When they met examples of structures that did not fit their existing models they found it difficult to develop beyond them.

It is part of the role of teachers and textbooks to present scientific ideas *as models*, but as the research literature suggests teachers are often not communicating this perspective to their students (Grosslight et al. 1991), even when they do hold such a view themselves (Justi & Gilbert, 2002a, b; van Driel & Verloop, 2002).

Analytical Lens: Learning Impediments

Meaningful Learning, Chunking and Scaffolding

The second perspective that will be presented as an example is developed in part from Ausubel's well known and widely accepted idea of 'meaningful' learning which highlights the role of the students' existing knowledge as a key interpreter and organiser of new learning (Ausubel & Robinson, 1971; Ausubel, 2000).

This perspective on learning implies that successful teaching needs to be designed so that it builds effectively on prior learning. This is a key principle of 'constructivist' approaches to teaching and learning (e.g. Driver & Oldham, 1986). In effect, the teacher needs to plan the presentation of new material with the students' current knowledge structures in mind. From this perspective, many 'failures' of teaching can be explained in terms of the relationship between the hypothetical cognitive structure, assumed by the teacher, that *would* make sense of new material, and the way students interpret the new material through their *actual* cognitive structure.

This principle – of effective teaching needing to be matched to the learners' existing conceptual structure – has been used to develop a typology of 'learning impediments' (described in Taber, 2001b – for the latest versions of the typology, see the website 'Science Learning Doctors - using diagnostic assessment in the science classroom', at http://people.pwf.cam.ac.uk/kst24/ScienceLearningDoctors.html).

The primary distinction is between situations when the intended learning does not take place because the learner cannot make sense of the presented material in terms of existing ideas ('null learning impediments'); and situations where the learner inappropriately (from a curriculum perspective) interprets the new material in terms of existing ideas ('substantive learning impediments'). Null learning impediments may occur because the assumed prior learning has not taken place, or because the learner does not bring it to mind when the new material is met. Substantive learning impediments may be acquired from various sources (but for pragmatic reasons those that derive from previous teaching are put into a separate category).

In the present analysis this typology was applied, and students' learning difficulties (Taber, 2002a, b) were identified in terms of a number of types of 'mismatches' between the teaching and the assumed prior knowledge (Taber, 2004):

- deficits in assumed prior learning (such as not being aware of the classical physics which would make 'planetary' atoms unstable);
- failures to make expected connections with prior learning (such as not recognising the relevance of work on spectral lines undertaken in physics);
- interpreting new information in terms of alternative conceptions (such as considering that a force can be too small to have any effect);
- interpreting new information in terms of oversimplified models from prior teaching (such as adopting the term 'orbital' as a synonym for electron shells).

These and other examples (see Table 1) suggest that mismatches between the material presented in teaching and the learners' prior knowledge structures may contribute significantly to the 'learning demand' (Leach & Scott, 1995) of this topic. It is important to point out that this approach does not imply that learners are 'to blame' for the mismatches – it is the teacher's responsibility to plan teaching to connect with the students' actual prior knowledge, and to undertake the necessary diagnostic assessment and any 'remedial' instruction that may be needed (Taber, 2002c),

Table 1. Classification of some student learning difficulties

Type of learning impediment	Learning difficulty
deficiency impediment (no relevant material held in existing cognitive structure)	D1. no knowledge of radiation of energy by electrical oscillators D2. no knowledge of centripetal acceleration and centripetal force D3. no direct experience available to understand wave-particle duality D4. no knowledge of the scientific concept of angular momentum
fragmentation impediment (learner does not see relevance of material held in cognitive structure to presented material)	F1. work on spectral lines stored as physics knowledge F2. areas of electron density above and below the benzene carbon skeleton not related to the 'spare' electrons not used in σ-bonding
ontological impediment (presented material inconsistent with intuitive ideas about the world held in cognitive structure)	O1. forces can be too small to have an effect O2. spin is a property of rotating bodies
pedagogic impediment (presented material inconsistent with ideas in cognitive structure deriving from prior teaching)	P1. electrons are found in concentric shells P2. electrons move around nuclei in planetary orbits P3. inertia is sufficient to explain planetary motion P4. energy input leads to (all) particles in system moving faster P5. heating metals till they glow causes emission [taken as emitted electrons] P6. bonds are between two atomic centres and are represented as lines

This perspective draws upon ideas about the conceptual (i.e. content) *and* cognitive (i.e. process) aspects of learning: where prior learning includes 'alternative frameworks' then new material may be interpreted accordingly, but not all 'faults in the system' are due to missing prerequisite knowledge or the presence of alternative conceptions (cf. Taber, in press).

It is not enough for a learner to hold relevant prior knowledge in memory, this relevance needs to be recognised and the prior learning brought to mind if prerequisite knowledge is to act as a substrate for new learning. There are at least two good reasons why this may not occur. Sometimes the new learning does not act as a sufficient cue to trigger the prior learning. This was seen in one of the examples listed above, where the student did not bring to mind the experiment she had undertaken in *physics* to measure the wavelengths of spectral lines. When the student was first asked whether it was possible for an atom to be 'excited' in electrical terms, she did not think it was. However, after being reminded of an experiment she had undertaken in her physics class "with a spectrometer, and you...measure angles, and work out the wavelengths of colours of light?" she changed her mind and was then able to explain an atom could be excited, when "you promote an electron to a higher energy level. And then it falls back and gives out the energy". Memory research suggests that recall is often context specific (Baddeley, 1990; Anderson, 1995; Parkin, 1987) and an interview *about chemistry* may have not been an ideal context to activate knowledge seen as *physics* (Taber, 1998b).

Another potential factor concerns the capacity of what is normally termed working memory, where "temporary holding and manipulation of information during the performance of a range of cognitive tasks such as comprehension, learning and reasoning" takes place (Baddeley, 1986, p.34). For new information to be considered alongside prior learning it is necessary to bring both to mind at once. Studies show that working memory has a very limited capacity, so that most people can only 'juggle' with about seven 'items' at once (Miller, 1968). Now it appears that these 'items of thought' are not fixed quanta. Rather it is possible that a single 'item' can sometimes be a rather complex entity, because people are able to 'chunk', that is, to see *familiar* complexes as single items of information, i.e. as gestalts (Calvin, 1997; Kellogg, 1995; Parkin, 1987; Sousa, 2001).

However this level of familiarity depends upon having already consolidated the information into long-term memory. In the context of science classes it is possible to suggest a wide range of symbolic information that can be recognised *by experienced teachers* as a single familiar pattern, but which may seem to be very complex information *to the novice*.

So a biology teacher will recognise, for example, a schematic of a typical insect body plan in terms of well-consolidated prior learning, but the detail may be too much for a student to take in all at once. In physics the teacher may recognise a graph as showing an exponential decay, and knowing the key characteristics of such a function, be able to take in all the essential information to reproduce the graph: but the student may not recognise the general form or identify the critical features needed to reproduce the graph.

In chemistry the teacher may recognise a formula, even a quite complex structural formula, or a familiar reaction scheme as single item of information. However, from the students' resolution these schemes may be too complex to take in at once (Taber, 2002c).

When relating new information to prior learning the situation is compounded. Not only will new information likely seem much more complicated to the student than to the teacher, but if prior knowledge is not yet fully consolidated it may be stored in a form that can not yet be effectively chunked. This is a significant issue when studies of memory suggest that this process of consolidation typically takes place over months and even years (Carter, 1998; Greenfield, 1997; Sousa, 2001). This limitation was considered to contribute to some of the learning difficulties identified in the source study of college students' understanding of quantum models of matter.

The research found that students readily adopted *terms* such as sub-shell and energy level, and the general pattern of orbital labels (Taber, 2002a). However, the students made a variety of errors in conflating or confusing this range of distinct concepts (referring to sub-shells as orbitals; shells as sub-shells; sub-shells as shells; shells as energy levels and so forth) and talked of electrons in orbitals that they gave invalid designations such as 1p and 1d.

From a *learning impediments* perspective these types of learning difficulties, perhaps best considered as 'confusing' similar ideas, can be understood when the information that students were being expected to learn in a relative short period is considered from 'the resolution' of the learner. Students begin their college course having a model of the atom with concentric shells of electrons, but the quantum model of electronic configuration, and the relationship of this to energy levels (important to match the atomic model to the periodic system) introduce significant complications.

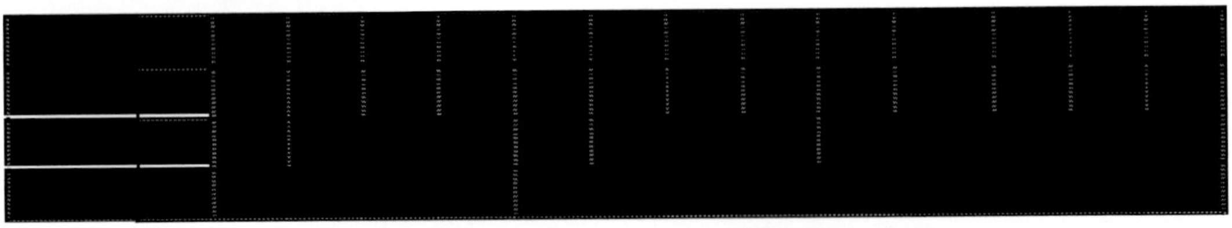

Figure 3. Orbitals, sub-shells and shells for n=1 to n=3.

Some characteristics of the electronic orbitals present in the hydrogen atom (and used as a model for other atoms) may be deduced from some simple mathematical relationships between the quantum numbers, and the pattern for the first three electron shells are shown in figure 3. However the energy level associated with an electron is not determined in any straightforward way from the combination of quantum numbers alone (Scerri, 1998), so that a range of considerations apply, as shown in figure 4. (This figure applies to comparisons within a single atomic system. Comparisons between orbitals in atoms with *different* nuclear charge provide an additional complication).

principal quantum number, n	$n=1<n=2<n=3\ldots$			
azimuthal quantum number, l				s=p=d=f...
magnetic quantum number, m_l				degeneracy razed (various permutations possible)
spin quantum number, m_s	(parallel in same orbital excluded)	spin-paired (anti-parallel) > singly occupied	parallel with electrons in *other* orbitals < antiparallel	

(key: < lower energy than; = same energy as; > higher energy than)

Figure 4. Relationship between quantum numbers and electronic energy levels for a specific atomic system.

Clearly this material is both abstract and complex, although to the teacher, to whom it is well consolidated and the basis of understanding electronic configurations and the periodic table, it will be meaningful, ordered and readily applied.

When these ideas are introduced by experienced teachers many students *will* understand the scheme, because the skilful teacher will in effect work as an 'add-on-memory unit' - or what Bruner calls acting as a vicarious form of consciousness (Hennessy, 1993, p.13) - whilst the student is mastering a new skill or concept. Experienced teachers become skilled at facilitating the construction of knowledge in the classroom (Edwards & Mercer, 1987; Ogborn, Kress, Martins, & McGillicuddy, 1996).

The teacher manages the information being considered at any moment so that the learner's working memory is not overloaded (Gallimore & Tharp, 1990). This is *analogous* to the process by which we are able to successfully undertake calculations that are too complex to 'do in our head'. We use paper to plan out, and keep track of, the stages of the calculation, and focus on each step in turn. In introducing complex conceptual material, the teacher manages the process so that the learner only has to be concerned with understanding each individual step, whilst the teacher monitors the process and navigates through the overall scheme. In this way the learning process will be scaffolded for the learner through the teacher's exposition (Scott, 1998).

However, consolidation of memory is a long-term process, so even if the material is initially understood, and can be recalled, it does not immediately become 'chunked' in long-term memory, and cannot be accessed alongside other information without overloading working memory.

It was also found that when students discussed electrons in molecules they commonly referred to the bonding electrons being in *atomic* orbitals rather than *molecular* orbitals (Taber, 2002a). Even when there was clear evidence that students appreciated the scheme they had been taught about for forming molecular orbitals from atomic orbitals ('linear combination of atomic orbitals'), so they 'knew' that the bonds were formed by combining atomic orbitals to form molecular orbitals, they still made this 'error'.

The scheme for labelling ground state atomic orbitals acts as pre-requisite knowledge for understanding the hybridisation model (which gives alternative sets of orbitals with energetic and geometric properties more suitable for forming bonds), and the geometry of the hybrid atomic orbitals is prerequisite knowledge for appreciating the schemes for forming molecular orbitals that are introduced to students. Even if the learner is able to understand the *components* of the scheme for forming bonds by orbital overlap (as presented at this level, i.e. see figure 5), they may still be unable to keep the whole scheme clearly in mind.

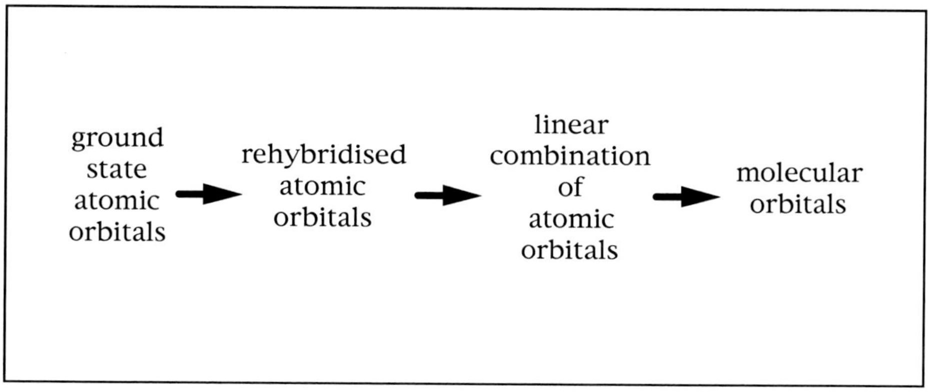

Figure 5. A scheme for conceptualising the formation of molecular orbitals.

Chemistry teachers are well aware that these aspects of college chemistry are abstract, and difficult for students to understand. What perhaps is less obvious, but is highlighted in the study considered here, is that even when students *understand* the ideas, and are able to *recall* them correctly in response to a direct question, they may still demonstrate difficulties applying the ideas in appropriate contexts (cf. Bloom, 1964/1968).

The importance of *understanding* concepts is usually well recognised by science teachers, but the research considered here highlights the need for students to be given time to consolidate new learning before it can be accessed and applied effectively (Taber, 2004). In Piagetian nomenclature, assimilation is not enough - accommodation is also needed (cf. Caravita & Halldén, 1994). Only after the new knowledge is well integrated into conceptual schemes (i.e. after equilibration has occurred), and can be conceived as a single pattern (or gestalt), can the material be processed as a single chunk of information.

The principle of memory consolidation over time is well established (Parkin, 1987; Greenfield, 1997; Carter, 1998), and is reflected in the common teaching principle of regular review of important concepts. Yet, this aspect of learning in science has received relatively little research attention (but see Gauld, 1989; Taber, 2003a), perhaps because of the logistics of collecting the necessary data. The analysis presented here offers one explanation for some of the learning difficulties uncovered in the source study (Taber, 2004). This leads to the testable hypothesis that restructuring the teaching scheme to allow more time between the introductions of new concepts that build upon each other may help students achieve the intended learning.

This perspective focuses on aspects of learners' cognitive structures such as 'missing' pre-requisite learning, or alternative frameworks. However, the approach also emphasises that it is not sufficient to know 'what' a student knows: it is also important to know how that information is organised and integrated within cognitive *structure* as this determines how readily it may be accessed, which contexts encourage it being activated, and the extent to which bringing it to mind will load working memory.

Analytical Lens: Student Ontologies

Exploring Student Ontologies

One aspect of cognitive structure is the way learners classify their concepts in terms of their basic categories for the types of entities in the world - their ontologies. Mariani and Ogborn comment that "the world appears to be understood as fundamentally constituted of objects which are real and have some permanence, and to which one can effect changes by exercising actions within a spatio-temporal framework" (1991: 84). Ogborn and co-workers (Mariani & Ogborn, 1991, 1995; Nicholls & Ogborn, 1993) attempted to explore the way people understand aspects of their worlds in terms of the notion of an 'ontological space', and they sought to identify the dimensions used to organise concepts within that space (cf. Kelly's 1963 notion of bipolar constructs).

Previous work (Taber, 1998a, 2001a) suggests that students' ontologies may well provide a fertile perspective for exploring their learning difficulties when studying structure concepts in chemistry. One example concerns the way that secondary (junior high school) students commonly conceptualise chemical bonding as a dichotomy of two distinct bonding types (see Table 2).

Table 2. A dichotomous bonding ontology

Type of bond	Covalent	Ionic
Formed between	Non-metal & non-metal	Metal & non-metal
Formed by	Electron sharing	Electron transfer

Students seem to strongly adopt this dichotomy and see bonding as *either* covalent *or* ionic (Taber, 2002c). One consequence of this is that when polar bonding is introduced at college (senior high school) level (i.e. Table 3) students do not readily appreciate how it can be an intermediate form of bonding, and tend to see it as *a type of* covalent bonding (i.e. Table 4).

Table 3. A continuous bonding ontology

Type of bond	Covalent...........polar...............Ionic
Formed when...electronegativity difference	none...high
Formed by	electrical interactions

Table 4. A modified bonding dichotomy

Type of bond (sub-type)	Covalent pure covalent	polar covalent	Ionic
Formed by	equal electron sharing	unequal electron sharing	electron transfer

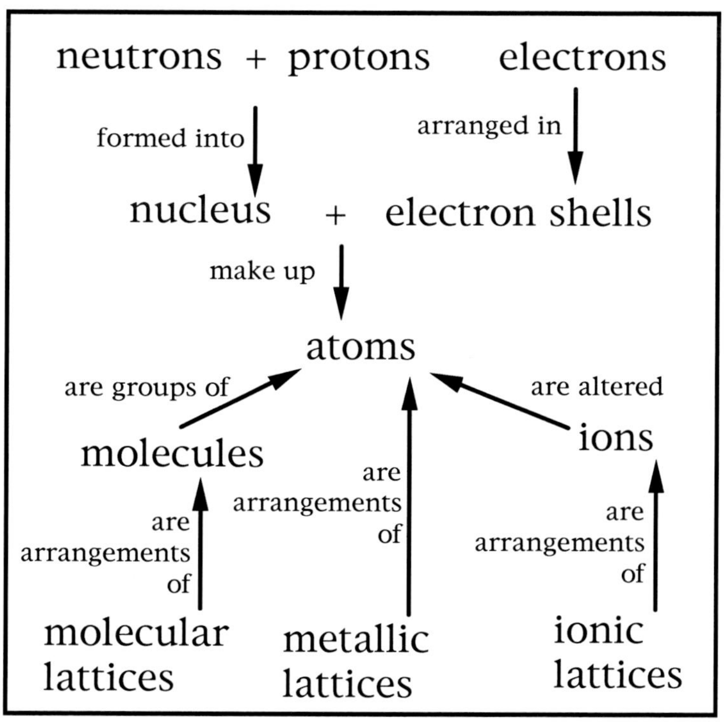

Figure 6. An atomic ontology (Taber, 2001a).

The previous research also shows that students at secondary (junior high) level commonly learn about the molecular world in terms of an 'atomic ontology', i.e. that the basic units of matter are atoms (Taber, 2003b), and other entities are constituent part of atoms (e.g. electrons), altered atoms (i.e. ions), or combinations of atoms (e.g. molecules). This is shown in figure 6:

Whilst this approach certainly represents one way of conceptualising the molecular world, it is not the only possibility. So figure 7 represent an alternative way of thinking about matter at this scale that is more consistent with the range of concepts introduced at college (senior high) level study.

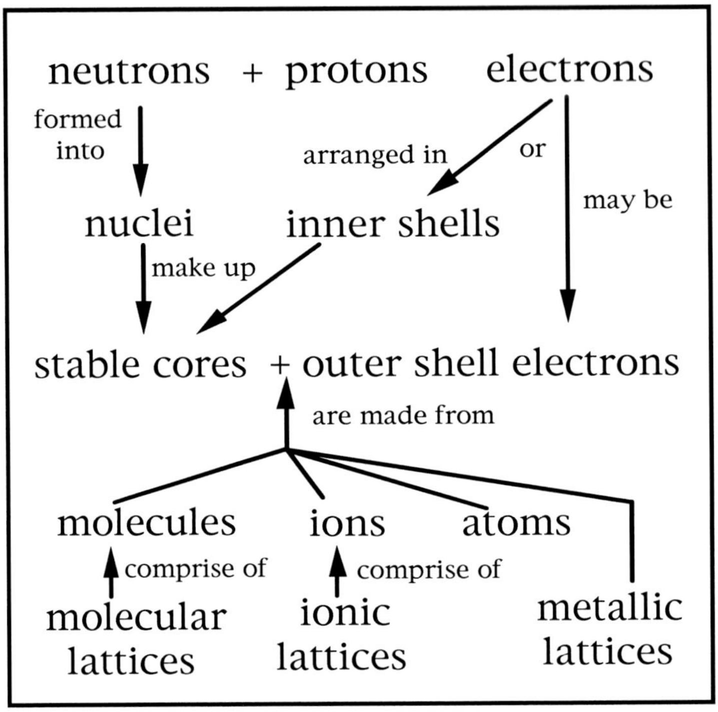

Figure 7. An alternative ontology of the molecular realm (Taber, 2001a).

This research suggested that some student learning difficulties in college chemistry derive from rigidly conceptualising the molecular model of the world in terms of an atomic ontology (figure 6) which channels students' thinking in particular ways. However, the 'atomic ontology', seems to have intrinsic psychological appeal (presumably because it is able to offer '*a* building block' of matter), and once acquired seems to act as an impediment to learning about some aspects of college chemistry (Taber, 2003b). Once the ontological classification is in place, it seems to be difficult to amend.

Modelling Student Ontologies

One way of modelling how individuals represent concepts within cognitive structure is in terms of tree diagrams, where subordinate concepts are shown as branches of superordinate concepts. Within such a model changes in cognitive structure may be represented as additions or deletions to a tree, or branches moving from one tree to another.

Thagard (1992) has used a computer model of this type, where the evolution of conceptual structures is governed by increases in explanatory coherence. Thagard suggested that major reorganisations of cognitive structure, conceptual revolutions, could not take place by a sudden radical restructuring, but rather that alternative structures may develop in parallel, so that a revolution involved the point where the habitually used conceptual framework no longer had as much explanatory coherence as an alternative which had been developing 'in the background'.

Although storing multiple versions of reality may initially seem inefficient, there are good reasons to think otherwise. Much of human cognition is based upon developing knowledge by metaphor and analogy (Lakoff & Johnson, 1980) and Karmiloff-Smith (1994) has argued that a major advance in human mental evolution took place when our ancestors developed the ability to make copies of mental representations from one domain and apply them elsewhere ('representational redescription'). Calvin's (1997) notion of cognition involving the cloning of spatio-temporal patterns of brain activity across the 'patchwork quilt' of the cortex also makes use of a similar principle.

Thagard's model uses a branching tree model of conceptual structure. The work of Chi and co-workers (1992, 1994) uses this type of branching representation within distinct conceptual trees to explore the idea that some of the alternative conceptions elicited from learners may be considered as categorisation errors. For Chi, conceptual change means a change in categorical status - i.e. a change in the branching of cognitive structure.

Chi suggests that there are fundamentally distinct concept trees for different basic ontological categories, so that, for example, matter, processes and mental states need to be considered as different trees. According to Chi, some common learning difficulties in science may be explained in terms of learners assigning a concept to the wrong basic ontological category, what Chi (1992: 133) refers to as "a mismatch between the intrinsic and the psychological ontology". Chi suggests that when learners are introduced to such concepts as heat and electrical current they class them (in their 'psychological ontology') with material substances. From the scientific perspective (the 'intrinsic ontology') these concepts are actually very different types of entity to material substance - Chi uses the term 'constraint-based events' – so learners draw inappropriate inferences about the nature and properties of the entities.

Chi and her colleagues argue that whereas conceptual change *within* an ontological tree may be more or less difficult, conceptual change across different trees does not occur. Rather, another version of the concept has to develop independently - thus perhaps explaining why 'life-world' notions should be retained when the scientific alternatives have been acquired (Solomon, 1993).

As in Thagard's model, Chi's analysis suggests that "new conceptions on an ontologically distinct tree can be developed *gradually*, and yet the final outcome of the development (the shift) may appear to occur *abruptly*" (1992: 134). Thus, according to Chi and colleagues, learners have particular difficulty learning the scientific version of a concept if it has a different ontological status to their own initial categorisation (Chi *et al.*, 1994).

Forming an Orbital Concept from the Existing Ontology

In the research discussed here it was found that learners readily acquired the concept label of 'orbital' although they initially saw this as an electron trajectory (Taber, 2002a). Students entered study at this level with an existing concept of electron shells, and developing the notion that shells are divided into sub-shells, which are divided into orbitals, can be modelled as adding branching to the existing concept tree. However, where a students' existing concept of electron shell consisted of a set of electrons orbiting the nucleus, then any notion of 'orbital' understood as a branch from the existing 'shell' concept would be a very different sort of entity to the 'orbital' target concept. (This example is discussed in more detail below.)

The Analogy of Spin

One of the ways that quanticles such as electrons are different from the particles of everyday experience is that they can have *intrinsic* angular momentum, a property which is known as quantum-mechanical spin (or just 'spin'). This property of spin is important at college level in chemistry, so - according to the Pauli exclusion principle - two electrons can only occupy the same orbital if they are 'spin-paired' (i.e. have 'opposite' spin).

It was found in the source research that students tended to transfer associations of movement to the term spin (Taber, 2002a). So that one of the students, Edward had read that electrons were "spinning on their axes", and he *assumed* that the electron spin direction meant,

> "that an electron moves about this volume of space that's called an orbital in *one particular direction*, whereas the other moves in *the opposite direction*"

Another student, Quorat, explained that she thought that the spin was caused by the electrical repulsion,

> "because they're all going to be repelling each other and circling, always trying [sic] to get as far apart, 'cause that's why they're always spinning."

These students associated the term 'spin' with the macroscopic phenomenon of that name, of which a key feature is movement, rather than recognising spin as "the intrinsic angular momentum of a subatomic particle, nucleus, atom, or molecule, which continues to exist even when the particle comes to rest" (Lafferty and Rowe, 1994, p.556).

This finding may be represented in terms of ontological trees, for the scientist sees the spin of an electron as an *inherent property* (like mass), whereas the students see it, more akin to velocity, a *contingent property*, as the consequence of a process the electron is currently undergoing.

So figure 8 shows how the electron is presented in the curriculum, where the (rest) mass, charge and spin are fundamental properties of an electron, and where the energy assigned to an electron is contingent upon the location of the electron (i.e. the energy does not reside in the electron itself, but is associated with the configuration).

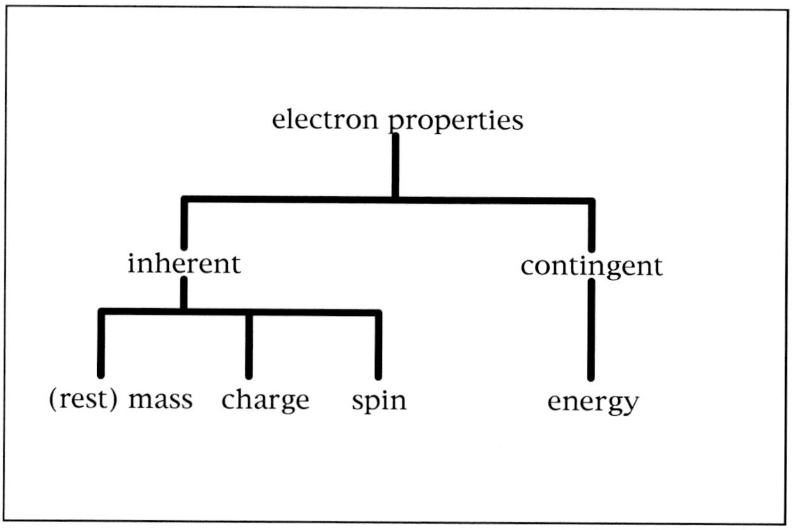

Figure 8. Properties of electrons.

However, students may conceptualise spin as something that is contingent upon the electron's motion in an orbital, rather than being a fundamental property of the electron itself (see figure 9).

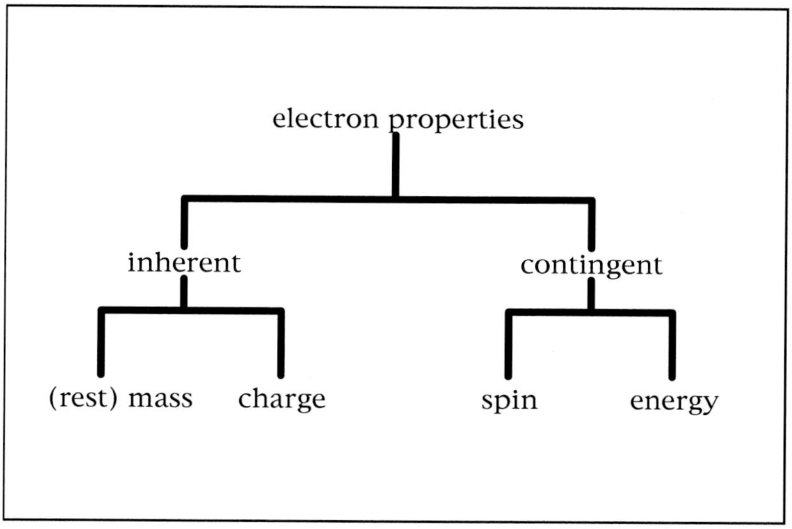

Figure 9. Student perception of properties of electrons.

This misclassification would seem to be similar (albeit in an inverse sense) to the situation where students identify potential energy as located in an object, seeing it as a inherent property of the object rather than better associated with a system configuration. So, for example, fuels are often thought to 'contain' energy (e.g. Brook, 1985).

The shift required here (moving 'spin' from the *contingent* to the *inherent* branch of the figure) does not seem very drastic when represented in simple diagrams, but, if these different types of property are ontologically very different, conceptual change may be very difficult. In terms of Chi's work, an intrinsic property of the electron would seem to be part of the 'matter' tree, where a contingent property would seem to be tied to a 'process'. Chi's (and Thagard's) research would suggest that the teacher's response to this learning difficulty would need to involve building up a new concept (of quantum-mechanical spin), rather than attempting to *shift* the learner's existing meaning for electron spin. More classroom-based research is needed to explore the importance of such 'mis-classifications', and the efficacy of strategies informed by the student ontology perspective.

Converging Lenses: Complementary Perspectives

The three perspectives discussed above provide insights into the learning difficulties faced by students when meeting the orbital model of atomic and molecular structure. Such 'analytical pluralism' is considered appropriate where the analytical lenses applied derive from consistent theoretical perspectives. Given this, the different perspectives can offer *complementary* ways of understanding research data. For example, the three analytical lenses applied here would all suggest reasons why students might have difficulty developing an appropriate understanding of polar bonding (an example from previous research discussed above).

To appreciate polar bonding as an intermediate form of bonding between pure covalent and pure ionic bonding the student has to understand the covalent—ionic distinction as a different sort of classifying entity: a continuum rather than a dichotomy. This is difficult where the student has established prior learning that acts as a learning impediment. Had the existing concepts of covalent and ionic bonding been understood as models – partial, simplified, of limited range of application, etc. – this transition might not be too challenging. However, the research shows that students often begin college (senior high) level studying with strong commitments to 'knowledge' that *bonding is covalent or ionic*, and that covalent bonding *is* electron sharing and ionic bonding *is* electron transfer. This is reflected in the way learners often attempt to make sense of metallic bonding in terms of covalent and ionic prototypes (Taber, 2003c).

Converging Lenses on the Orbital Model

The analytical lenses discussed can offer complementary insights in the present study. The analytical lens of student ontologies suggests that quantum-mechanical spin is something *fundamentally* different to 'life-world' spin, suggesting a possible explanation for the finding that students did indeed struggle to acquire the scientific concept, and so suggests a specific focus for future research into improving teaching practise.

This is not the only area where the notion of student ontologies might illuminate learning difficulties in this topic. Just as the shift in the atom concept (from an immutable, fundamental and indivisible particle of matter, to a system of particles that can be changed in interactions with other particles) is potentially problematic (Taber, 2003b), so is the transition from electron orbits, to orbitals.

In the model learnt in school a discrete particle moves around the nucleus along a particular trajectory. With the introduction of the quantum model of the atom, the electrons cease to be particles in the familiar sense, and can no longer even be considered to have a definite position at any moment in time. The quantum model of the electron in an atom describes a very *different sort of entity* to the notion of an electron that is familiar to students from their secondary education.

In the source study being considered here it was found that some of the students readily adopted the term ‘orbital’, but initially applied this to their existing notion of electrons in shells around the atomic nucleus (Taber, 2002a). Electrons were said to “go round, like in orbitals, or in spherical things” and an orbital was described as “the *path* the electron takes around the nucleus” as it “*circles* the nucleus in a sphere”. Acquiring the new vocabulary did not necessarily imply acquiring the intended new meaning. That the term *orbital* is so similar to *orbit* could well be significant for the meaning learners adopt (Schmidt, 1991) – i.e. a type of substantive learning impediment grounded in a linguistic cue (see http://people.pwf.cam.ac.uk/kst24/ScienceLearningDoctors.html).

The tendency for students to continue to think about atoms in terms of a ‘planetary-type’ orbit model has also been found in other studies (e.g. Cros et al. 1986, 1988; Mashhadi, 1994; Petri & Niedderer, 1998). That this is a common difficulty is not surprising in view of the analytical perspectives considered in this chapter. All three of the analytical lenses discussed would suggest reasons for this phenomenon.

Prior learning (‘e.g. electron orbits’) can act as an impediment to new learning (cf. Tsaparlis 1997). Students are asked to adopt a new model of the atom, which in many ways is inconsistent with the model they have already learnt, and which includes features that are ontologically very different from the familiar model. If the orbital concept is acquired as a subdivision of an existing simplistic notion of the electron shell then the acquired concept will be ontologically unlike the intended target concept (i.e. electrons should no longer to be seen as particles in orbit, but should be seen as having wave characteristics, and being considered either in terms of probability, or modelled as smeared-out into clouds of charge density).

Not only this, but this new very different model is used to complement the existing one: i.e. even after they have been taught about the quantum model of the atom there will be many contexts when electrons will continue to be discussed as if classical particles (e.g. where electron movement is represented with curly arrows in reaction mechanisms). Clearly, this is an area where a sophisticated appreciation of the nature of models in science would be helpful (cf. Jones, 1991).

Discussion - Analytical Plurality versus Methodological Purity

At the outset of this chapter it was suggested that the analysis of data relating to complex phenomena, such as learning in science, might benefit from the apparently eclectic approach of applying a range of analytical lenses as perspectives to interrogate the data. The science education research literature offers a range of potential insightful perspectives, and appropriate analytical frameworks may even be selected *in the light of* the research data, as long as the approaches used are congruous with the philosophical assumptions underlying the research, and with each other.

‘Qualitative research’, that is, interpretive research based upon the collection and analysis of primarily qualitative forms of data (Taber, 2007), is judged upon different criteria to 'quantitative research' (Eybe & Schmidt, 2001). In quantitative research it is normally considered good practice to design a research study so that features of data collection (sampling procedure, sample size) are chosen on the basis of the particular statistical tests to be applied: and specific hypotheses, and confidence levels (e.g. $p \leq 0.05$ for statistical significance) are part of the research design. Qualitative research often follows a more iterative procedure, where some methodological decisions may justifiably be made as the research progresses (e.g. Taber, 2000b).

As a narrative and rhetorical device, the term 'battery of tests' was introduced - a notion that was familiar from science, medicine and psychology, and might be considered by readers to signify an approach that is methodologically sound. However, this begs the question of whether the different tests said to work together in a battery *can* be considered to be part of a consistent approach in this sense. Within analytical chemistry, for example, it is clear that the techniques commonly used *are* congruous, as they are all underpinned by a coherent and consistent framework of theory. So, for example, there is no inconsistency between the theory that links aspects of chemical structure to (a) chromatographic properties and (b) spectroscopic properties.

If the approach discussed in the present chapter is to be seen as justified in the same way, then it is necessary for it to meet the same type of criteria. In others words it is important that any analytical tools applied to data are seen to be consistent with the methodological assumptions guiding the study.

In this chapter three different tools have been selected from the expanding repertoire of analytical frameworks that have been adopted or developed for use within science education research. Data from a published research study (Taber 2002a, b) were subjected to inspection from each of these perspectives, and the present chapter presents examples of how these different approaches can illuminate the nature and origins of some of the students' learning difficulties reported in the source research.

It is argued that these three approaches are consistent, both among themselves and also with the conceptual framework unpinning the study. The research was undertaken within the 'constructivist' tradition (Taber, 2006a), informed by notions such as the importance to learning of existing cognitive structure; the unique and personal nature of each individual's conceptual frameworks; and the potential significance of learners' alternative conceptions.

Although an information-processing perspective, drawing upon research from cognitive science, would normally be seen as a separate tradition, there is a clear link between notions of working memory capacity, and of consolidation of learning allowing chunking, with Ausubel's theory of meaningful learning (e.g. Ausubel, 2000), and this has informed the typology of learning impediments (Taber, 2001b, 2005). The link between these ideas and Piagetian notions of accommodation and equilibration was also noted above.

Part of the research into learners' ideas deriving from the constructivist tradition has led to studies into conceptual change and the status of learners' ideas (e.g. Hewson & Hennessey, 1992; Strike & Posner, 1992; Taber, 2001d), and that may be seen as one source of the current interest in students' understanding of models. In science we are looking for intelligible and feasible ideas - rather than articles of faith. Yet learners often take teaching models as facts. There is a connection here with neoPiagetian ideas about post-formal thinking which enables people to accept views of the world as incomplete - as partial models that can be complemented by alternative perspectives (Arlin, 1975; Castro & Fernández,

1987), and in particular Perry's (1970) scheme of intellectual development among college students.

The work on students' understanding of models also links with Chi's ideas about ontological category errors. Just as students may mis-classify processes (e.g. heating) as objects, or intrinsic properties as contingent properties - as in the example of electron spin discussed here - they may also mis-classify scientific models as realistic representations.

Of course the types of representation of cognitive structure, as a hierarchical system, used by some workers, such as Chi and Thagard, may not be an ideal model (cf. Taber, in press). Ogborn's approach also offers useful insights, as does Kelly's (1963) personal construct system. Seeing cognitive structure as a kind of semantic network (diSessa & Sherin, 1998) also has advantages. Each of these ways of conceptualising aspects of learners' ideas may present difficulties when we wish to accommodate certain reported phenomena that we might consider of significance: phenomena such as alternative frameworks; Vygotsky's (1986/1934) notion of the interaction of spontaneous and scientific concepts; gestalts (Andersson, 1986); phenomenological primitives (Hammer, 1996), etc.

This brings the argument full circle. Phenomena such as cognitive structure and student learning are obviously much too complex, too multifaceted, to be comprehensively described by simple models. Research into *students'* understanding of scientific models provides a limited perspective on student difficulties in learning. A typology of learning impediments will include gross simplifications and may force complex learning difficulties into a straightjacket of a limited system of categories. Hierarchical branching trees cannot surely *fully reflect* the way concepts are represented in our minds.

Yet this is not a reason to dismiss these perspectives. When faced with complex phenomena we turn to complementary partial models that can illuminate aspects of the phenomena being studied. The key question is not whether these particular tools provide a *complete explanation* - for example of student difficulties in learning about quantum models of matter - they clearly do not. The important questions are (a) whether these perspectives are congruous in the sense of being consistent with the theoretical position underlying the research, and (b) whether they have useful heuristic value.

The curriculum focus discussed - quantum models of atoms and molecules - is known to be a problematic one that would benefit from a better informed pedagogy. It is recognised that this topic is highly abstract, divorced from life-world experience (Knight, 2002), and this clearly makes it difficult for many students. Such an observation is, however, just a starting point for informing specific teaching approaches designed to help students understand the topic. A more detailed investigation of exactly where and why student learning 'breaks-down', or 'goes astray', is needed to advise teachers how to respond to the inherent difficulty of the topic.

The present chapter provides interpretations that can contribute to such a programme:

- *A better understanding of the nature and role of models in science* might help students develop more flexible conceptualisations that are more readily modified or augmented. In particular, teachers could put more emphasis on the partial and provisional nature of the models and representations used (Taber, 2006c).
- *The typology of learning impediments* has been used to identify some specific aspects of the topic where learners interpret teaching through alternative conceptual frameworks; and where the students do not have, or do not recognise the relevance

of, prerequisite knowledge (or where prior learning is not sufficiently well consolidated to be applied in new contexts without overloading working memory). One possible response is to reorganise teaching schemes to provide students with the time needed to reinforce and consolidate each new concept before it is to act as the basis of further learning.

- *The 'ontological discontinuities' that can restrict progression in developing understanding* - i.e. the way that concepts such as electronic orbital and electron spin are different kinds of entities to those already represented in the learner's cognitive structure that are most likely to be used as the anchoring prior knowledge (e.g. electron 'orbits', life-world spin that *entails* movement). The work of Chi (1992) and Thagard (1992) would suggest that here teachers may sometimes need to build-up new understandings rather than allow students to develop their existing ideas.

Conclusion

This chapter does not claim to present a comprehensive account of learning difficulties in this topic. I have drawn upon a single source study that provided suitable data to demonstrate how analytical pluralism might be applied in science education. The source study provided the in-depth data needed for this type of analysis, but only reported findings from a specific group of students in a particular educational context. Any attempt to provide a more definitive analysis of students' difficulties in the topic would certainly need to draw upon a larger and more representative sample, and could surely benefit from the application of a broader battery of analytical lenses. Nevertheless, the present chapter demonstrates how a range of complementary analytical lenses that draw upon congruous theoretical perspectives can be applied to interrogate a data set, leading to recommendations for developing pedagogy at the level of a science topic.

These recommendations are specific enough to inform teaching. They should, of course, be seen as hypotheses to be tested in practice – as the source of research questions for a further cycle of research. The ability to draw such conclusions demonstrates the heuristic value of the analytical tools used in this study. Just as Bohr's mermaid, his model of the atom, was open to criticism and refinement, the qualitative analytical tools currently available to science education researchers (such as the examples selected for this study) are not able to provide a complete and incontrovertible analysis of students' learning difficulties. They provide limited and provisional perspectives, but - like Bohr's mermaid - they provide avenues that can be tested and explored in further research.

References

Allinger, N. L., Cava, M. P., de Jongh, D. C., Johnson, C. R., Lebel, N. A. & Stevens, C. L. (1976) *Organic Chemistry* (2nd edition), New York: Worth Publishers.

Anastasi, A. (1982) *Psychological Testing* (5th edition), New York: Macmillan Publishing.

Anderson, J. R. (1995) *Learning and Memory: an integrated approach*, New York: John Wiley & Sons.

Andersson, B. (1986) The experiential gestalt of causation: a common core to pupils' preconceptions in science, *European Journal of Science Education*, **8** (2), pp.155-171.

Andrew, J., & Rispoli, P. (1999) Chemistry in Focus (2nd Edition), London: Hodder & Stoughton.

Arlin, P. K. (1975) Cognitive development in adulthood: a fifth stage?, *Developmental Psychology*, **11** (5), pp.602-606.

Atkins, P.W. (1974) *Quanta: a handbook of concepts*, Oxford: Oxford University Press.

Ausubel, D. P. & Robinson F. G. (1971) *School Learning: An Introduction to Educational Psychology*, London: Holt International Edition.

Ausubel, D. P. (2000) *The Acquisition and Retention of Knowledge: a cognitive view*, Dordrecht: Kluwer Academic Publishers.

Baddeley, A. (1986) *Working Memory*, Oxford: Oxford University Press.

Baddeley, A. (1990) *Human Memory: Theory and practice*, Hove, East Sussex: Lawrence Erlbaum Associates.

Beck, W. S., Liem, K. F. & Simpson, G. G. (1991) *Life: an introduction to Biology* (3rd edition), New York: HarperCollins.

Bloom, B. S. (1964) The cognitive domain, in Clark, L. H. (1968) *Strategies and Tactics in Secondary School Teaching: a book of readings*, London: MacMillan, pp.49-55.

Brook, A. (1985), Children's understanding of ideas about energy: a review of the literature, in Driver, R. & Millar, R., *Energy Matters*, Leeds, U.K.: Centre for Studies in Science and Mathematics Education.

Buddle, M., Niedderer, H., Scott, P. & Leach, J. (2002) 'Electronium': a quantum atomic teaching model, *Physics Education*, **37** (3), pp.197-203.

Buddle, M., Niedderer, H., Scott, P. & Leach, J. (2002) The quantum atomic model 'Electronium': a successful teaching tool, *Physics Education*, **37** (3), pp.204-210.

Calvin, W. H. (1997) *How Brains Think: evolving intelligence, then & now*, London: Weidenfeld & Nicolson.

Caravita, S. & Halldén, O. (1994) Re-framing the problem of conceptual change, *Learning and Instruction*, **4**, pp.89-111.

Carr, M. (1984) Model confusion in chemistry, *Research in Science Education*, **14**, 1984, pp.97-103.

Castro, E. A. & Fernández, F. M. (1987) Intellectual development beyond formal operations, *International Journal of Science Education*, **9** (4), pp.441-447.

Carter, R. (1998) *Mapping the Mind*, London: Orion Books.

Cervellati, R. & Perugini, D. (1981) The Understanding of the Atomic Orbital Concept by Italian High School Students, *Journal of Chemical Education*, **58** (7), July 1981, pp. 568-569.

Chi, M. T. H. (1992) Conceptual change within and across ontological categories: examples from learning and discovery in science, in R. N. Giere, (Ed.) (1992) *Cognitive Models in Science*, Minneapolis: University of Minnesota Press, pp.129-186.

Chi, M. T. H., Slotta, J. D. & de Leeuw, N. (1994) From things to processes; a theory of conceptual change for learning science concepts, *Learning and Instruction*, **4**, pp.27-43.

Clugston, M., & Flemming, R. (2000) *Advanced Chemistry*, Oxford: Oxford University Press

Cohen, L., Manion, L. & Morrison, K. (2000) *Research Methods in Education* (5th Edition), London: RoutledgeFalmer.

Cohen, R. J., Swerdlik, M. E. & Philips, S. M. (1996) *Psychological Testing and Assessment: an introduction to tests and measurement* (3rd edition), Mountain View, California: Mayfield Publishing.

Cros, D., Amouroux, R., Chastrette, M., Fayol, M., Leber, J. and Maurin, M. (1986) Conceptions of first year university students of the constitution of matter and the notions of acids and bases, *European Journal of Science Education*, **8** (3), pp.305-313.

Cros, D., Chastrette, M. and Fayol, M. (1988) Conceptions of second year university students of some fundamental notions in chemistry, *International Journal of Science Education*, **10** (3), pp.331-336.

Dawson, M. R. W. (1998) *Understanding Cognitive Science*, Oxford: Blackwell Publishers.

diSessa, A. A. & Sherin, B. L. (1998) What changes in conceptual change? *International Journal of Science Education*, **20** (10), pp.1155-1191.

Driver, R. (1989) Students' conceptions and the learning of science, *International Journal of Science Education*, **11** (special issue), pp.481-490.

Driver, R., Leach, J., Millar, R., & Scott, P. (1996) *Young People's Images of Science*, Buckingham: Open University Press.

Driver, R., Squires, A., Rushworth, P. & Wood-Robinson, V. (1994) *Making Sense of Secondary Science: research into children's ideas*, London: Routledge.

Driver, R. & Oldham, V. (1986) A constructivist approach to curriculum development in science, *Studies in Science Education*, **13**, pp.105-122.

Duit, R. (1991) *Students' conceptual frameworks: consequences for learning science*, Chapter 4 in Glynn, S.M., Yeany, R.H. & Britton, B.K., *The Psychology of Learning Science*, Hillsdale, New Jersey: Lawrence Erlbaum Associates, pp.65-85.

Duit, R. (2007) *Bibliography - Students' and Teachers' Conceptions and Science Education*, available from http://www.ipn.uni-kiel.de/aktuell/stcse/stcse.html. accessed 28/03/2007

Feynman, R. P. (1985) *QED: The strange theory of light and matter*, Princeton: Princeton University Press.

Edwards, D. & Mercer, N. (1987) Common Knowledge: The development of understanding in the classroom, London: Routledge.

Elliott, J. (1991) *Action Research for Educational Change*, Milton Keynes: Open University Press.

Ellis, H. C. & Hunt, R. R. (1989) *Fundamentals of Human Memory and Cognition* (4th edition), Dubuquw, Iowa: William C. Brown Publishers.

Ely, M., Anzul, M., Friedman, T., Gardner, D. & Steinmetz, A. M. (1991) *Doing Qualitative Research: Circles within circles*, London: RoutledgeFalmer.

Erickson, G. (2000) Research programmes and the student science learning literature. In R. Millar, J. Leach & J. Osborne (Eds.), *Improving Science Education: the contribution of research*, Buckingham, UK: Open University Press.

Eybe, H. & Schmidt, H.-J. (2001). Quality criteria and exemplary papers in chemistry education research, *International Journal of Science Education*, **23** (2), pp.209-225.

Feyerabend, P. (1988) *Against Method* (Revised edition), London: Verso.

Finar, I. L. (1973) *Organic Chemistry, Volume 1: The Fundamental Principles* (6th edition), London: Longman.

Fischler, H. & Lichtfeldt M. (1992) Modern physics and students' conceptions, *International Journal of Science Education*, **14** (2), pp.181-190.

Gauld, C. (1989) A study of pupils' responses to empirical evidence, chapter 3 of R. Millar, (Ed.), *Doing Science: images of science in science education*, London: The Falmer Press, pp.62-82.

Gallimore, R. & Tharp, R. (1990) Teaching mind in society: teaching, schooling, and literate discourse, in Moll, J. C. (Eds.), *Vygotsky and Education: instructional implications and applications of sociohistorical psychology*, Cambridge: Cambridge University Press, pp.175-205.

Gergen, K. J. (1999) *An Invitation to Social Construction*, London: SAGE Publications.

Gilbert, J. K. (1998) Explaining with models, in Ratcliffe M. (ed.) *ASE Guide to secondary science education* (London: Stanley Thornes), pp.159-166.

Gilbert, J. K., Boutler, C. & Rutherford, M. (1998) Models in explanations, Part 1: Horses for courses?, *International Journal of Science Education*, **20** (1), pp.83-97.

Gilbert, J. K. and Watts, D. M. (1983) Concepts, misconceptions and alternative conceptions: changing perspectives in science education, *Studies in Science Education,* **10**, pp.61-98.

Glaser, B. G. & Strauss, A. L. (1967) *The Discovery of Grounded Theory: strategies for qualitative research*, New York: Aldine de Gruyter.

Greenfield, S. (1997), *The Human Brain: A Guided Tour*, London: Weidenfeld & Nicolson.

Grosslight, L ., Unger, C., Jay, E. & Smith, C. L. (1991) Understanding models and their use in science: conceptions of middle and high school students and experts, *Journal of Research in Science Teaching* **28** (9), pp.799-822.

Hammer, D. (1996) Misconceptions or p-prims: how might alternative perspectives of cognitive structure influence instructional perceptions and intentions?, paper presented at the conference of the *American Educational Research Association*, April 1996.

Harrison, A. G. & Treagust, D. F. (1996) Secondary students' mental models of atoms and molecules: implications for teaching chemistry, *Science Education*, **80** (5), pp.509-534.

Harrison, A. G. & Treagust, D. F. (2000a) A typology of school science models, *International Journal of Science Education*, **22** (9), pp.1011-1026.

Harrison, A. G. & Treagust, D. F. (2000b) Learning about atoms, molecules, and chemical bonds: a case study of multiple-model use in grade 11 chemistry, *Science Education*, **84**, pp.352-381.

Hennessy, S. (1993) Situated cognition and cognitive apprenticeship: implications for classroom learning, *Studies in Science Education*, **22**, pp.1-41.

Hewson, M. G. A'B. (1985) The role of intellectual environment in the origins of conceptions: an exploratory study, Chapter 10 in West, L. H. T. & Pines, A. L. (eds.), *Cognitive Structure and Conceptual Change*, London: Academic Press Inc., pp.153-161.

Hewson, P. W. & Hennessey, M. G. (1992) making status explicit: a case study of conceptual change, in Duit, R., Goldberg, F. & Niedderer, H., (Eds.) (1992) *Research in Physics Learning: theoretical issues and empirical studies*, Kiel: I.P.N. (Institut für die Pädagogik der Naturwissenschraften), pp.176-187.

Hitchcock, G., & Hughes, D. (1989) *Research and the Teacher: a qualitative introduction to school-based research*, London: Routledge, 1989.

Hopkins, D. Harris, A., Watling, R. & Beresford, J. (1999) From inspection to school improvement?: Evaluating the accelerated inspection programme in Waltham Forest, *British Educational Research Journal*, **25** (2), pp.679-690.

Jensen, W. B. (1995) *Logic, History and the Teaching of Chemistry*, text of the Keynote Lectures, given at the 57th Annual Summer Conference of the New England Association of Chemistry Teachers, Sacred Heart University, Fairfield, Connecticut.

Johnson, P. (1998) Progression in children's understanding of a 'basic' particle theory: a longitudinal study, *International Journal of Science Education*, **20** (4), pp.393-412.

Johnson-Laird, P. N. (1983) *Mental Models*, Cambridge: Cambridge University Press.

Johnston, K. & Driver, R. (1991) *A Case Study of Teaching and Learning about Particle Theory: a constructivist reaching scheme in action*, Children's Learning in Science Project, Leeds: Centre for Studies in Science and Mathematics Education, University of Leeds.

Johnstone, A. H. (1989) Some messages for teachers and examiners: an information processing model, *Research in Assessment VII: Assessment of Chemistry in Schools*, London: Royal Society of Chemistry Education Division, pp.23-39.

Johnstone, A. H. (1991) Why is science difficult to learn? Things are seldom what they seem, *Journal of Computer Assisted Learning*, **7**, pp.75-83.

Jones, D. G. C. (1991) Teaching modern physics - misconceptions of the photon that can damage understanding, *Physics Education*, **26** (2), pp.92-98.

Justi, R. & Gilbert, J. (2000) History and philosophy of science through models: some challenges in the case of 'the atom', *International Journal of Science Education*, **22** (9), pp.993-1009.

Justi, R. S. & Gilbert, J. K. (2002a) Modelling, teachers' views on the nature of modelling, and implication for the education of modellers, *International Journal of Science Education*, **24** (4), pp.369-387.

Justi, R. S. & Gilbert, J. K. (2002b) Science teachers' knowledge about and attitudes towards the use of models and modelling in learning science, *International Journal of Science Education*, **24** (12), pp.1273-1292.

Karmiloff-Smith, A. (1994) Précis of Beyond Modularity: A developmental perspective on cognitive science, *Behaviour and brain Sciences*, **17**, pp.693-745.

Kelly, G. (1963) *A Theory of Personality: The Psychology of Personal Constructs*, New York: W. W. Norton & Company.

Kellogg, R. T. (1995) *Cognitive Psychology*, London: Sage.

Kind, V. & Taber, K. S (2005) *Science: Teaching School Subjects 11-19*, London: Routledge.

Knight, R. D. (2002) *Five Easy Lessons: Strategies for successful physics teaching*, San Francisco, CA: Addison Wesley.

Lakatos, I. (1970) Falsification and the methodology of scientific research programmes, in Lakatos, I. & Musgrove, A. (eds.), *Criticism and the Growth of Knowledge*, Cambridge: Cambridge University Press, pp.91-196.

Lakoff, G. & Johnson, M. (1980) The metaphorical structure of the human conceptual system, *Cognitive Science*, **4,** pp.195-208.

Lafferty, P. & Rowe, J. (eds.) (1994) *The Hutchinson Dictionary of Science* (Revised Edition), Oxford: Helicon Publishing Ltd.

Leach, J. & Scott, P. (1995) The demands of learning science concepts - issues of theory and practice, *School Science Review*, **76** (277), pp.47-51.

Lewis, E., & Berry, M. (2000) *AS and A Level Chemistry*, Harlow, U.K.: Prearson Education.

Lijnse, P. L., Licht, P, de Vos, W. & Waarlo, A. J. (1990) *Relating Macroscopic Phenomena to Microscopic Particles: a central problem in secondary science education*, Utrecht: Centre for Science and Mathematics Education, University of Utrecht: CD-ß Press.

Lovelock, J. E. (1987) *Gaia: a new look at life on earth*, Oxford: Oxford University Press (first published, 1979).

Mariani, M. C. & Ogborn, J. (1991) Towards an ontology of common-sense reasoning, *International Journal of Science Education*, **13** (1), pp.69-85.

Mariani, M. C. & Ogborn, J. (1995) The ontology of physical events: a comparison of two groups, *International Journal of Science Education*, **17** (5), pp.643-661.

Mashhadi, A., Advanced level Physics students' understanding of quantum physics, paper presented to the *British Educational Research Association* Annual Conference, Oxford, September 1994.

Matthews, M. R. (1994) *Science Teaching: The role of history and philosophy of science*, London: Routledge.

McNiff, Jean (1992) *Action Research: principles and practice*, London: Routledge.

Miller, G. A. (1968) The magical number seven, plus or minus two: some limits on our capacity for processing information, in Miller, G. A., *The Psychology of Communication: seven essays*, Harmondsworth: Penguin, pp.21-50.

Niaz, M. (1998) From cathode rays to alpha particles to quantum of action: a rational reconstruction of structure of the atom and its implications for chemistry textbooks, *Science Education*, **82** (5), pp.527-552.

Nicholls, G. & Ogborn, J. (1993) Dimensions of children's conceptions of energy, *International Journal of Science Education*, **15** (1), pp.73-81.

Nola, R. (1998) Constructivism in Science and Science Education: A philosophical critique, in M. R. Matthews (ed.), *Constructivism in Science Education*, Dordrecht: Kluwer Academic Publishers.

OFSTED (2003) *How we inspect state schools*, Office for Standards in Education, http://www.ofsted.gov.uk/, accessed February 2003.

Ogborn, J., Kress, G., Martins, I. & McGillicuddy, K. (1996), *Explaining Science in the Classroom*, Buckingham: Open University Press.

Parkin, A. J. (1987) *Memory and Amnesia: An Introduction*, Oxford: Basil Blackwell.

Perry, W. G. (1970) *Forms of Intellectual and Ethical Development in the College Years: A Scheme*, New York: Holt, Rinehart, and Winston.

Pescar, S. C. & Nelson, C. A. (1994) *The Wordsworth Medical Companion*, Ware: Wordsworth Editions Ltd.

Petri, J. & Niedderer, H. (1998). A learning pathway in high-school level quantum atomic physics, *International Journal of Science Education*, **20**, pp.1075-1088.

Petruccioli, S. (1993) *Atoms, Metaphors and Paradoxes: Niels Bohr and the construction of a new physics*, Cambridge: Cambridge University Press.

Pfundt, H., and Duit, R. (1994) *Bibliography: Students' alternative frameworks and science education*, IPN Reports in Brief, Germany: University of Kiel.

Platts, C. V. (1968) Structure models in chemistry teaching, *School Science Review*, **50** (171), pp.289-297.

Ponomarev, L. I. (1993) *The Quantum Dice*, London: Institute of Physics.

Pope, M. & Denicolo, P. (1986) Intuitive theories - a researcher's dilemma: some practical methodological implications, *British Educational Research Journal*, **12** (2), pp.153-166.

Pope, M.& Watts, M.(1988) Constructivist goggles: implications for process in teaching and learning physics, *European Journal of Physics*, **9**, pp.101-109.

Popper, K. R. (1989) *Conjectures and Refutations: The Growth of Scientific Knowledge*, 5th edition, London: Routledge.

Sattaur, O. (1987) Cuckoo in the nest, *New Scientist*, 24th December.

Sánchez Gómez, P. J. & Martín, F. (2003) Quantum versus 'classical' chemistry in university chemistry education: A case study of the role of history in thinking the curriculum, *Chemistry Education: Research & Practice*, **4** (2), pp.131-148.

Scerri, E. R. (1989) Transition metal configurations and limitations of the orbital approximation, *Journal of Chemical Education*, **66** (6), pp.481-483.

Scerri, E. R. (1991) The electronic configuration model, quantum mechanics and reduction, *British Journal of Philosophy of Science*, **42**, pp.309-325.

Scerri, E. R. (1998) How good is the quantum mechanical explanation of the periodic system?, *Journal of Chemical Education*, **75** (11), pp.1384-1385.

Scerri, E. R. (2003) Philosophical confusion in Chemical Education Research, *Journal of Chemical Education*, **80** (20), pp.468-474.

Schmidt, H.-J. (1991) A label as a hidden persuader: chemists' neutralization concept, *International Journal of Science Education*, **13** (4), pp.459-471.

Schwandt, T. A. (2001) *Dictionary of Qualitative Inquiry* (2nd edition), Thousand Oaks, California: Sage Publications.

Scott, P. (1998) Teacher talk and meaning making in science classrooms: a review of studies from a Vygotskian perspective, *Studies in Science Education*, **32**, pp.45-80.

Shiland, T. W., Quantum Mechanics and conceptual change in high school chemistry textbooks, *Journal of Research in Science Teaching*, **34** (5), 1997, pp.535-545.

Solomon, Joan (1993a) Four frames for a field, Chapter 1, in Black P.J. & Lucas, A. M. (eds.), *Children's Informal Ideas in Science*, London: Routledge, pp.1-19.

Solomon, J. (1993b) The social construction of children's scientific knowledge, in Black P. J. & Lucas, A. M. (eds.), *Children's Informal Ideas in Science*, London: Routledge, pp.85-101.

Sousa, D. A. (2001) *How the Brain Learns* (2nd edition), Thousand Oaks, California: Corwin Press.

Strike, K. A. & Posner, G. J. (1992) A revisionist theory of conceptual change, in Duschl, R. A. & Hamilton, R. J. (1992) *Philosophy of Science, Cognitive Psychology, and Educational Theory and Practice*, Albany, N.Y.: State University of New York Press.

Sykes, P. (1986) *A Guidebook to Mechanism in Organic Chemistry* (6th edition), Harlow: Longman.

Taber, K. S. (1998a) An alternative conceptual framework from chemistry education, *International Journal of Science Education*, **20** (5), pp.597-608.

Taber, K. S. (1998b) The sharing-out of nuclear attraction: or I can't think about Physics in Chemistry, *International Journal of Science Education*, **20** (8), pp.1001-1014.

Taber, K. S. (2000a) Multiple frameworks?: Evidence of manifold conceptions in individual cognitive structure, *International Journal of Science Education*, **22** (4), pp.399-417.

Taber, K. S. (2000b) Case studies and generalisability - grounded theory and research in science education, *International Journal of Science Education*, **22** (5), pp.469-487.

Taber, K. S. (2000c) *Molar and molecular conceptions of research into learning chemistry: towards a synthesis,* The Chemical Education Research Group Lecture 2000, available at http://www.rsc.org/Membership/Networking/InterestGroups/ChemicalEducationResearch/Lectures.asp (accessed 14th April 2007)

Taber, K. S. (2001a) Building the structural concepts of chemistry: some considerations from educational research, *Chemistry Education: Research and Practice in Europe*, **2** (2), pp.123-158.

Taber, K. S. (2001b) The mismatch between assumed prior knowledge and the learner's conceptions: a typology of learning impediments, *Educational Studies*, **27** (2), 159-171.

Taber, K. S. (2001c) When the analogy breaks down: modelling the atom on the solar system, *Physics Education*, **36** (3), 222-226.

Taber, K. S. (2001d) Shifting sands: a case study of conceptual development as competition between alternative conceptions, *International Journal of Science Education*, **23** (7), 731-753.

Taber, K. S. (2002a) Conceptualizing quanta - illuminating the ground state of student understanding of atomic orbitals, *Chemistry Education: Research and Practice in Europe*, **3** (2), pp.145-158.

Taber, K. S. (2002b) Compounding quanta - probing the frontiers of student understanding of molecular orbitals, *Chemistry Education: Research and Practice in Europe*, **3** (2), pp.159-173.

Taber, K. S. (2002c) *Chemical misconceptions - prevention, diagnosis and cure* (2 volumes), London: Royal Society of Chemistry

Taber, K. S. (2003a) Lost without trace or not brought to mind? - a case study of remembering and forgetting of college science, *Chemistry Education: Research and Practice*, **4** (3), pp.249-277.

Taber, K. S. (2003b) The atom in the chemistry curriculum: fundamental concept, teaching model or epistemological obstacle?, *Foundations of Chemistry*, **5** (1), pp.43-84.

Taber, K. S. (2003c) Mediating mental models of metals: acknowledging the priority of the learner's prior learning, *Science Education*, **87**, pp.732-758.

Taber, K. S. (2004) Learning quanta: barriers to stimulating transitions in student understanding of orbital ideas, *Science Education*, **89** (1), pp.94-116.

Taber, K. S. (2006a) Beyond Constructivism: the Progressive Research Programme into Learning Science, *Studies in Science Education*, **42**, pp.125-184.

Taber, K. S. (2006b) Constructivism's new clothes: the trivial, the contingent, and a progressive research programme into the learning of science, *Foundations of Chemistry*, **8** (2), pp.189-219.

Taber, K. S. (2006c) Towards a Curricular Model of the Nature of Science, *Science & Education* (published on-line first, 10.1007/s11191-006-9056-4, 40 pages.)

Taber, K. S. (2007) *Classroom-based Research and Evidence-based Practice: A Guide for Teachers*, London: SAGE Publications.

Taber, K. S. (in press) Conceptual resources for learning science: issues of transience and grain-size in cognition and cognitive structure, *International Journal of Science Education*.

Thagard, P. (1992) *Conceptual Revolutions*, Oxford: Princeton University Press.

Treagust, D. F., Chittleborough, G. & Mamialia, T. L. (2002) Students' understanding of the role of scientific models in learning science, *International Journal of Science Education*, **24** (4), pp.357-368.

Tsaparlis, G. (1997). Atomic orbitals, molecular orbitals and related concepts: conceptual difficulties among chemistry students. *Research in Science Education,* **27**, 271-287.

Tsaparlis, G. & Papaphotis, G. (2002) Quantum-mechanical concepts: are they suitable for secondary students?, *Chemistry Education: Research and Practice in Europe*, **3** (2), pp.129-144.

Wheatley, P. J. (1968) *The Determination of Molecular Structure* (2nd edition), Oxford, Oxford University Press.

Williams, D. H. & Fleming, I. (1973) *Spectroscopic Methods in Organic Chemistry* (2nd edition), Maidenhead: McGraw-Hill.

van Driel, J. H. & Verloop, N. (2002) Experienced teachers' knowledge of teaching and learning of models and modelling in science education, *International Journal of Science Education*, **24** (12), pp.1255-1272.

von Glasersfeld, E. (1989) Cognition, construction of knowledge, and teaching, *Synthese*, **80**, pp.121-140.

Vygotsky, L. (1986) *Thought and Language*, London: MIT Press (first published in Russian in 1934).

Ziman, John (1968) *Public Knowledge: an essay concerning the social dimension of science*, Cambridge: Cambridge University Press.

In: Science Education in the 21st Century
Editor: Ingrid V. Eriksson, pp. 107-135

ISBN: 1-60021-951-9

Chapter 3

LEARNING PHYSICS VIA MODEL CONSTRUCTION: ISSUES AND EXPERIMENTAL RESULTS

R.M. Sperandeo-Mineo and C. Fazio
Department of Physics and Related Technologies, University of Palermo, Italy

Abstract

In recent times a considerable interest has been developed, in science education community, in fostering model-based reasoning at all level of schooling. Furthermore, research reports and projects recommended shifting the focus of science education from facts to overarching themes and toward the increasing relevance of models in teaching and learning science. We describe an approach to teaching/learning physics focused at developing pupil reasoning aimed at making sense of their own physical experiences and at evaluating information reported by others. We analyze how modeling procedures are relevant to the process of knowledge construction as well as to the epistemology of physics. Moreover, since models are representations of reality, we describe how to project and develop pedagogical tools able to scaffold the process of constructing different kinds of models: descriptive as well as interpretative and explicative models that concretize abstract ideas, simplify and clarify complex phenomena, predict trends and explain mechanism and processes. We also report the design process and the experimentation of two teaching-learning sequences focusing on the central issue of the relationships between observable phenomena, like macroscopic behaviors of matter, and their interpretation and/or explanation in terms of microscopic structure models. The design process with respect to the general framework of the *Educational Reconstruction Model* and the used pedagogical tools are described. Results, coming from teaching/learning experiments involving high school pupils, are also reported. The main focus is on the pupils' representations of phenomena and on the cognitive strategies put in action in order to modify or support their descriptive and interpretative mental models. Results are discussed from two different points of views: 1) by pointing out the efficacy of strategies focusing on the process of constructing predictive conceptual models and by identifying the concept of 'level of analysis' as different ways to look at the same phenomenon; 2) by analyzing how different kinds of modeling environments can stimulate pupil abilities in building mental models about the dynamical evolution of simple physical systems and improve their formalization.

Introduction

Models are a central topic in contemporary science education with debates focused on including a modeling perspective in science curricula and pragmatic strategies for designing classroom activities that enable pupils to learn about science as a modeling endeavor [1, 2, 3, 4]. Modeling is intended as the process of constructing and using models [5]. Learning physics via model construction should, therefore, involve the development of modeling strategies making pupils able to figure out their own physical experiences and evaluate information reported by others.

At the core of the debates about a modeling approach to learn science, and physics in particular, are questions connected with the different ways in which the term model is intended and used: in particular, how models and modeling procedures are involved in the individual process of knowledge construction as well as how mental models are related to physics models and to their mathematical representations.

In the next section, we analyze as researches in cognitive science, physics and physics education look at the role of models and modeling in the process of knowledge construction. In the second section, we describe a pedagogical approach to modeling for high school physics teaching and, following, we give two examples of experimented teaching/learning sequences in the field of mechanical wave propagation (Section 3) and heat transfer process (Section 4). The last two sections report the main results of our pedagogical experimentation in some high school classrooms and draw some relevant conclusion.

1. Models and Modelling

Many researches on cognitive science identified the relevant role of mental models in the process of individual knowledge construction. Moreover the epistemology of physics identified models as fundamental constructs of the discipline and ascribed to them a relevant role in the process of scientific knowledge construction. This should involve that a modeling approach can support the process of individual construction of scientific knowledge. We analyze this statement by trying to answer to the three questions reported in the following.

1.1. Is Modelling Relevant to the Process of Knowledge Construction?

Cognitive science research has shown that people tend to organize their experience into patterns or mental models [6-7]. Mental model researchers investigated the significant role of models in humans reasoning concerning many and often conflicting purposes: to communicate experiences, to provide descriptions of observed facts and phenomena, to predict what happens under prescribed circumstances, and sometimes to explain why things happen and in which way they do.

Cognitive questions about the nature of mental modeling and its role in human reasoning have been initially proposed by Kenneth Craik [8] that related people reasoning to the carrying out of thought experiments on internal models. The centrality of mental modeling to cognition is, actually, a hypothesis under investigation in many domains: from thinking about causality in physical systems [9] to reasoning with representations of domain knowledge [10], to analogical reasoning [11], to deductive inferring [6].

Although several distinct theoretical accounts of mental models exist, it is nowadays widely accepted the Johnson-Laird's [6] account about mental model as a '*structural analog*' or 'surrogate' of a real world or imaginary situation, event, or process that a person constructs to reason with in is mind. Two features of the Johnson-Laird's [6] mental model definition are relevant for the scientific discourse: the first one is the characteristic of structural analogy that embodies a representation of spatial and temporal relations as well as the causal structure connecting the events and entities depicted; the second one involves the selective abstractions of reality that we create and then carry out in our head. We have models of various aspect of reality and we simulate these models in order to '*make meaning*' out of what we are experiencing and to arrive at decisions that inform our actions. This involves that each model is constructed in a particular context, by selecting certain aspects that are perceived as relevant for the context itself, i.e. we develop the model for that experience, or system. A relevant research problem is the framing of transfer among different contexts in order to extract generalizations, categories of behaviors and explanations.

1.2. Is Modelling Relevant to Physics?

It is widely accepted that the general aim of physics is to interpret and predict the physical world. The epistemological analysis of physics evolution has pointed out that these processes of interpretation and prediction imply the construction of an understanding of the physical world in a modeling process that involves three different levels (all necessary in the functioning of the physics knowledge and constantly interacting): theory, model, experimental field of reference. The theory contains the explanatory system: the set of beliefs, basic principles (conservation, symmetry, etc...), laws, recognized values and techniques (shared by the members of a given group of physicists) [12]. The model is not an imitation of phenomena; it represents only some properties of reality [13]. The experimental field of reference corresponds to the experimental situations, which belong to the domain of validity of the theoretical construction (theory + model) brought into play in modeling.

Physicists frequently rely on modeling and model-based reasoning to concretize abstract ideas, to simplify and clarify complex phenomena, to predict trends, and to explain mechanism and processes. Model-based reasoning is strongly based upon different kinds of hypotheses [14]: *empirical law hypotheses*, that are summaries of perceived patterns in observations and *explanatory model hypotheses* that introduce imaginable models at a theoretical level and often contains currently unobservable entities. Molecules, field, waves are not simply condensed summaries of empirical observations, but rather are inventions that contribute new theoretical terms and images that are part of the scientist's view of the world and that are not 'given' in the data.

The Campbell [14] distinction has been supported by other researchers [15] and it involves, for example, that merely being able to make predictions from the empirical gas law ($PV=nRT$) is not equivalent to understanding the explanation for gas behavior in terms of an imaginable model of billiard-ball-like-molecules in motion. Unlike the empirical law, the model provides a description of a hidden process that explains how the gas work and answers '*why*' questions about where observable changes in temperature and pressure come from.

Moreover, the model cannot only add significant explanatory power to one's knowledge, but can also suggest questions that stimulate the growth of the theory.

All explanatory models are formed by the process of analogy, that is the seeking of similarities and differences between a *source* (which is something perceived to be somewhat like the phenomenon under study) and the phenomenon itself, which is the *target* [16-17]. In Nagel's [18] terms such analogue models help scientists to '*make the unfamiliar familiar*'. This suggests that analogical reasoning may be an important no inductive source for generating such hypothetical models.

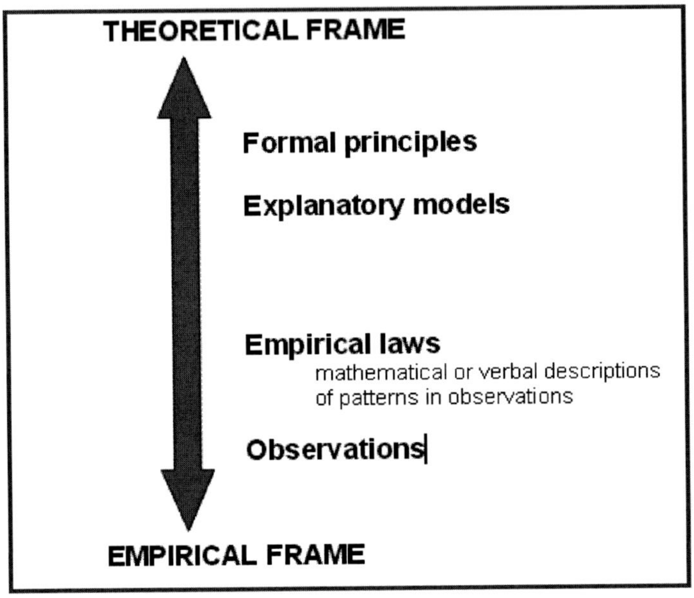

Figure 1. The relationship between the empirical and the theoretical worlds in science.

The double ended arrow reported in figure 1, which synthesizes the kinds of knowledge used in science, puts in evidence that the physics discourse involves a continuous two way analysis correlating theoretical and empirical frames.

1.3. Is Modelling Relevant to Physics Education?

The analysis of cognitive processes as well as the epistemology of physics evolution show that a modeling approach can constitute a good framing in order to design teaching-learning situations both faithful to the discipline and relevant for the learner. However, many pedagogical problems need clarifications and unambiguous choices.

The double-ended arrow, depicted in figure 1, indicates two possible ways of introducing the '*physics to be taught*': the first as a deductive reasoning process from general principles to specific empirical predictions, the second as a process of constructing scientific knowledge by modeling (at different level of abstraction) signals coming from the experiential field. The first approach is usually followed in traditional teaching, where particular systems and contexts (already formalized or easily to be formalized) are, at first, presented and then confronted with the real world phenomena. On the contrary, the modeling approach focuses on the real world where pupils live and on their experiences that are considered the starting points for the construction of a scientific knowledge.

Scientific models are usually very different from pupils' personal views of the world, the spontaneous models [10], and some transformations of the scientific models are necessary in order to fill this gap by gradually adapting pupils' conceptions to scientific models. This approach implies that the physics content structure for teaching is not mainly, or even solely, oriented to physics issues, but it will have to include educational issues and pupils' conceptions. This framework has been adopted in many research studies, founded on '*constructivist epistemology*' [19] and based on the main assumption that there is not a single content structure in a particular content area, but indeed different content structures can be performed, according to the specific (implicit or explicit) aims of the authors.

This process of reconstruction of the physics content to be taught raises some pedagogical problems that need a deep analysis involving a careful design of innovative approach/sequences as well as of the competences involved in the modeling processes and in the supporting tools.

Many new approaches to physics teaching define as their main objective the pupils' awareness of the *'scientific method'*, and most of them assume a data-driven model of discovery. Scientific reasoning includes not only observation and interpretation of data, but also reasoning with the theoretical or explanatory models, which underlie events or phenomena. Because they are not directly observable, their constituent entities and constructs cannot be acquired by induction from observations and/or laboratory activities. Very soon, explanatory models require analogical reasoning and/or the ability to construct complex mental models and to perform *'thought experiments'* [20]. Sometime, experimentation can be presented as an activity whose purpose is to develop, interpret, or evaluate explanatory models of the phenomena being investigated.

Moreover, many physics explanatory models are represented by mathematical equations correlating properties of abstract entities. It is well known that this involves many learning difficulties, since verbal and mathematical communications not always give productive results: very often, teachers are of the opinion that pupils' understanding is very different from that they supposed to explain and communicate. Pupils that are unable to make sense of teacher explanations usually take refuge in memorization.

In order to foster a realistic view of science at all level of schooling, we should find ways to foster both pupil model-building and experimental skills. It is well known that the use of technology, in the form of microcomputer-based laboratory to aid experimentation, is really effective in classroom, in order to stimulate interest and learning. Still, computers can play a relevant role in helping pupils to learn model building and testing, by supplying the necessary scaffolding for the process of using, integrating and constructing several models of knowledge (common and scientific).

Simulations, micro-worlds and graphical/analysis tools can supply different kinds of support, as for example:

- to allow the visualization of structure and evolution of models (explaining mistakes, animating physic processes and illustrating connections);
- to point out models making physics accessible (connecting to what pupils know or are supposed to know);

- to lighten the load of math difficulties (models of complex system can be more broadly accessible because they can be set up without dealing with the formalism of calculus);
- to focus on qualitative reasoning with models (the form of reasoning that most likely will continue to play an important part in pupil's life).

Many researches have pointed out evidences for the effective relevance of computer for learning, although the characteristics of commercially available software are very different and sometime contrasting. However, it is a shared idea that the use of modeling software, integrated with laboratory work, can supply the relevant supports in scaffolding the pupils' construction of scientific knowledge.

2. A Teaching/Learning Approach to Modelling

To redesign, implement and evaluate the structure of teaching/learning environments for science education need a new kind of *'developmental research'* that involves theoretical reflection and conceptual analysis of the physics content as well as classroom research on the interaction of teaching-learning processes [21]. Some relevant trends of such a research approach have been analyzed in recent papers (see for example [22]) that correlate teaching and learning with a deep analysis of the discipline and the cognitive dimension.

Many today researches are based on the *Model of Educational Reconstruction* [23-24], that points out the need of a detailed design stage, based on hypotheses of learning pathways coming out from a twofold analysis of the content structure and the cognitive requirements. The first analysis involves concepts, ways of reasoning and processes of thinking and the second one the results of empirical studies about pupils' conceptions and representations. These two issues, pupils' models and ways of reasoning as well as statements and procedures of the scientific knowledge, are therefore accepted to be of the same relevance and treated as resources for science education. In this way the science content to be taught is reconstructed in order to realize the main goal: to allow pupils to gain a fruitful knowledge of the physical world.

We describe an approach to physics education aimed at acquiring expertise in physics that can be transferred to different contexts. The approach supplies a unitary presentation (in method and structure) of the discipline and the accomplishment of a teaching/learning process based on relevant research results. In fact, it gives a representation of the properties and structure of phenomenological systems and processes that correlate descriptions on different organization levels of matter (macroscopic and microscopic). In the following, we describe the approach to the physic content and the involved pedagogical strategies.

2.1. The Pedagogical Approach to the Physics Content

Many patterns in the world are usually the result of interactions among large numbers of smaller pieces that somehow combine in different ways to create the large-scale pattern. Such macro-patterns, which arise out of the interactions of numerous interacting micro-agents, are called *emergent phenomena*, that is, phenomena emerging from interactions at a lower level or scale [25].

Our approach proposes a phenomenological understanding through an analysis performed according to levels, built on the basis of the identification of different structures, in which every level has its own elements and laws of interaction. Behaviors at a higher level originate from behaviors which are very often totally different at a lower level. This approach, which is considered especially interesting in the field of science education [26], should provide competences and methods of analysis that can be transferred to the broad area of the complex systems of multi-disciplinary interest.

In order to clarify the meaning of our approach, we give two examples: in a sea wave we see water moving, but it is energy that travels, not the water molecules; similarly, with a wave travelling along a rope, the pieces of the rope move up and down, but the wave moves along the length of the rope. In the process of diffusion (for example of the perfume coming from a bottle posed in a corner of a room), we perceive the perfume flow as if the perfume molecules alone are 'moving into' the whole room. So, the flow pattern arises from a consideration of the collective (or net) effect of the constituent level interactions (i.e., the bouncing around and into each other) of both or all of the perfume and air molecules, for each successive instant of time [27]. This underlying mechanism, relating the constituent component level behavior to the pattern level behavior, can be described as one of emergence, in the sense that the pattern is the summative (or net) outcome of the random interactions of all the molecules (both perfume and air), and not just due to the behavior of one class of molecules (such as the perfume ones).

In many physics processes (such as diffusion), neither the macro level components themselves nor their constituents (agents or molecules) are directly causing the global pattern to occur. Instead, the macroscopic properties and behaviors must be explained in terms of the collective interactive outcomes of all the constituent components (agents or molecules), such that neither an individual component nor a group of individual constituent components cause(s) the global pattern [27]. Therefore, such kinds of mechanisms (as the flow in a diffusion kind of process) are not direct and can be referred to as emergent.

It has been shown that people find difficult to understand this emergent sense of levels [26-27], although they are able to use appropriate reasoning resources in some contexts [28]. In fact, it can be easily shown that many pupils have directly experienced the idea of emergent levels by participating in *human waves* at sports stadiums. Individual people simply stand up and sit down, but the wave moves around the stadium. This is a simple example of the way behaviours at one level arise out of very different behaviours at another level.

In our approach, the macroscopic properties observed and measured will be connected with microscopic models whose characteristics and behaviors will be studied through computer simulations. As a consequence, the physics content to be taught is reconstructed through an analysis at different levels which involves a progress from the analysis of phenomena to the models for their description and interpretation.

2.2. The Teaching/Learning Strategies

In our approach, the process of modeling is intended as a set of procedures (thinking and doing) to find explanations of phenomena. An explanation has been defined as '*an answer sought or provided to a specific question*' [29, p. 85]. Scientific questions involved in the modeling of the natural world supply different kinds of explanations. It is possible to make evident [29] a pattern of relationships stemming from the interactions between the nature of questions and the explanations elicited: questions like '*how does the phenomenon behave?*' or '*why does the phenomenon behave as it does?*' or '*how might it behave under other conditions?*' are perceived differently by pupils on the basis of their different models. As a consequence, our teaching strategy supports pupil learning by using different kinds of questions aimed at stimulating pupils to elicit their representations and, mainly, to make explicit how their models are useful to describe and/or explain phenomena.

Questions support the process of knowledge construction by appropriately shaping it. Our shaping procedure involves a slight modification of the Predict-Observe-Explain (POE) format [30] into a Predict-Explain-Observe-Explain (PEOE) format [31]: questions require pupils to make predictions about observed pathways in physical phenomena or results of experiments/simulations and to explain their prediction (on the basis of their representations). Then, observations are performed and new kinds of questions are supplied in order to encourage pupils to think about the viability of their own representations, to construct new ideas and to negotiate shared meaning with their partners (classmates and teacher). The introduction of two moments connected with the search for explanations is aimed at the construction of interpretative and predictive models [29] by making explicit the representations of entities and their relationships [32].

Our modeling procedures use simulation software in parallel with laboratory experiments for qualitative demonstrations and quantitative analyses. Two kinds of simulations have been prepared: visualizations and quantitative simulations. The first ones are aimed at visualizing mechanisms of functioning, by making evident how the observable perceptual global phenomenon emerges from intrinsically distinct underlying local interacting processes. Their objective is in scaffolding the development of a *sense of mechanism*, helping pupils to *run* their mental models. We use the word *mechanism* by referring to processes that we visualize in terms of ordinary sense experience, as it happens in many cases of microscopic effects (as for example gas pressure or diffusion) in terms of familiar behavior of macroscopic objects. The quantitative simulations are aimed at visualizing the underlying mechanisms as well as at stimulating patterns of thinking, common with the laboratory activity, such as:

- to perform continuous iterative loops between models and experimental observations, each refining the other;
- to move from qualitative observation to quantitative measurement, as the phenomenon becomes clearer.

The prepared simulations are highly interactive and allow the variation of parameter values to test and extend hypotheses, to answer new questions and so on. In this sense, we call this way to use simulations '*to experiment with models*'.

Following, we describe two Teaching Learning Sequences (TLSs) [22] based on the same approach and aimed at connecting system macroscopic properties/behaviors with characteristics and structures of microscopic models able to describe and interpret them on the basis of physical theories. Each sequence is divided into steps involving different activities. Each step has been designed with defined learning outcomes, by taking into account novices' and pupils' mental models reported in literature and also pointed out by our research. These constitute the starting point (i.e. the pupils' mental models that can be supposed as initial conceptions drawn from everyday experience) orienting the pathways toward the scientific representations (i.e. more adequate mental models whose construction was supported by the learning strategy). The two TLSs have been experimented in some high school classes as described in Section 5.

3. The TLS about Modelling Mechanical Wave Propagation (TLS_MMWP)

The TLS here described is part of a wider project aimed at the introduction of wave propagation (mechanical and electromagnetic waves) with a unified approach, looking at the signal generation (mechanical perturbation, electromagnetic signal), its propagation (where, how,...) and its detection (receivers, comparison between original and received signals,...). Here, we concentrate on the sound wave propagation and, in particular, on the role played by the physical characteristics of media where sound waves propagate. We refer to procedures to put in action in order to stimulate pupils in the construction of mental models productive for the understanding of scientific models.

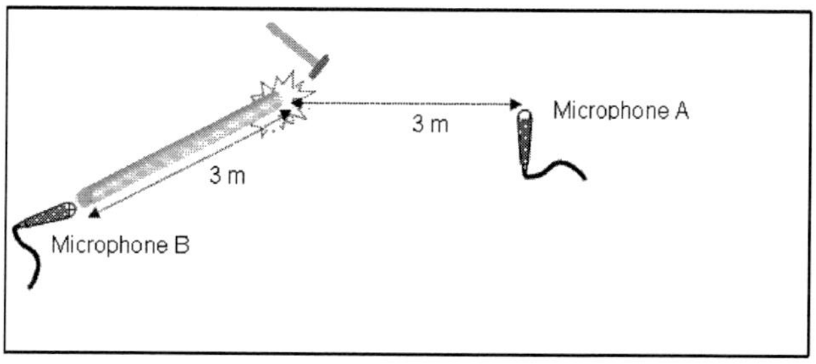

Figure 2. Sketch of the experimental setting. The figure represents a 3 meters long metallic rod fixed to the lab desk by means of vices. The rod is hit by means of a heavy hammer. The produced sound is detected by two microphones, A and B, placed as in figure. Pupils are required to predict if both microphones detect the sound at the same time or will one of them detect sound first. They are also required to explain the reasons of their predictions.

Many studies report that sound and mechanical wave propagation is an area in which pupils display many difficulties in understanding. Several studies [33, 34, 35, 36] suggest that a naïve mental model is associated with sound propagation and this model is a common denominator for various alternative ideas which represent sound traveling as a particle-like object or through the translation of particles travelling in successive pulses. These authors used the term *entity model* to describe a pupils' idea of the sound as being carried by individual medium molecules and passing from one molecule to another, and/or the terms *sound particle model* as the materialization of the supply, a mixture of energy, intensity and speed, given by the sound source to the medium. Sound is considered as an entity that may or may not be material and it may or may not need the medium to propagate. Pupils may call this *sound entity* not only a *sound particle* but also a *sound wave*, a *disturbance* or a *vibration*. The latter of these are scientifically acceptable terms.

The starting point of our TLS has been the analysis of the experimental situation schematized in figure 1.

The analysis of pupils' predictions and explanations evidenced two main categories of mental models, classified as novice mental models in Table I. These models assign to the medium two different contrasting roles: in fact, a relevant number of pupils assigned to the medium the role of obstructing and slowing down sound propagation and an almost equivalent number of pupils thought that density of the medium determines the distance a molecule can move and then transfer sound (or force, or energy).

The TLS has been designed in order to scaffold the construction of mental models productive for the understanding (at qualitative as well at quantitative levels) of the scientific model described in Table I.

Table I.

Novice Mental Models	Scientific Mental Models
Sound velocity is higher in more dense media (lower molecular separation facilitates the sound transfer). Sound velocity is lower in denser media (bigger/closer molecules obstruct sound transfer).	Propagation velocity c of mechanical wave depends on density and elasticity of the medium: $c \propto \sqrt{(\tilde{Y}/\rho)}$ (with Y indicating the elastic compressibility of the medium and ρ its density).

The TLS has been divided in different steps and its methodological approach required the pupils to perform activities in groups, write descriptive and explanatory answers to each investigation or problem, and discuss them individually and in small groups with the tutors (during the work) and the teacher (usually at the end of the investigation). A final class discussion, where pupils were requested to shortly describe activities and results of their investigations, completed each class period.

I) Observation of everyday phenomena.

This phase was aimed at pointing out connections between spontaneous pupils' models and everyday phenomena. Observations, using easy available materials (ropes, slinky,...) and analysis of videos involving wave propagation were performed. Pupils were stimulated in finding the involved physical variables, inferring their relationships and predicting behaviors.

II) Experiments and data analysis.

Pupils were involved in performing quantitative and semi-quantitative experiments and data analysis, using commercial sensors (with computer interfaces), for open-ended investigations. Experimental work involved direct comparisons of sound speed in solid rods and air by performing the experiment shown in figure 2. The microphones A and B (see figure 2) were connected to a computer data logger. The time the sharp sound, produced by hitting the metallic rod end, takes to travel the 3 m distance between the sound source and the two microphones was measured.

Figure 3 reports the signals detected by microphones A and B: a delay between the onset of the two signals is evident. The points were collected with a data sampling rate of 5000 points s^{-1}. Since the difference between the two peaks corresponds to 38 points, this involves a delay time of approximately 8 10^{-3} s.

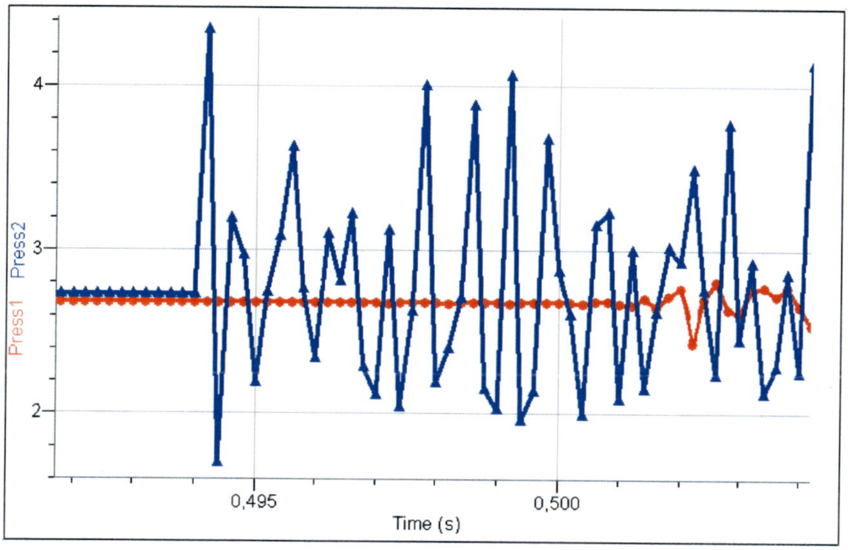

Figure 3. Signals detected by the two microphones A and B placed 3 m distant from the sound source.

Later on, a measure of the sound speed in air was also performed. Figure 4 shows the experimental apparatus: a couple of microphones, placed at a fixed distance, L, were connected to a computer data logger and used to measure the time it takes a sharp sound to travel the distance between them. The ratio between the microphone distance and the time delay between the start of each detected signal gives the speed it takes to sound to go from one microphone to the other one.

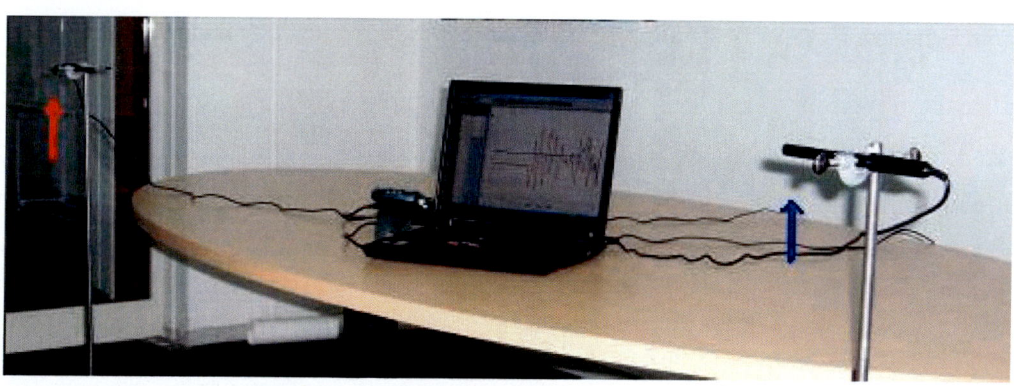

Figure 4. Experimental apparatus to measure the delay time between the sound pulses registered by two microphones placed 2 m apart.

Figure 5 shows the signals detected by two microphones placed 2 m apart. The points were registered with a data sampling rate of 5000 points s^{-1}. Since the difference between the two peaks corresponds to 28 points, this involves a delay time of 5.6 10^{-3} s. Measurements were performed several times with microphones placed at different relative distances; the mean value of the sound speed in air, obtained with this method, at a room temperature of 22 °C, was V=360 ± 30 m/s.

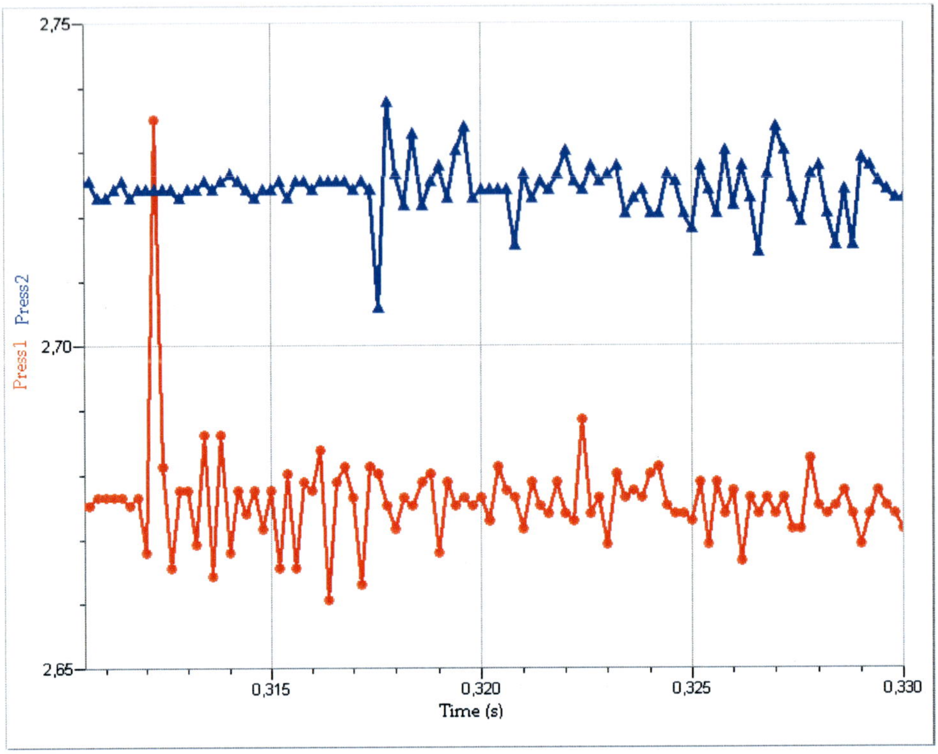

Figure 5. Signals detected by two microphones placed 2 m apart when a sharp sound gets to the nearest one and then to the other.

By recalling the previous result about the delay time between two sound pulses travelling in air and inside a brass rod, the sound speed in brass can be estimated. By averaging among the calculation performed by the different groups of pupils a mean value V = 3500 ± 200 m/s was obtained.

III) Modeling

In the third step, pupils were involved in activities aimed at building physical models able to describe and interpret observations and experiments as well as at finding relationships between experimental data and hypotheses supporting the dynamical models explaining the analyzed processes. They were directly involved in dynamical modeling by using appropriate software (simulations and spreadsheets) and in the re-analysis of the various experiments, explanations as well as predictions, on the basis of features of the proposed models. Modeling activities were developed in order to formalize the dependence of propagation speed from the medium properties.

Simplified models of solids, constituted by linear chains of masses and springs (see figure 6) were analyzed. Pupils were stimulated in finding analogies between their mental models of solid and the proposed model, by looking at the relationships between spring elastic constants and solid compressibility, as well as particle masses and solid densities.

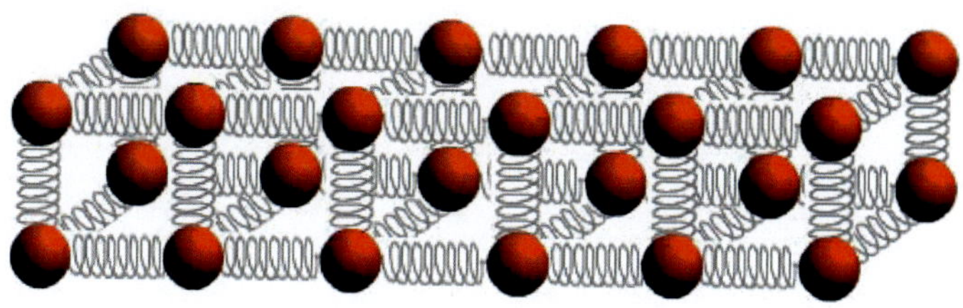

Figure 6. Model of solid structure, represented as a lattice of masses interacting each other by means of springs.

In order to formalize the dependence of propagation speed from the medium characteristics, an oversimplified model has been analyzed in more details by means of a linear chain simulation (some masses connected with mass-less springs, see fig. 7). The propagation of longitudinal pulses was analyzed in the simulated model.

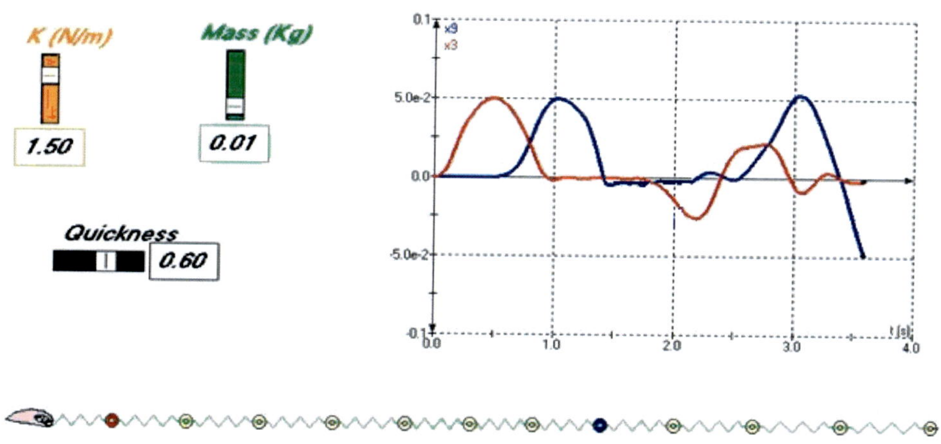

Figure 7. Pulse propagating in a linear chain of masses connected by springs. Displacements of particles red and blue are shown as a function of time.

The simulation allows the user to modify the values of masses and elastic constants and to analyze variations of the pulse propagation speed. The analysis can be performed at a qualitative level, by visualizing the delay times among the motions of successive particle, as well as at a quantitative level in order to find a relationship between the propagation speed and the mass values and/or the spring stiffness (see figures 8 and 9). Through a fitting procedure, the mathematical equations connecting the pulse speed with the chain mass and spring constant values can be found. Analogies between the simulated model and the qualitative models of solids were developed, by looking at the relationships between spring elastic constants and solid compressibility, as well as at particle masses and solid densities [37] (see Table I). The obtained relationships were, then, confronted with those previewed by the theory and with experimental data taken from textbooks.

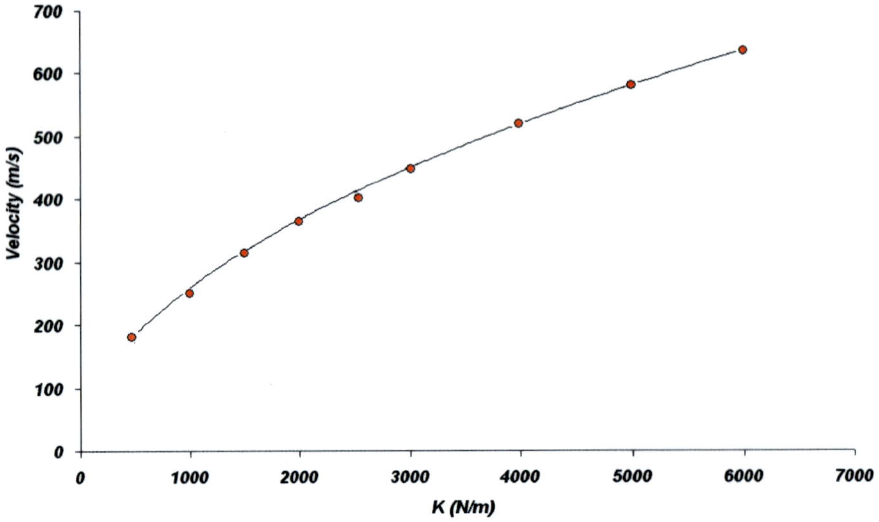

Figure 8. Graph of the pulse propagation speed vs. spring elastic constant values.

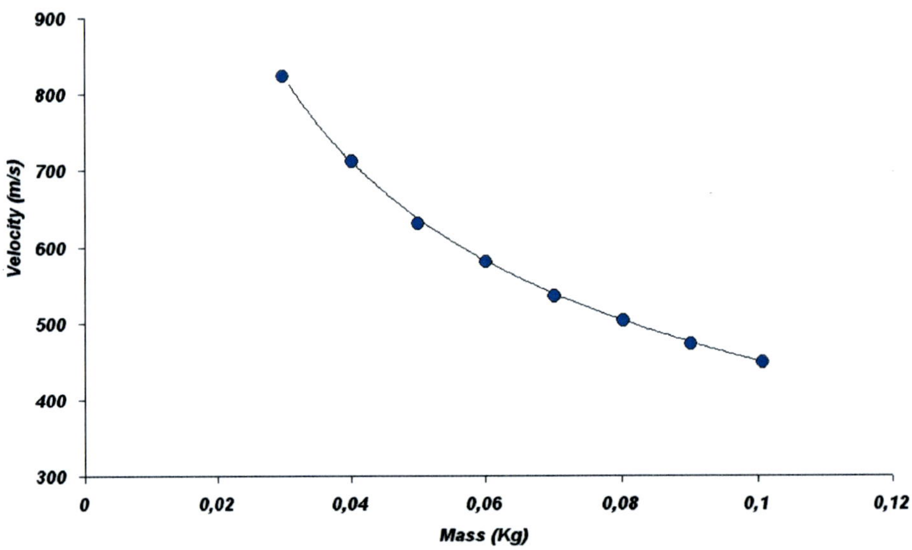

Figure 9. Graph of the pulse propagation speed vs. chain mass values.

At the end of this step, pupils were supposed to be able to describe mechanisms for sound propagation in solids and to identify variables affecting wave speed.

Modeling involved also the analysis of simulations visualizing sound pulse propagation in a column of air-particles. Pupils were stimulated to observe pulses generated by a piston moving at the end of a long column of gas particle and to find analogies with longitudinal waves generated at the end of a long spring. It was analyzed how a forward movement of the piston (see figure 10) produces an increase of density and pressure in the initial region, which propagates along the column with a velocity that can be calculated as a function of gas pressure and gas density. This increase is registered as a force on the spring posed at the end of the column.

Simulations were supposed to supply representations of air motion where a displacement from density or pressure equilibrium is propagating. By looking at the motion of particle layers, pupils were stimulated to connect particle motion with generator motion and point out how the air (or medium) properties influence the propagation speed.

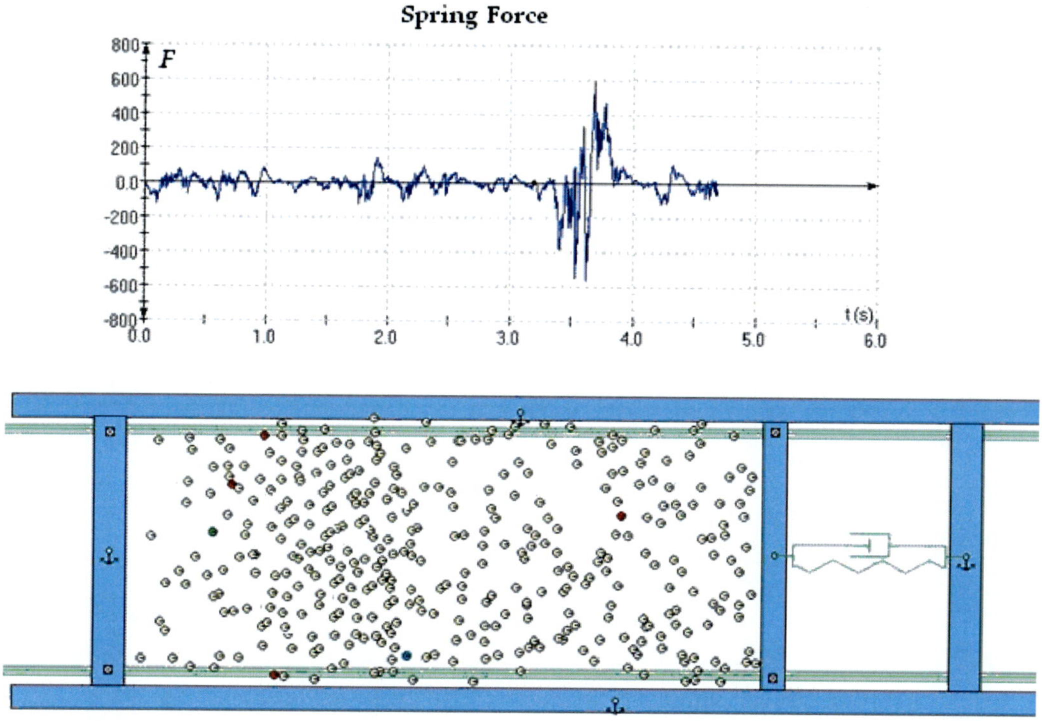

Figure 10. Pulse propagating in a gas of elastic ball particles. The spring on the right registers the variation of pressure after a delay time registered in the graphical display.

Our teaching strategy supported pupils' learning by using different kinds of questions aimed at stimulating them to elicit their representations and mainly to make explicit how the different models are useful to describe and/or explain the observed phenomena. A detailed analysis of pupils' learning pathways is reported elsewhere [31]. In Section 5, we point out the main results involved in pupils' learning of physical concepts as well as of modeling procedures.

4. The TLS about Modelling the Heat Transfer Process (TLS_MHTP)

The TLS here described is part of a wider project aimed at the educational reconstruction of thermal interaction between bodies by analyzing the category of phenomena that is possible to describe -in first approximation- only in terms of thermal interaction. These phenomena are divided into three subcategories: interaction where the bodies only change their temperatures, interaction between liquids and gasses with mixing of substances and interaction where bodies change their state (liquefying, boiling,…). Physics theory describes these phenomena through heat transfer between the two interacting systems. We here focus on modeling the heat transfer in solids.

The educational research has pointed out that pupils' explanations of thermal phenomena are usually different from scientific models. In fact, the scientific view of thermal processes is subtle and difficult: their description in terms of microscopic structure of matter, or more simply in terms of the particulate nature of matter, shows many learning problems. On the other hand, to analyze these phenomena only at a macroscopic level makes difficult a meaningful understanding, since it involves a deep understanding of thermodynamics principles, which is usually out of high school pupils' grasp. Our approach, by providing qualitative and semi-quantitative models of microscopic behaviors can supply a sense of mechanism that helps pupils in the construction of adequate mental models.

In our educational approach, empirical qualitative and/or quantitative experiences are strictly integrated with the search for explanations through modeling procedures. The main point is in maintaining coherence between the models to be taught and the corresponding experimental field, which provides the experiential basis for meaning making. We focuses on the construction of physics models representing adequate and plausible representations of phenomena and helping pupils to integrate their different unrelated understandings: model properties have to be at the same level of analysis as pupils' observations of the natural world. In fact, temporary and pragmatic qualitative models are also adopted [38, 39], as an intermediate phase, in order to gradually induce pupils to understand more comprehensive and predictive physical models.

Although many obstacles in the learning of particulate structure of matter have been put into evidence [40, 41], we think that the understanding of microscopic models of matter is an important objective for high school physics teaching. The reason is twofold: the epistemological features of physics models (including their character of hypothetical models sometimes based on theoretical more than on empirical foundations) and the need to realize, co-operatively with the other disciplines (as for example chemistry), possible unitary descriptions of systems and phenomena.

According to our model of the educational reconstruction, our approach grounds on some well known pupils' mental models about thermal processes and relationships between the concepts of heat and temperature. Some researches [42, 43, 44] have pointed out some relevant ideas/models about thermal processes that pupils construct in their everyday interaction with thermal phenomena. These can be synthesized as following:

i) the experimental fact that all objects in prolonged contact reach the same final temperature is not obvious to pupils;
ii) the ability to be heated is considered a '*natural*' property of particular substances;
iii) the substances are divided by pupils in two categories, '*hot*' and '*cold*', but their coldness depends on the peculiar situations in which they are used.

The common pupils' interpretation of thermal phenomena seems to take origin from two specific ideas: the conception of heat as a form of substance and/or energy stored in the bodies and its partial or total coincidence with the concept of temperature (usually intended as a measure of the heat 'contained' in the body).

The TLS, we here describe, is divided in different steps involving various pupils' activities aimed at characterizing, as the previous one, the processes of observing, experimenting and modeling.

I) Observation of everyday phenomena

In this first step, pupils were stimulated to perform observations about common thermal phenomena or to recall their previous experiences with them. For example, observation were performed involving different objects that have been maintained in the same environment for long time, or cups of different materials (metal, plastic, styrofoam,...) each containing almost equal quantity of hot coffee and all cooling in the same environment, or objects at different temperatures placed in thermal contact,...... In all the cases, pupils were requested to predict the behaviors of the different systems and to explain their predictions. A final class discussion was aimed at pointing out the connections with the pupils' spontaneous models as well as the relevant characteristics of phenomena to be pointed out by identifying the involved variables.

II) Experiments and data analysis

The second step was devoted at performing experiments aimed at controlling predictions and behaviors. Pupils worked in groups of two or three to perform quantitative and semi-quantitative experiments, by using cheap and easily available materials and commercial sensors (with computer interfaces). Sensors allow to reduce the time available for each experiment (for example by reducing the quantities of substances to be heated or cooled) and to perform more experiments in a class period. Moreover, the real time graphical display enabled pupils to modify variables in real time. Figure 11 gives an example of a cooling experiment: two small cylinders of copper and brass were warmed by placing them in an oven at about 75°C. Then they were placed on the lab table and cooled by forcing the environment air, at about 15°C, around them.

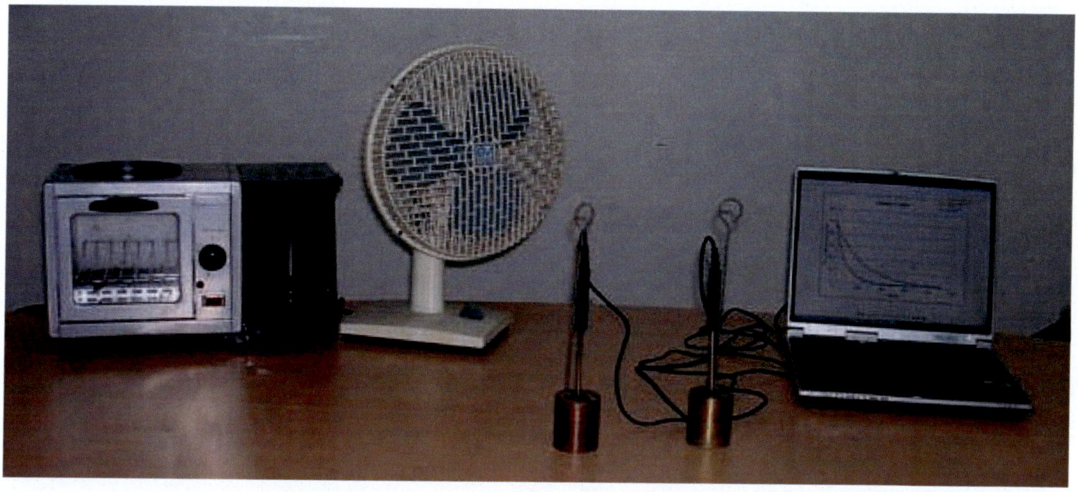

Figure 11. The experimental apparatus. Two sensors are placed at the centre of cylinders and register their temperature.

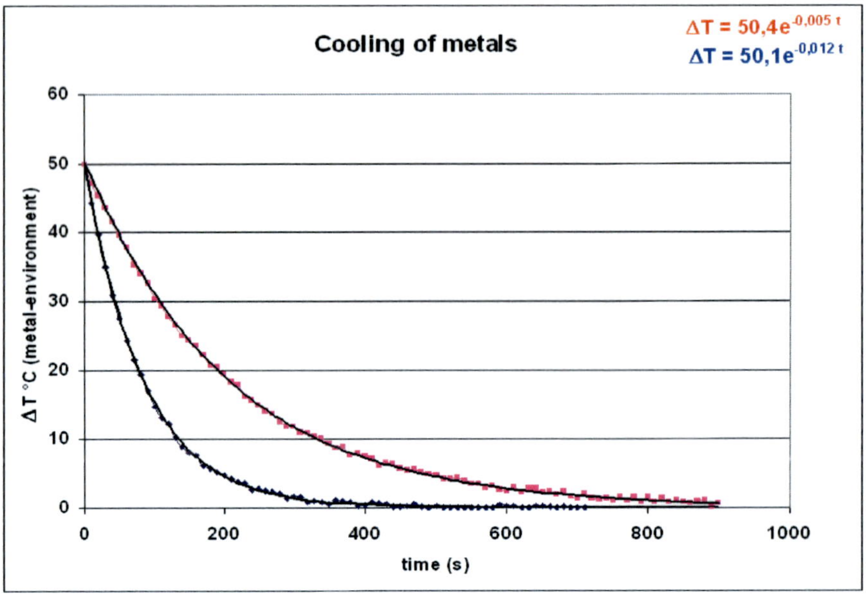

Figure 12. Graphical display of the time dependence of temperature in the process of cooling of the two cylinders represented in figure 10. The copper cylinder cools down more quickly than the brass one.

Figure 12 reports the graphical display of the cooling experiment shown in figure 11. Through a best fitting procedure, the quantitative relationship of the temperature difference between the metal cylinder and the environment (ΔT) vs. time (t) can be found. The dynamic of the two cooling curves gives evidence of different coefficients that are related to the different thermal conductivity of the two metals.

III) Modeling

An interpretation of matter behaviors, observed at the macroscopic level, can be found through a change of the analysis level. Simple microscopic models can be constructed by hypothesizing the existence of well defined elements constituting matter and of equally well defined interactions between them, and between them and the environment. Then, we have to check if, and in what extent, this model can explain the macroscopically emerging thermal properties of matter. Such kinds of models should help us to explain, for example, in what a hot body is different from a cold one, or what happens when a body reaches the thermal equilibrium with another body or with the environment. We give an example of a modeling approach that can describe and explain the experimental data reported in figure 12.

We simulate metals by using a crystalline solid structure (a regular structure where atoms are positioned in a lattice and repeat almost exactly a fundamental structure in all the space). The atoms are supposed to interact so that they can oscillate with respect to their respective equilibrium positions. A simple way to represent this interaction is to do it with springs connecting all the atoms in the lattice.

We describe two models used in the TLS in order to simulate:

a. the thermal interaction of a solid with the environment,
b. the thermal interaction of two solids initially at different temperature.

These models are implemented by using the Net Logo [45] simulation environment. Figure 12 shows a two-dimensional display of the model[1].

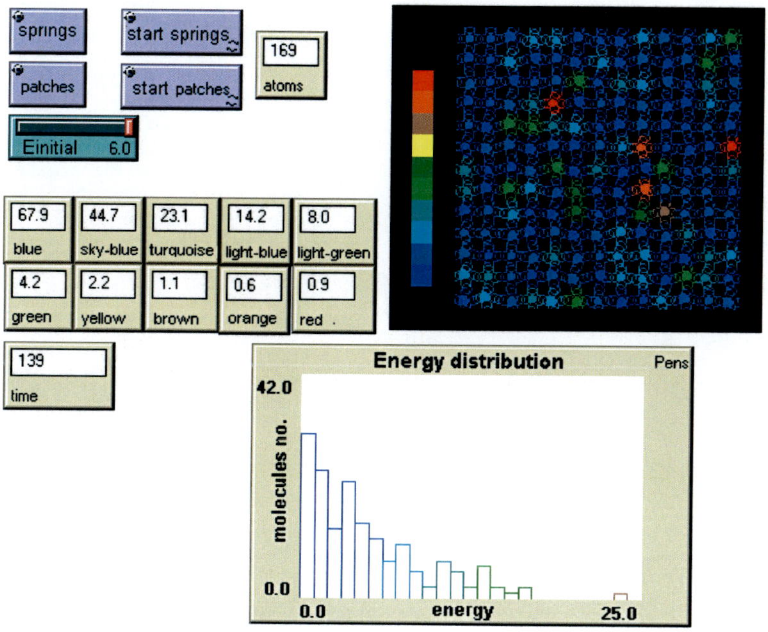

Figure 13. The Net Logo screenshot of the simulation of the energy distribution in a thermally insulated solid.

The solid temperature is linked to the atomic oscillation energy, in the sense that in two identical solids at different temperatures, the atoms of the solid at a higher temperature possess higher oscillation energy. An initial velocity, identical in module, but randomly directed in the plane, is given to each sphere (modeling an atom). Since the springs are initially in the equilibrium positions, spheres have initially the same (kinetic) energy. Through the use of various colors, Net Logo allows us to visualize the energy of each sphere. The total energy is divided in ten intervals, represented by ten different colors. Initially, then, all the spheres have the same color. Starting the simulation, each sphere can interact with the surrounding by exchanging energy. As a consequence, one can observe a variation of the sphere individual energies, described by a window reporting the number of spheres with a well defined energy (i.e. color) that updates while times goes on.

The same basic model is used to simulate the thermal interaction of a solid with the environment (see figure 14) and the thermal interaction between two solids initially at different temperatures (see figure 15).

In the first simulation the hot solid molecules interact among them and with the air molecules by exchanging energy. The graph of solid temperature vs. time shows how the solid molecules diminish their energy and how the equilibrium is reached.

---

1 The model can be deepened at different levels: as a classical model of masses oscillating around their equilibrium positions or as the solid Einstein model, where quantization of energy is considered. However, also the classical model allows us to describe the heat flux in a solid and the establishing of thermal equilibrium.

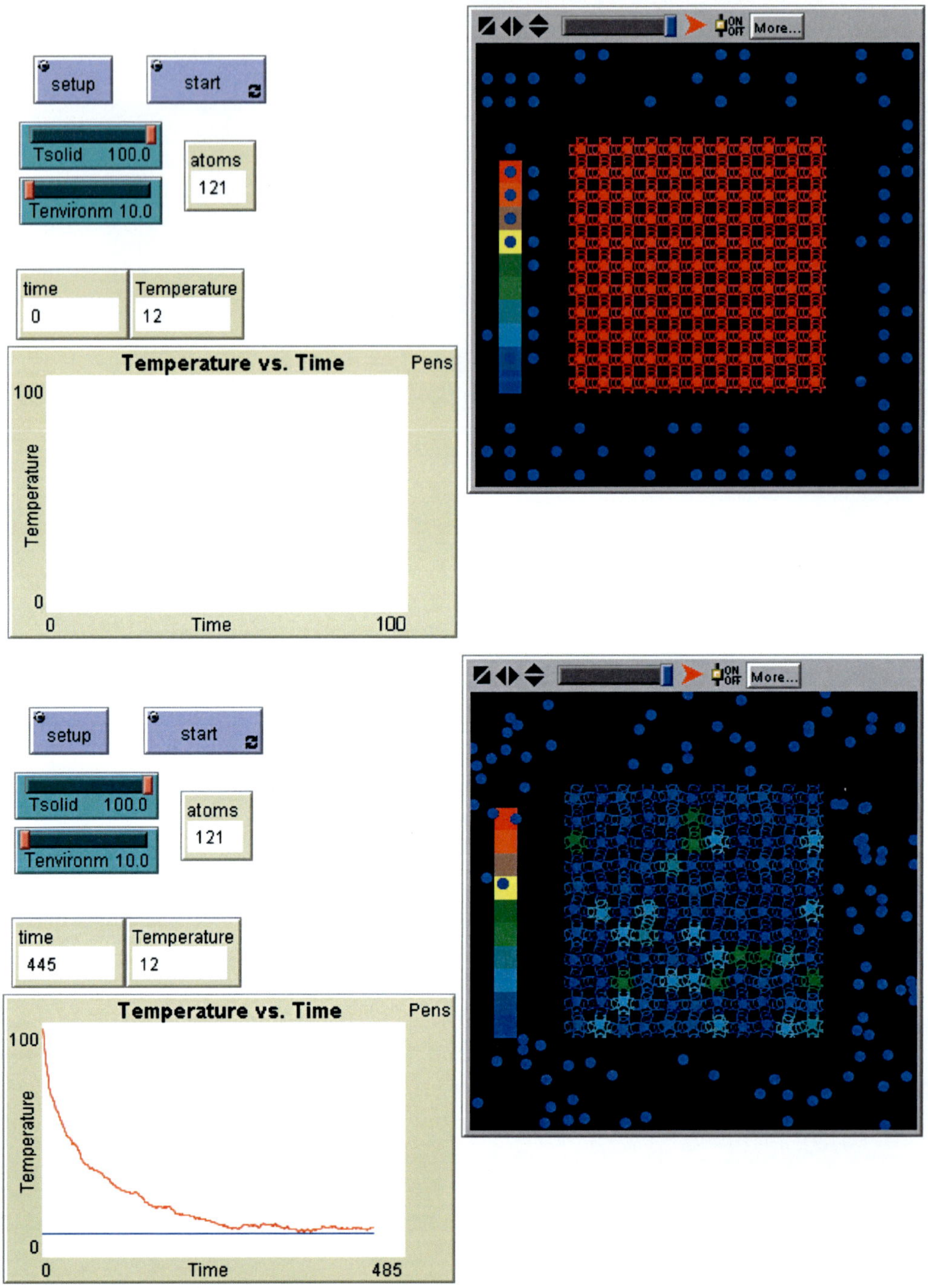

Figure 14. The Net Logo screenshot of the simulation of a solid interacting with the environment. On the left: the initial situation, where the hot solid (red particles) is placed in a colder gaseous environment (blue particles). On the right: the equilibrium situation, when the solid is at the same temperature of the environment. Note the Temperature-Time graph, indicating the reaching of thermal equilibrium.

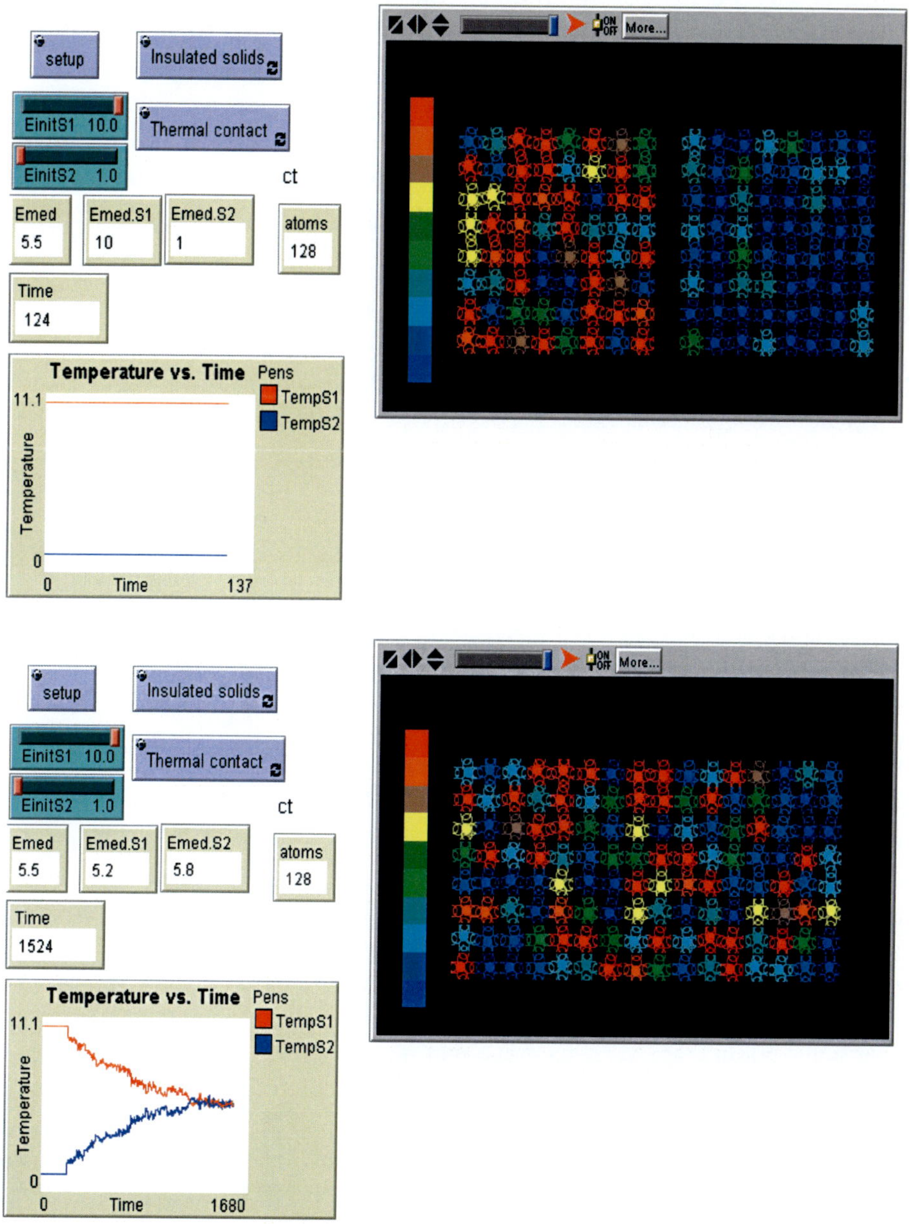

Figure 15. The Net Logo screenshot of the simulation of two solids thermally interacting. On the left: the solids have different temperatures. On the right: the equilibrium configuration, when the solids are at the same temperature. Note the Temperature-Time graph, indicating the reaching of thermal equilibrium.

In the second simulation, two solid blocks of different temperatures (see the left side of figure 15) are brought into contact: the solid molecules of the contact region start to lower or increase their energy allowing a variation of the energy of the two entire blocks. The equilibrium at an intermediate temperature is reached, as show by the temperature vs. time graph of the two bodies (see figure 15, right side).

Both the simulations visualize, through the variation of the sphere color and the calculation of the system mean energy, the evolution toward the equilibrium state. They supply a visualization of possible microscopic behaviors that can help pupils in constructing mental models representing mechanisms of functioning, that emerge in a behavior of the systems as a whole analogous to the observed and experimented (see the analogue cooling graphs). The macroscopic characteristics of heat transfer in a solid can be interpreted at a qualitative level and, sometime, at a quantitative one.

5. Main Results of the Teaching/Learning Experiments

The described TLSs has been experimented in four classes (fourth-year of Science Oriented Secondary School) by a teacher that previously participated to the implementation of the pedagogical materials. The experimentation involved 92 pupils (age range 16-17 years). Two researchers took part in the teaching/learning experiment; they cooperated the teacher in tutoring the different groups of pupils, in order to register details of their learning processes.

The TLS_MMWP has been developed in four class periods (each one involving almost 2 hours) and the TLS_MHTP in five class periods. In both cases, home work was also requested. Each class period involved interactive lectures (for less than the 30% of the class period) and active working sessions, where pupils performed experiments and simulations. Two more hours have been devoted to pre-test and post-test administration.

Pupils worked in small groups (two or three pupils) and discussion with peers was an essential method of learning. We call lectures '*interactive*', as the teacher asked questions and the pupils were free to ask questions and make comments. Whole class discussions, as well as small group discussions, were considered important strategies to stimulate learning and some of them have been audio recorded for a successive analysis. Data, obtained from pupils' individual work and group interactions, have been analyzed in order to produce a description of the setting complexity as well as of the knowledge development of the whole class. The teacher and researcher logbooks, registering the critical episodes and some relevant points of the discussions, gave also useful insights about the adequacy of the individual steps for pupils' learning. Pupils' answers to tests and transcriptions of discussions have been processed through triangulation, in order to obtain a comprehensive analysis of pupils' participation in the work from several perspectives.

Three papers [31, 46, 47] report a detailed contextual analysis of some events and conditions, with the goal of seeking to understand how the pupils' behaviors and understanding could have been changed or affected by the designed teaching/learning environments. Here, we report some qualitative results that allowed us to portray pupil progress in learning, due to the various activities and peer interactions, their common learning difficulties in order to partially represent their learning evolution.

According to the *Model of Educational Reconstruction* [23, 24] our approach, in both TLSs, took into great account empirical studies about pupils' conceptions and representations. For this reason, both experimental designs provided for different situations where pupils could be stimulated to make explicit their mental models. The analysis of pupils' spontaneous models in the field of mechanical wave propagation as well as of the heat transfer, described in literature, reveals some peculiar characteristics. The conceptions displayed by pupils show different facets usually depending from the context. However, both categories of phenomena seem mainly explained on the basis of naïve models that usually are extrapolations from our everyday experience with the mechanical world and visible substances [48].

Our teaching/learning experiment pointed out some differences about the characteristics of pupils' representations, as well as about the ways learning was approached by pupils which showed different naïve mental models. For example, the discussions between pupils and tutors, as well as the analysis of pupils' worksheets showed relevant differences in the approach to find meanings to experiments and/or simulations between pupils thinking sound as an entity completely independent from the medium and those thinking sound as some thing that medium particles (atoms/molecules) transfer between each other. The two kinds of models stimulated different approaches in making sense of wave propagation: the first one induced pupils to look at the medium as something hindering propagation, whereas the second one influenced pupils in assigning to the medium a role, even if an incorrect one.

Our hypothesis was that learning of mechanical wave propagation involved for many pupils a relevant conceptual change, perhaps of the kind described by Chi et al. [49, 50, 51]: an ontological shift involving a kind of re-categorization of a concept from being one kind to being another kind, in which the kinds are fundamentally distinct, as well as '*entities*' are fundamentally different kind from '*processes*'. On the other hand, pupils showed to have a mastery of adequate thinking resources [28] in analyzing some processes, as human wave propagation in a sport stadium or the falling domino chain. In fact, in these contexts they were able to differentiate between the motion of individual agents and the emergent property in the system and to single out the thing propagating (information or a collision or an impulse and so on).

The same kind of concept re-categorization has been necessary for the heat concept: in the course of the teaching activities pupils gradually gave up expressions like '*the heat contained in the body*' or '*that body possess more heat*' and referred to hot bodies as '*bodies with more thermal energy*' or '*bodies where atoms move more quickly*'.

It has been pointed out [50] that conceptual change is difficult, not because it is hard to do in principle, but because it is not something that pupils are aware of. To stimulate this awareness was a main objective of many steps of our TLS_MMWP, where we focalized on the need to scaffold the construction of mental models that differentiate properties and behaviors of single elements (molecules, atoms, agents) of the systems where waves were propagating from those of the waves (the emergent property). This was also the main objective of our TLS_MHTP, where properties and behaviors of single atoms of solid bodies (oscillations with different energies) were differentiated from the emergent property of heat transfer between bodies at different temperatures.

Many of the contexts, presented in our TLSs, show that it is possible to differentiate between two kinds of processes: causal and emergent processes. The first one, that can be explained by identifiable actions, with temporal and spatial contiguity, and with an identifiable beginning and ending; the second one caused by the collective aggregation of multiple causal events. Our TLSs emphasize, through simulation and visualizations, the two kinds of processes at the two levels: causal processes at micro-level and emergent processes at macro-level. From the analysis of pupils' worksheets, as well as from many discussions, it can be pointed out that visualizations and simulations, depicting and confronting micro and macro behaviors of the systems, supplied to pupils the useful scaffolding, stimulating them to activate appropriate thinking resources.

The role of qualitative knowledge in understanding and problem solving [52] is well acknowledged. Experts engage in qualitative analysis of systems and/or problems prior to working with equations. This involves the construction of a running model of the system that takes into account the characteristics/variables perceived as relevant. In our framework, this research for qualitative functional relations between physical quantities constitutes the first level of modeling that can supply one/more representations of the selected aspects of a set of material situations. For each context of our TLSs, we begin our analysis by focusing on the qualitative reasoning in order to stimulate pupils into acquiring a sense of mechanism: i.e. a qualitative understanding of *'how things work'*, which is used as an explicatory framework. Quantitative analysis is successively faced by applying principles and laws of Newtonian mechanics and energy conservation.

Our detailed analysis of class discussions as well as of the post-test [31, 47] supplied evidence that the majority of our pupils gained the ability to make sense of wave phenomena, as well as of heat transfer process. In particular, concerning the ability to develop modeling procedures making pupils able to give sense to the observed phenomenological areas, our results show that the modeling procedures, made explicit in different steps of our approach, supplied pupil reasoning ways to face their naïve mental models and results of experiments and simulations. The PEOE strategy (see Section 3), used in the worksheets, gave pupils the opportunity to explicit the mental models constructed for their prediction, to face them with the task and to search for their revision or their further elaboration. Sometimes, pupils became able to reason about information, as if they were conducting a thought experiment [28]: '.....*if particles behave ... ... then it should happen that...'*

Our results give also support to the view that conceptual change is a slow evolution where pupils' learning proceeds from initial, scientifically more or less incorrect, views via some intermediate or *'transitional states'* to scientific views [53, 54]: pupils' conceptual change requires time and explicit attention to the development of concepts. However, our results also show that the TLSs were not successful in bringing about conceptual change in all the dimensions of the wave concept or the heat transfer process for a relatively small percentage of the pupils. In fact, these pupils did not achieve coherence in the transfer of concepts among different contexts. A possible interpretation is that only *'a peripheral conceptual change'* [55] took place, in this respect, for some pupils and this needs further investigations, by focusing on the individual times requested for a fully aware conceptual change and the enlargement of the experimental field as relevant factor to investigate.

At the end of the teaching experiment, the teacher and the two researchers, rethought all the TLS steps by trying to make evident the learning progress at each step and their connections, in order to identify the characteristics of the proposed teaching/learning environments that contributed to modify the disciplinary as well as the modeling competencies of our pupils. Their evaluation showed two characteristics as mainly affecting learning: the first one involving the adopted working method and the second one involving the pedagogical tools.

i. The audio recorded registration supplied us the evidence that learning mainly resulted from an interaction between the pupils and the tutors, that supplied support for the process of construction of meanings: thus, not a responsive tutoring but a problematic and scaffolding one. Another methodological general point involved the collaboration and co-construction of new ideas through peer cooperation in the working group setting. The continuous peer interaction involved the listening of new ideas produced by others as well as co-constructing new ideas with others. This stimulated reflection: i.e. comparing what one understands with what the other person is saying.

ii. Pedagogical tools played an important role in scaffolding pupils' understanding: simulations and visualizations allowed pupils to connect experimental results with models supplying explanations, by making evident how the observable perceptual global phenomena emerge from intrinsically distinct underlying local interacting processes. The tools allowed us to initially remove mathematics and to focus on clarifying the sub-concepts: it has been possible to reveal the non-perceptible parts of the concepts, by making these more concrete and by providing hands-on experiences situating the concepts in an everyday context through appropriate analogies. The dynamic computerized displays allowed an appropriate representation of such dynamic concepts, in contrast with the usual ways of describing and symbolizing phenomena that use static representations.

Interactions among peers and with tutors, as well as appropriate pedagogical tools, stimulated pupils in how making thinking visible by explicating their multiple views of the same phenomenon, sorting out ideas, entertaining new ideas and forming more cohesive and sophisticated views. Technological supports made scientific models visible to pupils and comparable with their mental models, by scaffolding the process of considering alternative explanations and consequently of singling out mistakes.

Conclusion

The reported study is concerned with a method to develop TLSs and experiment them in a classroom setting. Their design approach as well as their implementation involved various theoretical perspectives. In particular, we have reported:

- the articulation of '*local steps*' within each sequence to facilitate the pupils' understanding of the desired knowledge and to investigate the anticipated learning demands;
- some qualitative results emerging from observations of pupils' conceptual evolution during engagement in the proposed teaching/learning environments. More detailed analysis of the pupils' learning are reported elsewhere [31, 46, 47].

The results, drawn from the analysis of classroom discussion transcripts and pupils' worksheets, show that the majority of pupils' resultant conceptions was, very likely, the product of gradual conceptual change processes produced by the application of thinking resources appropriate for the level of analysis requested by the posed problems. Pupils showed to experience a form of conceptual change in which the status of the scientific conception rose at the expense of their intuitive conceptions. However, although the exact nature of this conceptual restructuring is unclear, this research pointed out some significant indicators of these changes.

Our research does not directly face the problem of motivation, although it recognizes its relevance in the construction of knowledge and in the process of conceptual change. In fact, our approach widely uses some teaching techniques such as eliciting pupils' views, multiple representations of scientific knowledge, hands-on activities and application to real life, that have been considered effective factors stimulating interest and motivation [56].

We believe that our research study provides a successful example of how to point out fruitful relationships between findings on conceptual change and instructional practice [57]. Moreover, the focus on details of both the instructional design principles and the learning demands of specific areas of the subject is, in our opinion, of considerable importance in the gradual development of effective teaching approaches. We agree with Tiberghien's remark [58] that the design of teaching situations for each domain of physics can be an endless task. However, teaching approaches and strategies that imply general characteristics of physics knowledge and process skills are also present in our research: these involve the focus on the process of constructing predictive conceptual models, the need to identify relevant model building procedures and the identification of the teaching/learning validity of the '*level of analysis*' concept as different ways to look at the same phenomenon.

References

[1] Gilbert, J.K.; Boulter, C. J. (Eds.) (2000). *Developing Models in Science Education.* Dordrecht: Kluwer.

[2] Gilbert, J. K. (Ed.) (2007). *Visualization in Science Education* (Models and Modeling in Science Education) London: Springer

[3] http://modeling.asu.edu/

[4] http://www.wcer.wisc.edu/NCISLA/MUSE/

[5] Boulter, C. J. & Buckley, C. K. (2000). Constructing a Typology of Models for Science Education. In Gilbert, J.K.; Boulter, C. J. (Eds.) *Developing Models in Science Education.* Dordrecht: Kluwer.

[6] Johnson-Laird, P. N. (1983). *Mental models: Towards a cognitive science of language, inference, and consciousness.* Cambridge, MA: Harvard University Press.

[7] Norman, D. A. (1983). *Some Observation on Mental Model.* In D. Gentner, & A. L. Stevens (Eds.), Mental Models (pp 7-14). Hillsdale, NJ: Lawrence Erlbaum Associates.

[8] Craik, K. (1943). The Nature of Explanation. Cambridge: Cambridge University Press.

[9] de Kleer, J. and Brown, J.S. (1983*). Assumptions and Ambiguities in Mechanistic Mental Models.* In: D. Gentner, D., & A. L. Stevens (Eds). Mental Models. (pp.15-34).Hillsdale, NJ: Lawrence Erlbaum Associates.

[10] Gentner, D., & Stevens, A. L. (Eds.). (1983). *Mental models.* Hillsdale, NJ: Lawrence Erlbaum Associates.

[11] Dunbar, K., & Fugelsang, J. (2005). Scientific thinking and reasoning. In: K. J. Holyoak & R. Morrison (Eds.). *Cambridge Handbook of Thinking & Reasoning.* (pp. 705-726). Cambridge: Cambridge Univ. Press..

[12] Greca, I. M., & Moreira, M. A. (2002). Mental, Physical, and Mathematical Models in the Teaching and Learning of Physics. *Science Education*, **86**(1), 106-121.

[13] Bachelard, G. (1938) La formation de l'esprit scientifique: contribution à une psychanalyse de la connaissance objective. Paris: Vrin. Translated (2006) Formation of the Scientific Mind. Manchster, U.K.: Clinamen Press Ltd.

[14] Campbell, D. (1920). Physics: The Elements. Cambridge: Cambridge University Press (Republished in 1957 As The Foundations Of Science. New York:Dover)

[15] Giere, R. (1990). Explaining Science: A Cognitive Approach. Chicago: University of Chicago Press.

[16] Harré, R. (1970). The Principles Of Scientific Thinking. Chicago: University Of Chicago Press.

[17] Gilbert , J.K. Boulter, C. (1998). Models In Explanations, Part1:Horses For Courses? *International Journal Of Science Education*, **20**, 83-97.

[18] Nagel, E. (1961). The Structure of Science. *Problems in the Logic of Scientific Explanations.* New York: Harcourt, Brace & World.

[19] von Glasenfeld, E. (1995). Radical constructivism: *A way of knowing and learning.* London: Falmer.

[20] Nersessian, N. J. (1999). Model-based reasoning in conceptual change. In: Magnani, L., Nersessian, N. J., & Thagard, P. (eds.) *Model-Based Reasoning in Scientific Discovery.* (pp. 5-22). New York: Kluwer Academic/Plenum Publishers.

[21] Lijnse, P. L. (1995). 'Developmental research' as a way to an empirically based 'didactical structure' of science. *Science Education*, **79**, 189-199.

[22] Méheut, M., & Psillos, D. (2004). Teaching learning sequences: aims and tools for science education research. *International Journal of Science Education*, **26**(5), 515-535.

[23] Duit, R., Komorek, M., & Wilbers, J. (1997). Studies on educational reconstruction of chaos theory. *Research in Science Education*, **27**, 339-357.

[24] Duit, R., Gropengieβer, H., & Kattmann, U. (2005).Toward science education research that is relevant for improving practice: The model of educational reconstruction. In H. E. Fisher (Ed) *Developing Standard in Research on Science Education.* (pp. 1- 9). London, UK: Taylor and Francis.

[25] Resnick, M., & Wilensky, U. (1998). Diving into Complexity: Developing Probabilistic Decentralized Thinking Through Role-Playing Activities. *Journal of the Learning Sciences*, **7**(2), 153-171.

[26] Wilensky, U., & Resnick, M. (1999). Thinking in Levels: A Dynamic Systems Perspective to Making Sense of the World. *Journal of Science Education and Technology*, **8**(1), 3-19.

[27] Chi, M. T. H. (2005). Commonsense Conceptions of Emergent Processes: Why Some Misconceptions Are Robust. *Journal of the Learning Science*, **14**, 161-199.

[28] Hammer, D. (2000). Student resources for learning physics. *American Journal of Physics*, **68**(S1), 552-559.

[29] Gilbert, J. K., Boulter, C., & Rutherford, M. (1998). Models in Explanations Part 1: Horses for Courses? *International Journal of Science Education*, **20**(1), 83-97.

[30] White, B., & Gunstone, R. F. (1992). Probing understanding. London: Falmer.

[31] Fazio C, Guastella I, Sperandeo-Mineo R.M., Tarantino G. (2007). Modelling Mechanical Wave Propagation: Guidelines and Experimentation of a Teaching Learning Sequence. *International Journal of Science Education.* in Press.

[32] Gilbert, J. (2004). Models and modelling: routes to more authentic science education. *International Journal of Science and Mathematics Education*, **2**, 115-130.

[33] Linder, C. J. (1993). University physics students' conceptualizations of factors affecting the speed of sound propagation. *International Journal of Science Education*, **15**(6), 655-662.

[34] Maurines, L. (1993). Spontaneous reasoning on the propagation of sound. In J. Novak (Ed.), *Proceedings of the Third International Seminar on Misconceptions and Educational Strategies in Science and Mathematics*. Cornell University, Ithaca, NY, USA (distributed electronically).

[35] Wittmann, M. C., Steinberg, R. N., & Redish, E. F. (1999). Making sense of how students make sense of mechanical waves. *The Physics Teacher*, **37**(1), 15-21.

[36] Wittmann, M. C., Steinberg, R. N., & Redish, E. F. (2003). Understanding and Affecting Student Reasoning about Sound. *International Journal of Science Education*, **25**(8), 991-1013.

[37] Fazio, C., Guastella, I. , Sperandeo-Mineo, R M, & Tarantino, G. (2006). Measuring longitudinal wave speed in solids: two methods and a half. *European Journal of Physics*, **27**, 687-702.

[38] Linn,M. C. & Songer, N. (1991). Teaching thermodynamics to middle school students: What are appropriate cognitive demands? *Journal of Research in Science Teaching*, **28**, 885-918.

[39] Lewis, L. & Linn, M. (1994). Heat energy and temperature concepts of adolescents, adults, and experts: Implications for curricular improvements. *Journal of Research in Science Teaching*, **31**, 657-677.

[40] Novick, S. and Nussbaum, J. (1978). Junior High School Pupils' Understanding of the Particulate Nature of Matter: *An Interview Study. Science Education*. **62**, 273-281.

[41] Andersson, B. R. (1990). Pupils' conceptions of matter and its transformations (age 12-16). *Studies in Science Education*, **18**, 53-85.

[42] Rozier, S and Viennot, L. (1991). Students' Reasoning in Thermodynamics. *International Journal of Science Education*, **13**, 159-170.

[43] Engel-Clough, E. and Driver, R. (1985). Secondary students' conceptions of the conduction of heat: bringing together scientific and personal views. *Physics Education*, **20**, 176-182.

[44] Carlton, K. (2000). Teaching about heat and temperature. *Physics Education*, **35**, 101-105.

[45] http://ccl.northwestern.edu/netlogo/

[46] Aiello, M. L., & Sperandeo-Mineo, R. M. (2000). Educational reconstruction of physics content to be taught and pre-service teacher training: a case study. *International Journal of Science Education*, **22**, 1085-1097.

[47] Sperandeo-Mineo R.M., Fazio C. & Tarantino G. (2006) Pedagogical Content Knowledge Development and Pre-Service Physics Teacher Education: A Case Study. *Research in Science Education*, **36**, 235-268.

[48] Reiner, M., Slotta, J. D., Chi, M. T. H., & Resnick, L. B. (2000). Naive physics reasoning: A commitment to substance-based conceptions. *Cognition and Instruction*, **18**(1), 1-34.

[49] Chi, M. T. H., Slotta, J. D., & De Leeuw, N. (1994). From Things to Process: a Theory of conceptual change for learning science concepts. *Learning and Instruction*, **4**, 27-43.

[50] Chi, M. T. H., & Roscoe, R. (2002). The processes and challenges of conceptual change. In M. Limon, & L. Mason (Eds.), *Reconsidering Conceptual Change: Issues in Theory and Practice* (pp. 3-27). Dordrecht, Netherlands: Kluwer Academic Publisher.

[51] Chi, M. T. H. (2005). Commonsense Conceptions of Emergent Processes: Why Some Misconceptions Are Robust. *Journal of the Learning Science*, **14**, 161-199.

[52] Chi, M. T. H., Feltovich, P. J., & Glaser, R. (1981). Categorization and Representation of Physics Problems by Experts and novices. *Cognitive Science*, **5**, 121-152.

[53] Vosniadou, S. (Ed.). (1994). Capturing and modelling the process of conceptual change. *Learning and Instruction*, **4**, 45-69.

[54] Vosniadou, S., & Ioannides, C. (1998). From conceptual development to science education: a psychological point of view. *International Journal of Science Education*, **20**, 1213-1230.

[55] Chinn, C. A., & Brewer, W .F. (1993). The role of anomalous data in knowledge acquisition: a theoretical framework and implications for science instruction. *Review of Educational Research*, **63**, 1-49.

[56] Palmer, D. (2005). A Motivational View of Constructivist Informed Teaching, *International Journal of Science Education*, **27**, 1853-1881.

[57] Duit, R., & Treagust, D. F. (2003). Conceptual change: a powerful framework for improving science teaching and learning. *International Journal of Science Education*, **25**, 671-688.

[58] Tiberghien, A. (2000). Designing teaching situations in the secondary school. In R. Millar, J. Leach, & J. Osborn (Eds), *Improving Science Education - The contribution of Research* (pp. 27-47). Buckingam, UK: Open University Press.

In: Science Education in the 21st Century
Editor: Ingrid V. Eriksson, pp. 137-164

ISBN: 1-60021-951-9

Chapter 4

The Professional Development of Secondary Education Science Teachers: A Case Study as Methodological Integration

Bartolomé Vázquez-Bernal[1,2], Roque Jiménez-Pérez[2] and Vicente Mellado[3]

[1] Secondary school physics and chemistry teacher, Huelva, Spain
[2] Dept. of Science Education and Philosophy, Faculty of Education, University of Huelva, Spain
[3] Dept. of Science and Mathematics Education, Faculty of Education, University of Extremadura, Badajoz, Spain

Abstract

The work that we describe here is a case study of a secondary education science teacher, whom we shall refer to as Ana, about the interaction between action-oriented reflection and action itself, and their influence on professional development. The study was carried out from two different perspectives —one with a qualitative orientation using diverse data collection and analysis instruments, and the other collaborative action-research to form the backbone of Ana's professional development. In our theoretical outline, we stress two concepts— that of reflection which sustains the theoretical-practical dialectic, and that of complexity as a progression hypothesis of central importance in which we distinguish the three dimensions of technique, practice, and criticism.

The results showed Ana to be in transition from a technical to a practical dimension, with her reflection and her classroom practice being closely integrated and in the process of becoming more complex. A central core of obstacles was observed that impeded her professional development.

Acknowledgements

This work was financed by Research Projects SEJ2006-04175 of the Ministry of Education and Science (Spain) and European Regional Development Fund (ERDF).

1. Professional Development as the Integration of Reflection and Practice: A model Based on Complexity

For many years the 'professional development' (PD) of science teachers focused almost exclusively on renovating and increasing knowledge of scientific content. In the last two decades, however, pædagogical content knowledge (Shulman, 1986) has also found a place in this development, and has been the motivation of numerous studies with science teachers (Gess-Newsome & Lederman, 1999). A number of educators have taken a broader sociocultural or social constructivist perspective in PD models (Bell & Gilbert, 1998; Elliot, 1993; Hewson *et al.*, 1999; Howe & Stubbs, 2003; Kepler, 1999; Luft, 1999, 2001; Marx *et al.*, 1998).

Nevertheless, there have been many approaches to PD. A synthetic description is given in Mellado et al. (2006), including the union of the social and the personal (Hargreaves, 1996), long-term sustained support (Sánchez & Valcárcel, 2000), the didactics of science as the axis of teacher education (Tobin & McRobbie, 1996; Tobin et al., 1997), and action-research (Lyons et al., 1997; Roth, 1998). The teacher is an integral part of the community of a school, and it is very difficult for change to be individually implemented, and even more so for it to be consolidated, against the current of that school's educational culture and socially accepted norms (Bell, 1998; Hargreaves, 1996; Mellado et al., 2006; Milicic et al., 2004; Sánchez & Valcárcel, 2000).

The present study of PD reformulates some of these proposals. For convenience, we divide PD into three 'stages' (Fig. 1) in the sense of 'levels of perception', by which we mean that the teacher's development involves progressing from observing events as examples of generic classes to seeing them in increasing detail and complexity.

In a first stage, we accept the view that teachers' development involves professional, social, and personal growth (Bell & Gilbert, 1994; Bell, 1998).

In a second stage, PD is linked more particularly to three spheres: (i) a sphere of knowledge – knowledge of the practice of the profession; (ii) a sphere of know-how – procedures for the development of the teacher and for curricular innovation through action-research (Stenhouse, 1987); and (iii) a sphere of being – of attitudes, of a search for identity as a member of a community of continually evolving learners.

The third stage is the most explicit, with complexity in both reflection and classroom practice. In this stage, we will inquire into some of the obstacles that teachers face in developing this complexity of reflection and of classroom practice, and their mutual integration.

Experienced teachers have very stable personal practical knowledge and beliefs. These have become consolidated throughout their careers, and are very resistant to the change. Moreover, there exist conditioning elements that reinforce traditional models, and are obstacles to changing them (Shwartz, Ben-Zvi & Hofstein, 2005; Tobin, 1998; Verjovsky & Waldegg, 2005).

As suggested by Lakatos's metaphor of competition between research programs (Lakatos, 1983), there is undoubtedly a central core in teachers' conceptions and practice which is very resistant to change (Mellado, 2003). If we can determine just what these core obstacles are, we will then have a real possibility of reconstructing more complex practical theories (Fourez, 1994; Niaz, 2002).

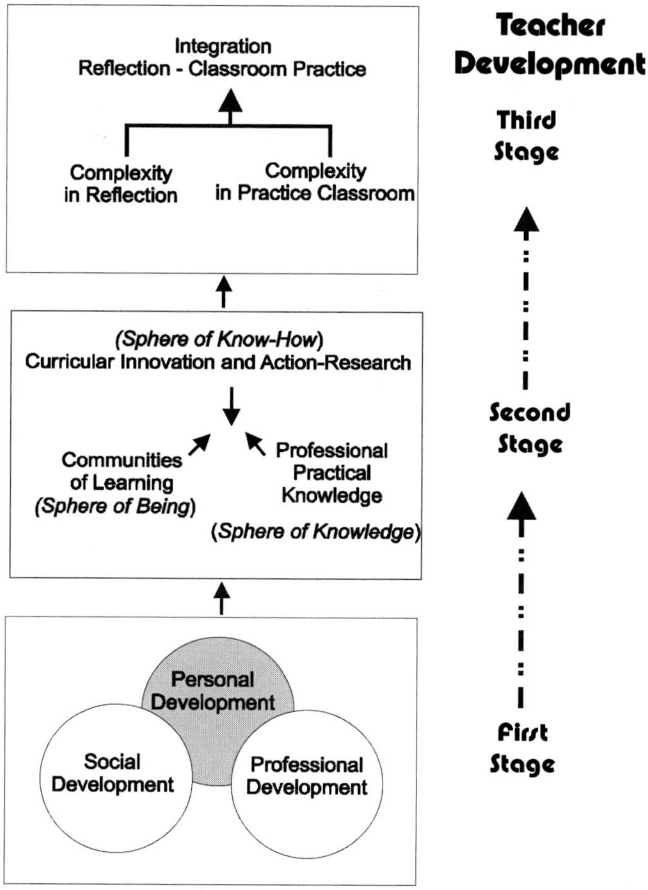

Figure 1. Stages in our proposal of Teacher Development.

2. The Complexity of Reflection and Practice

Advances in education research have led to growing recognition of the processes of reflection (Goodman, 1987; Kemmis, 1999; Loughran, 1996; Schön, 1983; Zee & Roberts, 2002; Zeichner, 1993; 1995). Reflection is interpreted as an act inherent to practice itself (Jiménez et al., 1999) or as a necessary catalyst for the development of professional knowledge (Bryan & Atwater, 2002). This opening to personal and shared dialogue is especially important for teacher development (Gunstone & Northfield, 1994), since change is usually continuous and gradual, and takes place in collaboration with colleagues (Mellado et al., 2006; Reyes et al., 2001). Our present interest is in particular to understand and interpret reflection and its relationship with practice (Baird et al., 1991; Gunstone et al., 1993, 1994; Gunstone, 1999; Windschtil, 2003).

We establish the hypothesis of complexity, the heir of earlier theoretical developments in the field of science education, above all of the different approaches to the evolution of the teacher-centred models of education. There then arises the search for personal models of education that are specific to each teacher. One result of such studies was the elaboration of the Hypothesis of Progression on Teachers' Professional Knowledge (Porlán & Rivero, 1998; Wamba, 2001; Jiménez & Wamba, 2003).

Although the literature mostly shows a preference for the existence of three levels associated with reflection (in chronological order: Hoyle, 1974; van Manen, 1977; Tom, 1984; Zimpher & Howey 1987; Carr & Kemmis, 1988; Louden, 1991; Schön, 1998; Elliot, 1999), studies of a complex nature require approaches of a diverse conceptual nature (Herrán, et al., 2005). On the basis of the foregoing contributions, we proposed a Hypothesis of Complexity (Vázquez-Bernal, 2006) in which we differentiated three dimensions in place of the three levels described by the aforementioned authors, but now associated with a somewhat different concept. Feldman (1994, 2002), in a reflection on the necessity to analyze how teachers reason, on their practices, and on the social and political context, suggested a metaphor of horizons — a teacher's view of his or her educational horizon is in some way like a physical horizon. In this same sense, we associate metaphorically the capacity for reflection with the concept of dimension. Thus, we distinguish three dimensions: technical, practical, and critical. In our hypothesis, each dimension becomes more complex, adding increasing complexity to reflection itself from purely instrumental interests in the technical dimension, through practical problem solving, to the social awareness and emancipating role of education in the critical dimension.

In the Hypothesis of Complexity, we recognize an increasing level of complexity in teachers' reflection linked to their classroom practice. Zeichner (1993) and Cachapuz (1995) reject the reduction of the process of reflection to the consideration of teaching skills and strategies, to the exclusion of the ends of education. We concur with the view of reflection as a promoter of knowledge described by Cochran-Smith & Lytle (2003), who distinguish between knowledge for practice which is generated in the university, knowledge in practice which is generated in everyday classroom practice, and knowledge of practice which is the merging of the former two.

Our basis is the necessary integration of teachers' classroom practice and the quality of their reflection as generator of knowledge. From our perspective, this has to result in a certain degree of complexity in their professional development. Figure 2 is a graphical representation of the Hypothesis of Complexity in which we emphasize the increasing degree of complexity as the teacher advances from one dimension to the next.

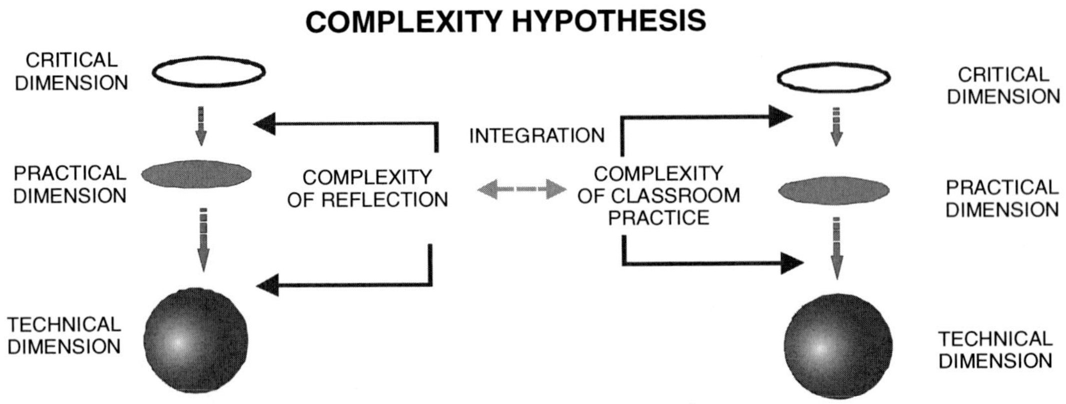

Figure 2. The integration of reflection and classroom practice.

How to actually develop and practise reflection is a teacher's own choice based on his or her context and personal history. This does not mean that we do not defend a complex and systemic view of a teacher's discourse, with the incorporation of different voices or models (Pozo, 1999). The notion of complexity, extensively developed by Morin (1995), is a value, thought, and action orienting ideological option, and a perspective on action and the construction of knowledge (Bonil & Pujol, 2005).

3. Case Studies as Methodological Integration

Various authors (Colás & Buendía, 1998; Latorre et al., 1996) have described the existence of three predominant philosophical currents in educational research — logical empiricism, interpretative theory, and critical theory. Indeed, these approaches coexist in educational research today. Nevertheless, given the complex and multidimensional nature of education itself, other authors (Herrán et al., 2005) consider it necessary to formulate a new research paradigm specific to education which they call 'Complexity', and which integrates theory and practice.

With respect to the professional development of science teachers, some authors have questioned whether the professional perspectives of these teachers have improved with the projects that were implemented three decades ago (Lesh & Lovitts, 2000). To respond as to how a PD project may have influenced teaching practice requires studies that extend over time and are undertaken in natural contexts. For that reason, researchers in education need to recognize the sociocultural, historical, systemic, political, cognitive, affective, and psychological aspects of teaching and learning (Nelly & Lesh, 2000). In sum, as indicated by Romberg & Collins (2000), there exists no single qualitative research program, but multiple approaches characterized by the levels at which the research is targeted: ontological, epistemological, methodological, and technical. The choice of method is conditioned by the nature of the research problem and its related questions. Nevertheless, it needs to be emphasized that no method can provide answers to all the questions that can arise in education research (Mestre, 2000). But they do share a common objective — to generate scientifoc-educational knowledge using a scientific methodology (Fox, 1981; Bartholomew, 1984; De la Orden, 1985; Parrilla, 1992; Latorre et al., 1996; Colás & Buendía, 1998; Sandín, 2003). This form of dealing with the subject requires medium and long term longitudinal studies, since the changes occur over long periods of time, and only longitudinal studies will show whether those changes are ephemeral or permanent (White & Arzi, 2005).

In our research in particular, we used a case study as a tool for methodological integration. In the terminology of Stake (1998), this would be a study of an instrumental nature as it was aimed at trying to understand the problematic situations that arise from the research itself, although it has a critical aspect (Keeve, 1998) in that it has the additional goal of inducing change, and an evaluational aspect in its objective of determining results.

4. Research Questions

Our work forms part of an action-research program carried out in a state secondary-education school in a town of 20 000 inhabitants in the province of Huelva, located in the southwest of

Spain. It was carried out by a teacher from the same school, with four other teachers from the school and university science education researchers. In another work (Vázquez et al., submitted for publication), we described the case of Marina, another of the participating teachers. In the present article we will centre on the case of a teacher we will refer to as Ana, one of the teachers participating in the research. Although the results are specific for each teacher, both are part of the same research group, so that the context, method, and presentation of the results have many aspects in common.

We summarize in the form of questions the problems that we set ourselves:

a) Methodologically, can we find a balance between methods following different paradigms without reducing the credibility of the research?
b) Is it possible to base professional development on the integration and complexity of reflection and classroom practice?
c) Is there a degree of convergence between reflection and classroom practice? How do they both evolve over time?
d) In which dimension of the Complexity Hypothesis is Ana situated, both for reflection and classroom practice?
e) What kind of obstacles hinder the integration of reflection and classroom practice, preventing the desirable development?
f) In what way can a program of action-research affect the professional development of teachers?

5. Research Methods

To put into operation the process followed with the teachers, we applied the Kemmis and McTaggart (1988) action-research model, whereby successive methodological cycles of planning, acting, observation, and reflection are established. The work was carried out during two consecutive school years, 2001-2002 and 2002-2003, with students of the 3rd year of Secondary Obligatory Education (S.O.E.: 14-15 year-olds). Ana, the pseudonym of our second case, is a chemistry graduate. On joining the working group in the 2001-2 school year, she had twelve years experience as a teacher. She has tenure, so that her administrative situation is one of stability. On occasions Ana describes herself as "the mother". Previous research has shown that the metaphors used by teachers allow one to discover the implicit referents that sustain the teacher and that have a powerful influence on his or her teaching behaviour in the classroom (McRobbie & Tobin, 1995; Mellado et al., 2006; Tobin et al., 1994).

Instruments of Data Collection and Analysis

We shall classify the instruments according to their methodological function: first order (data collection), second order (category systems, theoretical and taxonomic models), and third order (presentation and interpretation of the data).

The reflection data collection instruments were the teacher's diaries and memos, transcriptions of the work group meetings, questionnaires, and interviews. For the classroom practice, the ethnographic notes and extracts from the videotapes of the class sessions, and other documentary sources such as the programs and teaching units, or the work produced by the students (first-order). The fundamental instrument for data analysis, both for reflection and practice, was the Reflection Analysis Categories System, R.A.C.S. (second-order: Annex I). We also used third-order instruments for the representation, such as complexity spheres for reflection and practice and the reflection-practice integration horizon, which allowed the evolution of the teacher to be viewed over the course of the two school years that the study lasted, as well as giving an overall representation of the integration of reflection and practice.

The variety of instruments allowed us to undertake a triangulation from different perspectives. Thus, for reflection we distinguished three forms according to the context: introspective, inquiry oriented, and obtained from the diaries; interrogative, statement oriented, and obtained from the memos and interviews; and group, interaction oriented, and obtained from the records of the group meetings.

We considered six analytical frames to elaborate the R.A.C.S.: ideological, teacher education, psychological, contextual, epistemological, and curricular. We included in the ideological frame the influence of ideologies on the teaching environment and the relationship occupied by the teachers and their context. The teacher education frame included the education of teachers in general, and that specific to science education. The psychological frame included the way in which the teachers view school-level learning and their understanding of the obstacles in the teaching-learning process. In the contextual frame we included the interactive teacher-student relationships, and social and scheduling organization. In the epistemological frame, we included school-level knowledge and how the teachers perceive it, the criteria with which they select the sources of information, and the utility and construction of this kind of knowledge. Finally in the curricular frame we included the types of problems that are proposed in the classroom, the sequences of activities, the curricular material employed, and evaluation. To analyze the classroom practice we used the Practice Analysis Categories System. This consists of the psychological, contextual, epistemological, and curricular frames. The ideological and teacher education frames were excluded, not because they have no influence, but because they are not directly inferred from the data available.

We distinguished three dimensions in the Complexity Hypothesis —technique, practice, and critique— for each analysis structure or substructure that characterize each frame. Annex I gives the categories associated with the different frames. As an example, Table I includes the four categories associated with “Curricular Materials or Information Sources” (T. D. = technical dimension; P. D. = practical dimension; C. D. = critical dimension). Each of them has an associated code (TTEX, PSOU, and CSIM) formed from its meaning in English, and which we shall use in the representations of the analysis of the results.

Table I. Categories associated with the "Curricular Materials" structure.

FRAMES	STRUCTURES	SUBSTRUC-TURES	CATEGORIES	CODES
Curricular	Curricular materials or information sources	For the teacher / student	T.D.: Use of the textbook as principal source of information.	TTEX
			P.D.: Use of different sources of information.	PSOU
			C.D.: Socially important problems.	CSIM

The information was processed using the AQUAD computer program. Its application was mainly centred on coding, search options, calculation of the information unit frequency, associations in simple coding sequence formats, and the verification of links (Huber, Fernández & Herrera, 2001). The documents that were analyzed by means of AQUAD were the diaries, memos, minutes of the meetings, interviews, and ethnographic records. Besides the frequency calculations, the AQUAD program allowed us to obtain the linkages of the categories that appear clustered in grouped sequences of two or three. From the links between the codes, we obtained the core categories which stand out because of their high frequency rates, or the nucleating agents which also link positively with other categories.

6. Results

Frequency Analysis of Reflection Categories

Table II. Frequencies of reflection categories for Ana during the school year 2002/3.

	Technical dimension	Practical dimension	Critical dimension
Ideologies in the educational environment	TAPO (0)	POPT (0)	CADM (6), CENV (2), CINQ (1), CHIS (0), CIDE (0)
Interactive teacher-student relationships	TBOX (2)	PTWK (3)	CPRO (1)
Teacher education	TACC (2)	PPER (22)	CREF (10)
Science education	TRES (2), TDOM (3)	PREA (0), PPCK (14)	CEXP (2)
School-level learning	TMEM (7), TASS (10)	PMEA (1)	CLSC (0)
Obstacles in the teaching-learning process	TUND (37)	PDIF (54)	COBS (2)
Student motivation	TIND (22), TMTV (10)	PPAR (33)	CDIS (5)
Classroom atmosphere	TCON (5)	PATM (3)	CREG (1)
Social organization	TCOM (0)	PSWK (9)	CSOC (15)
Temporal organization	TTIM (8)	PADA (9)	CDYN (2)
School-level knowledge	TSCI (99)	PDLY (32), PAPD (1)	CEPI (0)
Selection criteria for the information sources	TEFF (21)	PANS (4)	CDEC (1)
Use of school-level knowledge	TPRG (1)	PSKI (11)	CCIT (4)
Construction of school-level knowledge	TSOL (1), TREI (17)	PINT (21)	CNEG (0)
Types of problems	TCAP (14)	POPR (12)	CRSP (5)
Sequence of activities	TRIG (3)	PFLE (15)	CDIV (2)
Curricular materials or information sources	TTEX (1)	PSOU (13)	CSIM (1)
Objective of evaluation	TOBJ (1)	PSUB (3), PIDE (2)	CCEN (1)
Purpose of evaluation	TSAN (0)	PSUM (4)	CDEV (2)
Participation in evaluation	TGUA (5)	PSTU (0)	CCOE (0)
Instruments of evaluation	TEXA (8)	PDSE (15)	CMET (0)

Reflection was analyzed from three perspectives: frequency analysis, linkage analysis, and content analysis. For the first two, we used the AQUAD computer program. Because of its length, we shall not show the content analysis in the present work. Table II gives in parentheses the frequencies of appearance of the code associated with a given category within each form of reflection for the 2002–3 school year, as the sum of the three forms of reflection.

The category that appears with greatest frequency (99) is scientific knowledge as the fundamental knowledge of the curricular content (TSCI), belonging to the technical dimension. This is followed by reflection on the difficulties of the students (PDIF), from the practical dimension (54). These two categories were also the most frequent with Marina, another of the participating teachers (Vázquez et al., submitted for publication). Next appear allusions to the students' lack of understanding (TUND: 37), the effort and participation of the student as extrinsic factors (PPAR: 33), reference to the daily experience of the student (PDLY: 32), the individual effort of the student as an intrinsic factor (TIND: 22), attention to the students' interests (PINT: 21), acritical efficacy and its role in improvement (TEFF: 17), the reinforcement of the ideas presented by the teacher (TREI: 17), support for the socially more needy (CSOC: 15), and the use of problems with closed answers (TCAP: 14). We would highlight the presence of categories of a technical nature in the teacher's reflections, even though in an emergent form: ranked second in her reflections is her interest in the students' difficulties.

Table III presents the evolution of the frequency analysis for Ana in the two school years studied. During the year 2002/3, we observed a tendency towards the practical and critical dimensions, although the most important increase in complexity towards the practical dimension occurred in the interrogative reflection (final interview). Group reflection remained the same in both years, ending at the same level as introspective reflection.

Table III. Evolution of Ana's reflection frequencies during the two school years.

Type of reflection	**Nº of Participants**	**Oriented towards**	**Analysis instruments 2002**	**Results for Ana 2001 (704 codes)**	**Results for Ana 2002 (662 codes)**
Introspective	One	Inquiry	Teacher's diary *(187 codes)*	Technical dim.: 87% Practical dim. 13% Critical dim.: 0%	Technical dim.: 56% Practical dim.: 43% Critical dim.: 1%
Interrogative	Two	Statement	Interview *(55 codes)*	Technical dim.: 53% Practical dim.: 35% Critical dim. 12%	Technical dim.: 36% Practical dim.: 32% Critical dim.: 32%
Grouped	More than two	Interaction	Record of meetings *(396 codes)*	Technical dim.: 53% Practical dim.: 44% Critical dim.: 3%	Technical dim.: 42% Practical dim.: 47% Critical dim.: 11%

Figure 3 shows this evolution graphically. One observes that the teacher is more technical when she is reflecting alone (diary) than when she is interacting with the others.

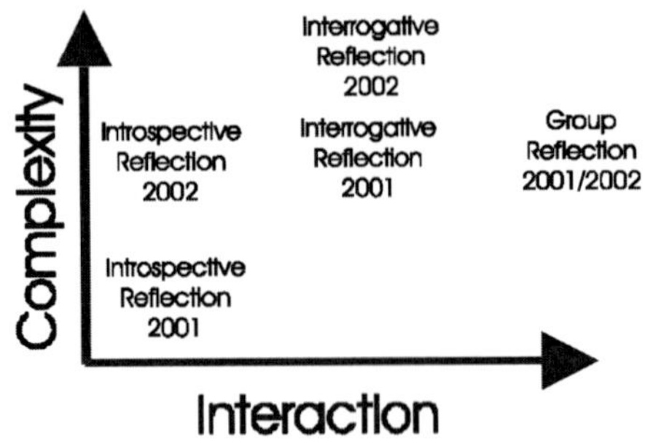

Figure 3. Complexity vs interaction.

Analysis of the Reflection Category Linkages

The linkage analysis allowed us to determine the associations between two and three categories that appeared over 10 consecutive lines in a segment of text, and to establish hypotheses of linkages between these categories. Selecting the most frequent linkages in each school year, we elaborated the diagrams shown in Figure 4 for the categories in the technical dimension.

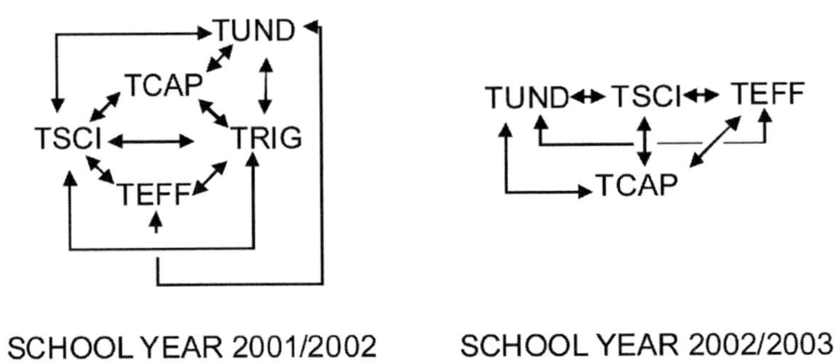

Figure 4. Most significant links between Ana's technical dimension categories in the two school years.

The stability of Ana's reflections is shown in the existence of similar linkages in the two study years. However, in that second school year, there disappeared the reflections with regard to the inflexibility of the activities (TRIG). Observing the linkages between categories in Figure 4, one can indicate that scientific knowledge as fundamental knowledge of the curricular content (TSCI), the students' lack of understanding (TUND), the use of closed-answer problems of purely scientific content (TCAP), as well as the use of an acritical and technological way of thinking based on criteria of efficiency (TEFF) form the irreductible core of Ana's explicit theories in the technical dimension.

For the practical dimension the most significant linkages are those represented in Figure 5. Observing the categories, one appreciates that they have little stability, i.e., they only appear in a single school year. Moreover, they are linked to technical categories, which indicates the influence that the latter exert on the teacher's thinking. We also would highlight that the practical categories are not linked to each other in either of the school years under study.

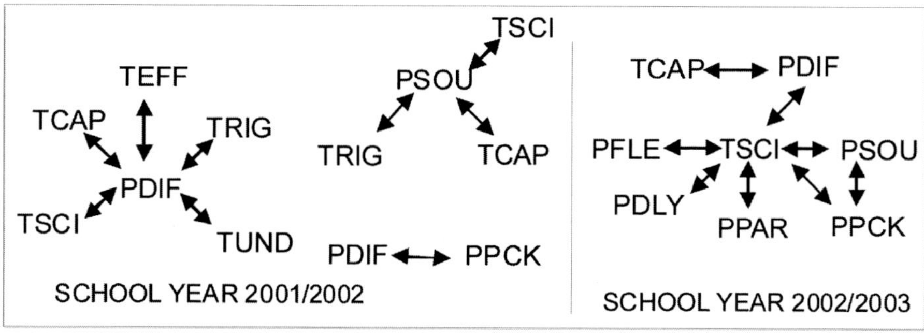

Figure 5. Most significant links between the categories of the practical dimension in the two school years analyzed.

In the school year 2002/2003 the central category is that of reflection on students' difficulties (PDIF). However, this category appears to be closely linked to categories of the technical dimension, being affected by that of efficiency criteria (TEFF), scientific knowledge as fundamental knowledge of the curricular content (TSCI), the students' lack of understanding (TUND), the use of closed-answer problems of purely scientific content (TCAP), as well as the use of rigid activities (TRIG). The irreductible core is much smaller than that analyzed previously for the technical dimension, above all because the absolute values at which the links appear are much lower in most cases. The practical categories of greatest frequency in Table II (PDIF, PPAR, PDLY, PINT) do not appear directly linked to each other.

The categories of the practical dimension that appear most linked in Figure 5 are reflection on the students' difficulties (PDIF), the use of different sources of information (PSOU), and pædagogical content knowledge (PPCK). These constitute the nucleating agents around which Ana's reflection evolves.

Figure 6 shows the most significant links for the critical dimension. In the first school year, one finds the critical category on the school's role in overcoming social inequalities (CINQ) to be associated with the practical category of the recognition of what is personal as starting point in the teacher's improvement (PPER).

In the second school year, the central critical category in the linkages is the use of problems of research into the social and natural environment (CRSP). Other critical categories that appear are the educational administration and its pressure as political fact (CADM) and the formation of citizens with critical capacity concerning advances in science and technology (CCIT). Their interest lies in that they are linked to categories of a practical nature. One observes, however, that the categories of a critical nature that appear are different in the two school years and have no linkage with each other, reflecting their instability in the teacher's reflections.

By way of synthesis, the analysis of the two school years indicates that the irreductible core of the explicit theories in the technical dimension is consolidated around certain categories. This core has many similarities with the case of Marina, another of the participating teachers (Vázquez et al., submitted for publication). The two teachers have a strong initial education in their discipline and very similar career histories. The categories are those of reflections involving the students' lack of motivation and understanding, the use of closed problems of purely scientific content, the use of an acritical and technological way of thinking based on criteria of efficiency and the inflexibility of the activities. The category of reflection on students' difficulties in the teaching-learning process is the nucleating agent of Ana's explicit theories, and a true zone of future development towards complexity.

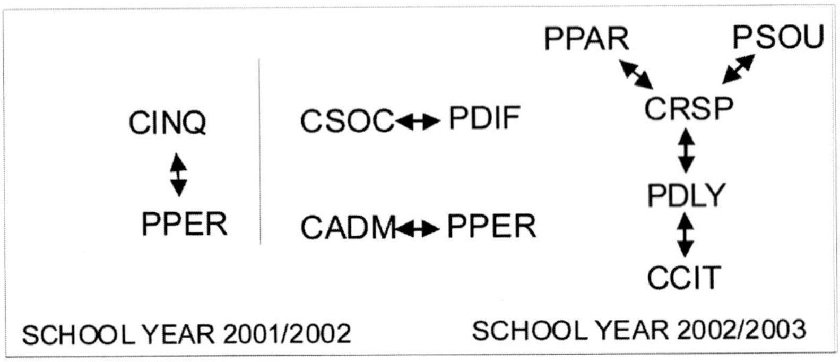

Figure 6. Most significant links between the critical dimension categories.

The Complexity Sphere for Ana's Reflection

Figure 7, the complexity sphere, is a synthesis of the results for Ana's reflection over the two school years of the study. This is a representation of our complexity hypothesis metaphor, displaying its different dimensions. We have put the technical dimension on the first inner ring (one-dimensional), the practical dimension on the outer ring simulating therefore a (two-dimensional) surface, and the critical dimension at a certain height above the two rings thus giving a spherical perspective (three-dimensional).

We have adopted a set of keys to interpretation so as to facilitate rapid access to the information. These keys are on two levels — codes and arrows. A code can be of three types — solid black (the code appears in both school years), open black with white fill (the code appears in a single school year), and solid white (the code does not appear in either school year). The arrows express movement towards increasing complexity. Thus, a solid black arrow indicates there has been a completed transition of complexity from an initial dimension to the next, a white arrow that there still exists no complexity, and a dashed arrow that the transition is in progress.

When the evolution of the teacher's reflection is observed overall, one perceives the existence of a process of undoubted complexity. This holistic vision, found in all the frames of analysis, shows the teacher to be in a situation of transition from the technical towards the practical dimension. A closer view of the complexity sphere shows that the structures that have most evolved towards complexity are the social organization of the classroom, the criteria for the selection of the information sources, and the purpose of evaluation. On the contrary, the teacher's understanding of how school-level learning takes place and who participates in evaluation shows structures that are less complex and more resistant to change.

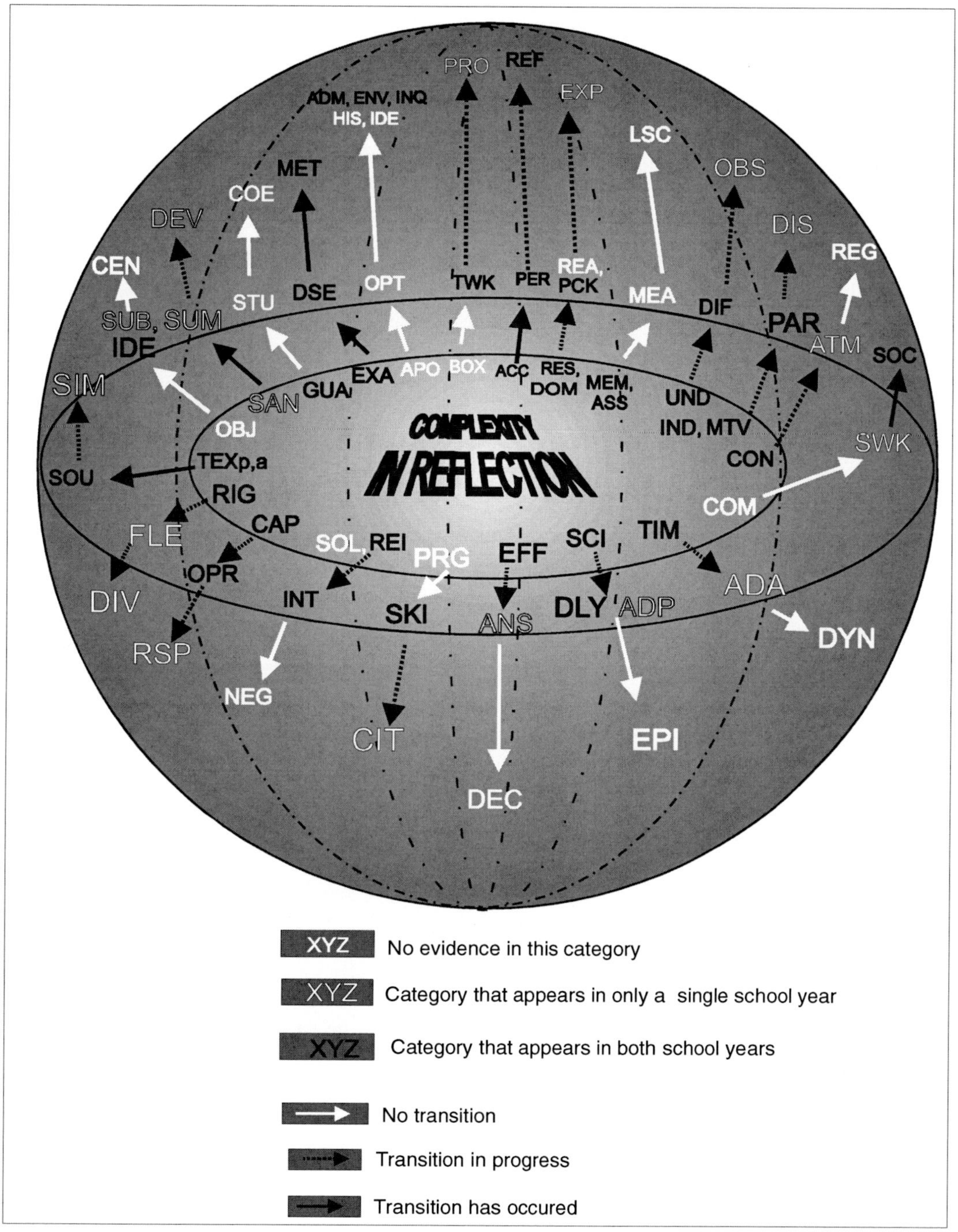

Figure 7. Complexity sphere for the evolution of Ana's reflection.

Analysis of Classroom Practice

As an example of the classroom practice results, we shall describe the structure of the intervention corresponding to "Student Curricular Materials or Information Sources".

Technical dimension: The analysis of the ethnographic notes indicates that the main source of information for the students is the set of notes (TTEX) prepared by the teacher. These consist of a considerably revised adaptation of textbooks based on program-guides of Hierrezuelo et al. (1998). The teaching unit was prepared anew, however, to adapt it to the new guidelines and agreements that came out of the work group. Corresponding to the first time that these notes were used is the following extract from the ethnographic notes (RET1–2002: ethnographic record 1 of school year 2002–3):

> "11:22 h. Gives specifications of how to recover the corresponding parts. Adds that, for this subject, photocopies are going to be given out.
> "11:23 h. Says to Gema that if she keeps up with her attitude being just as positive, her results will improve and that she will see it.
> "11:24 h. The teacher comments that last year they saw the complete unit. RET1–2002."

In sum, it is clear the importance that the teacher gives to the teaching unit notes as the main curricular material for her students. This constitutes a true obstacle in the development of her classroom practice.

Practical dimension: Among the diverse sources of information for the students (PSOU), there was the use of an educational documentary in video format on Molecular Kinetic Theory, followed by a questionnaire on the content of the documentary. The following is an extract from the ethnographic notes:

> "Adds that they will do the theory that comes next tomorrow. Says that now they are going to watch the MKT video, adding 'you have to take notice of what it says'. Antonio seems a bit unruly.
> "12:59 h. The teacher asks the students to come up closer to see the video. The students seem to be getting a bit jumpy, especially Antonio.
> "13:00 h. The teacher prepares the film. The students are talking among themselves. RET7–2002.
> "Projects a video documentary 'Journey to the Interior of Matter'. RET152002."

The students are also asked to look for information on food labels about the composition and concentration of certain foodstuffs, as is expressed in the following extract:

> "When you arrive home, look at the composition of some bottled foods (at least three, and at most five), and answer:
> "What are the components?
> "Which is the solvent and which the solute(s)?
> "What is the concentration of each? RET10–2002."

In sum, together with the use of the activity program-guides, the teacher uses other sources of information for the students such as video materials and documents that can be found in everyday life.

Critical dimension: Proposed as an important social problem (CSIM) was to investigate the 'sinking of a oil-carrying barge at Algeciras'. This was related to the 'Prestige' oil-spill catastrophe, a subject of burning interest in the mass media throughout the time at which the teachers were doing the practical development of the teaching unit, and was naturally considered important for treatment in class. The following shows the first time that the teacher told her students that they were going to use this social problem in their class:

"8:44 h. Expresses that they are going to make posters about the sinking of a barge carrying fuel in Algeciras. She says that oil is a solution with solvent and solutes, as they will see at the end of the topic. She requests them to bring material to do this tomorrow, since it is about approaching the problem of contamination. RET11–2002."

Another important social problem was the treatment of drug abuse and its effects on health, a topic of especial importance to the students in that class. The following extract from the notes refers to the first time that this topic was dealt with in class:

"8:25 h. They are going to begin reading page 11 on drug abuse. She designates Antonio to read, who protests. José volunteers and reads the text.
"8:26 h. The teacher says: 'Think if using them is worth it.' The teacher comments a little on it. Later she says that the substances from 'joints' remain in the urine for 2 days to 3 weeks. Several students show interest in that. RET19–2002".

For the structure 'Student Curricular Materials or Information Sources', Ana makes use of diverse sources without abandoning the prepared teaching unit. In this aspect, she is making her teaching practice more complex. Because of her use of important social and environmental problems from articles in newspapers, magazines, and written media in general, we think that she has initiated the transition towards the critical dimension. As an illustration of this spatio-temporal synthesis, Fig. 8 shows the result of the analysis of classroom practice for this structure.

Figure 8. The evolution of the complexity of Ana's classroom practice for the structure 'Student Curricular Materials or Information Sources'.

In Fig. 9, we show the complexity sphere for classroom practice obtained from the analysis of the two school years studied.

The overall analysis of the classroom practice shows a perceptible complexity from the technical dimension towards the practical. Areas belonging to the psychology and curricular frames, such as school-level learning and participation in evaluation, present technical positions. At the opposite end, such areas as the selection criteria for the information sources, obstacles in the teaching-learning process, and the purpose of evaluation, lie within the critical dimension, or are in transition towards it. These aspects represent practical theories that are firmly established in the teacher.

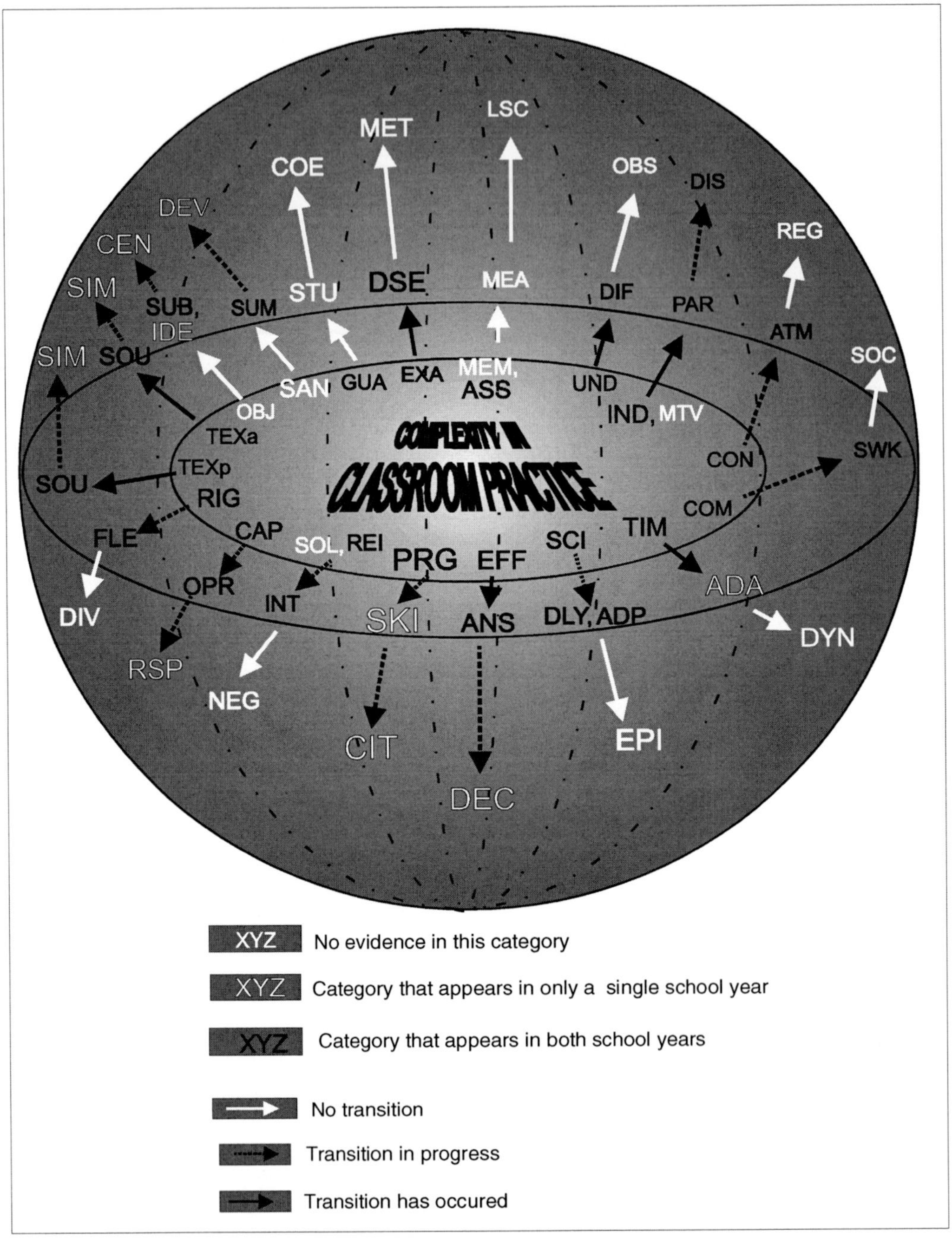

Figure 9. Complexity sphere for the evolution of Ana's classroom practice.

The Interaction between Reflection and Practice: The Integration Horizon

The analysis of the interaction between classroom reflection and practice requires the convergence of diverse sources of information: analysis of the reflection from the different proposed perspectives, analysis of the classroom practice, and comparison with the teacher's initial conceptions. Figure 10 is a synthesis of the results for the integration between classroom reflection and practice. This is therefore a third-order instrument which we call the integration horizon.

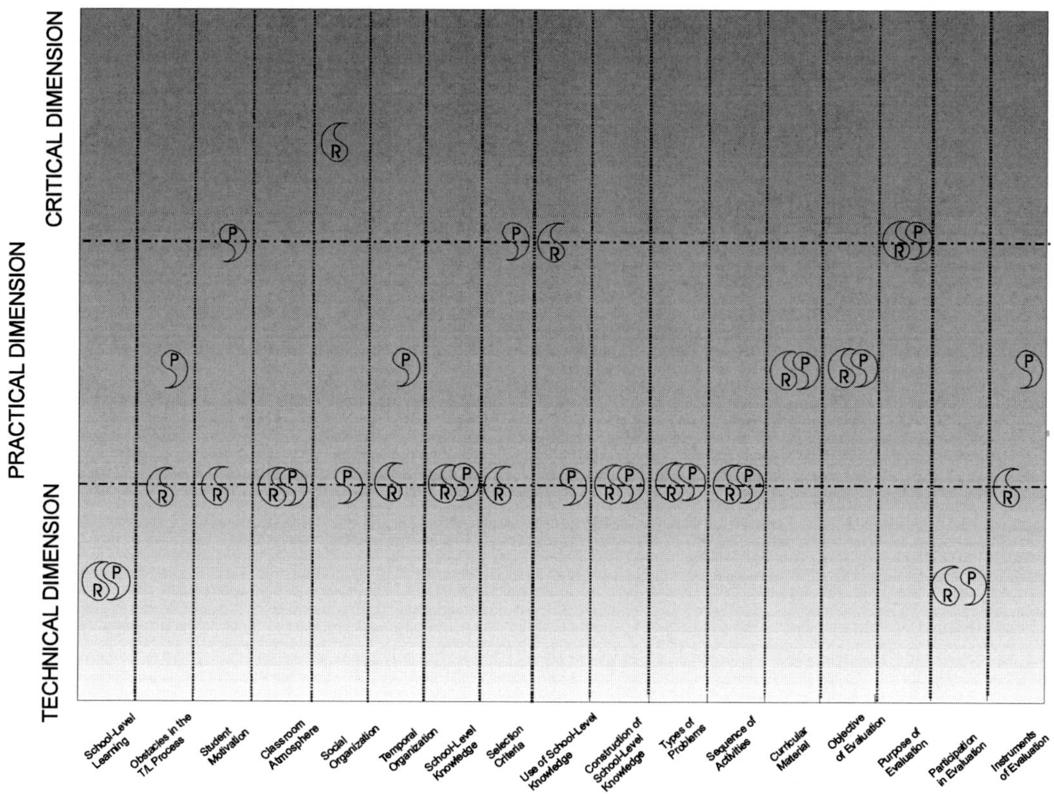

Figure 10. Ana's reflection-practice integration horizon.

All the structures and substructures common to Ana's reflection and classroom practice are distributed along the horizontal axis. The three zones corresponding to the critical, practical, and technical dimensions are located on the vertical axis, and between them there are two intermediate zones, represented by dash-dotted lines, which symbolize the transition towards the practical or critical dimensions. In the centres of each dimension are the zones of the dimensions themselves. These forms of representation express the degree of complexity obtained in each structure or substructure analyzed. At the same time, the possible integration or non-integration is represented by the corresponding symbol. Two paired halves of the sphere indicate integration, and unpaired halves non-integration. The left half corresponds to reflection, and the right to classroom practice.

The results of integration lead to us to make the observations presented in Table IV from the two analytical perspectives of integration and complexity.

Table IV. The integration and complexity perspectives in the case of Ana.

The integration perspective	The complexity perspective
• There exists a degree of total integration in the following study areas: school-level learning, classroom atmosphere, school-level knowledge, construction of school-level knowledge, type of problem, activity sequence, teacher and student curricular materials, objective of evaluation, purpose and participation in evaluation. They belong to four frames without distinction — psychological, contextual, epistemological, and curricular. • A degree of partial integration is observed in the obstacles in the teaching-learning process, the temporal organization, and evaluation instruments. • A lack of integration is noticeable, however, in the remaining areas: social organization (more complex in reflection), student motivation (more complex in classroom practice), selection criteria for the information sources (more complex in classroom practice), and use of school-level knowledge (more complex in reflection). These areas belong to the contextual and epistesmological frames, respectively.	• A transition zone is perceived between the technical and practical dimensions in which most of the structures or substructures analyzed are located. • In a second zone of the purely technical dimension, there are only three representations — school-level learning, and participation in evaluation. • Lastly, there is a transition zone between the practical and critical dimensions in which the integrated purpose of evaluation are located. The indication of the preceding results is that Ana is essentially in a phase of complexity from the technical towards the practical dimension.

7. Conclusions and Implications

In the following paragraphs, we will present our conclusions in terms of the questions that guided the study, with emphasis given to the main methodological aspects that underlay the case study.

a) The research instruments were conditioned by the theoretical and methodological framework of the 'Complexity Hypothesis'. This required the use of diverse and versatile instruments characteristic of different methodological approaches. Our choice of quantitative instruments was aimed at obtaining the information given by the count, comparison, and evolution of the frequencies of the codes (corresponding to the R.A.C.S. categories), and at the possible establishment of hypotheses of linkages between those categories using a Boolean method of quantitative comparison. We used a diversity of qualitative instruments — participant observation (ethnography), structured interviews ('Grounded Theory'), diaries (biography), video records (speech analysis), and conversation records (phenomenology). As characteristic of the critical methodological approach, we planned cycles of action-research with a clearly emancipatory conception of education. Finally, the model of complexity gave us an interactive and systemic perspective on education (García

Díaz, 2005). The case study has therefore shown itself to be a system that is capable of integrating diverse methodological approaches, and has allowed enrichment of the generation of new knowledge.

b) Our Complexity Hypothesis allowed us to establish In the following paragraphs, we will present our conclusions in terms of the questions that guided the study, with emphasis given to the main methodological aspects that underlay the case study.

c) The research instruments were conditioned by the theoretical and methodological framework of the 'Complexity Hypothesis'. This required the use of diverse and versatile instruments characteristic of different methodological approaches. Our choice of quantitative instruments was aimed at obtaining the information given by the count, comparison, and evolution of the frequencies of the codes (corresponding to the R.A.C.S. categories), and at the possible establishment of hypotheses of linkages between those categories using a Boolean method of quantitative comparison. We used a diversity of qualitative instruments — participant observation (ethnography), structured interviews ('Grounded Theory'), diaries (biography), video records (speech analysis), and conversation records (phenomenology). As characteristic of the critical methodological approach, we planned cycles of action-research with a clearly emancipatory conception of education. Finally, the model of complexity gave us an interactive and systemic perspective on education (García Díaz, 2005). The case study has therefore shown itself to be a system that is capable of integrating diverse methodological approaches, and has allowed enrichment of the generation of new knowledge.

d) Our Complexity Hypothesis led to the establishment of six frames of influence — ideological, teacher education, psychological, contextual, epistemological, and curricular— which proved to be effective filters through which to gain insight into Ana's reflections and actions. The establishment of levels of increasing complexity has permitted us to orient and make explicit the degree of development of the participating teachers.

e) From the results of the integration of reflection and classroom practice, the analysis confirmed that there existed a degree of full integration between reflection and classroom practice in 10 of the 17 areas studied. In other areas, however, there was only a degree of partial integration or even an evident lack of integration between reflection and practice. These results largely coincide with those reported in another work (Vázquez *et al.*, submitted for publication) for the case of Marina, one of the same team of teachers although they both maintain their own particular characteristics. While we can not claim that this fact is extendable to all science teachers, it does serve to make one reflect on the model of initial teacher education that these teachers experienced, and in general on the design of teacher education courses in Spain (Oliva, 2005).

Our results indicated that progress occurred in reflection and classroom practice, but many areas were frequently out of phase, although it could not be said in general terms that one of them was more advanced than the other. In some domains, there existed full integration between reflection and practice, but in others explicit reflection was more advanced than the teacher's actual practice, or, *vice versa*, practice was more advanced than reflection. This is coherent with our fundamental premise that reflection not only guides action, but is in turn guided by it in a process of mutual and convergent interaction, and that a program of professional development has to take both aspects into account. Previous research has shown that teachers' advanced concepts may not be converted into classroom practice if the teachers lack schemes of practical action that are coherent with their beliefs (Gess-Newsome & Lederman, 1993; Mellado, 1998).

f) The existence of a process of increasing complexity is clearly perceived when the evolution of Ana's reflection and classroom practice are observed overall, and, interpreted holistically, one sees how the teacher is in transition from the technical towards the practical dimension. It is clear that Ana's pædagogical model has not undergone a total change, but rather a gradual evolution with some aspects having evolved more than others. For us, particularly noteworthy was her conception of school-level learning, since previous research has shown that a fundamental factor that stimulates science teachers' change is becoming aware of the existence of the students' alternative ideas, and moving from a content-centred model to one that is more student centred (da Silva *et al.*, 2007; Hewson *et al.*, 1999).

g) As a result of our inquiry into the obstacles to professional development, we would highlight that Ana conceives of learning as a process of assimilation, not of construction of knowledge. The intrinsic character of motivation makes her think that if the students' ideas do not progress, it is because the students are not making the necessary effort. There persists the pressure of time in preventing diversification of the activities and attending to different rates of learning in the classroom. In her stated conceptions of the nature of science, she showed an empiricist outlook, imbued with a certain epistemological absolutism, and an accumulative view of scientific knowledge. The acritical efficacy, inflexibility in the activities, the use of closed problems, the punishment aspect of evaluation, and setting herself at the centre of most of the classroom activities, form part of the difficult to reduce core of obstacles hindering her professional development (Vázquez *et al.*, 2006a). These obstacles are closely related to the structures that have remained in the technical dimension, making it difficult for her to evolve. Such structures are strongly ingrained and consolidated by years of professional experience, and for them to evolve it is necessary to continue working with and providing support to teachers (Vázquez *et al.*, 2007). This claim is based on our having compared and contrasted the results of the different case studies that we have performed within the same group of teachers, since they all share a similar core of conceptions and practices that are refractory to change.

h) We believe that the action-research program has had a very positive impact on the professional development of all the participating teachers, in spite of their presenting cores of obstacles that are very resistant to change. At the time of writing, the group of teachers has dissolved for motives of professional mobility. This leads us to meditate on the difficulty of establishing stable work groups of teachers actively involved in their professional development and sympathetic to curricular innovation. Our research is currently continuing with one of the teachers of the group (Marina), focusing on the uses of the new information technologies, on the consequences of classroom practice for the students' learning, and on the importance of pædagogical content knowledge (Gárritz & Trinity-Velasco, 2004). All these facets, together with the other contributions from Science Education research, constitute the central axis for teachers' professional development (Vázquez *et al.*, 2006b).

Annex I

Reflection Analysis Categories System (R.A.C.S.)

FRAMES	STRUCTURES	SUBSTRUCTURES	CATEGORIES	CODES
Ideological	Ideologies in The Educational Environment		D.T.: No ideological referents exist, it being an apolitical activity.	TAPO
			P.D.: Ideology as personal option.	POPT
			C.D.: Educational administration and its pressure as political fact. Influence of the social environment. The school's role in overcoming social inequalities. History as conforming the current educational situation. The analysis of ideologies in the educational context.	CADM CENV CINQ CHIS CIDE
	Teacher-Environment Relationship		T.D.: Resistance to break with the sensation of the classroom as a black box.	TBOX
			P.D.: Teacher team work.	PTWK
			C.D.: Educational profession in continual interaction with the context.	CPRO
Educational	Teacher Education		T.D.: Teacher education as a mere accumulation of credits.	TACC
			P.D.: The personal as the starting point in the improvement of the teacher.	PPER
			C.D.: Recognition of reflection as guarantor of personal change.	CREF
	Science Teaching		T.D.: Resistance to reading pædagogical material. Insecurity in the curricular domain of the material being taught.	TRES TDOM
			P.D.: Interest in reading pædagogical material. Pædagogical content knowledge.	PREA PPCK
			C.D.: Belonging to a community of professionals in which science teaching experiences are exchanged.	CEXP
Psychological	School-Level Learning		T.D.: Role of memory as principal guarantor of learning. Assimilation as guarantor of learning.	TMEM TASS
			P.D.: Construction as guarantor of meaningful learning.	PMEA
			C.D.: Learning as social construction.	CLSC
	Obstacles in the Teaching-Learning Process		T.D.: Students' lack of understanding.	TUND
			P.D.: Reflection about the students' difficulties.	PDIF
			C.D.: Collective inquiry into the nature of the obstacles behind the difficulties.	COBS

Reflection Analysis Categories System (R.A.C.S.) (Continued)

FRAMES	STRUCTURES	SUBSTRUC-TURES	CATEGORIES	CODES
Contextual	Interactive Teacher-Student Relationships	Student Motivation	T.D.: Individual effort of the student as intrinsic factor. Motivation already present in the groups of students.	TIND TMTV
			P.D.: Student effort and participation as extrinsic factors.	PPAR
			C.D.: Positive discrimination towards students with special educational needs and/or social deprivation.	CDIS
		Classroom Atmosphere	T.D.: Discipline and control of the class.	TCON
			P.D.: Negotiation about the working atmosphere in the classroom.	PATM
			C.D.: Systematic use of contracts with the students to regularize classroom work.	CREG
	Social Organization		T.D.: Competitiveness as impulse to learning.	TCOM
			P.D.: Student team work.	PSWK
			C.D.: Support for the socially needy.	CSOC
	Temporal Organization		T.D.: The lack of time characterizes the form of classroom intervention.	TTIM
			P.D.: The time factor is made flexible and adapted to classroom intervention.	PADA
			C.D.: Time is dynamically adapted to the different rates of learning.	CDYN
Epistemological	School-Level Knowledge		T.D.: Scientific knowledge as fundamental knowledge of the curricular content.	TSCI
			P.D.: Reference to the daily experience of the student. Adaptation of the scientific content to the school environment.	PDLY PAPD
			C.D.: School science possesses its own epistemological status inherent to its social character.	CEPI
	Selection Criteria for the Information Sources		T.D.: Acritical efficacy and its role in improvement.	TEFF
			P.D.: Providing answers to open questions that are proposed.	PANS
			C.D.: Making decisions and acquiring commitments.	CDEC
	Use of School-Level Knowledge		T.D.: Concern to attain objectives and complete the programming.	TPRG
			P.D.: Acquisition of basic skills in solving problems	PSKI
			C.D.: The formation of citizens with critical capacity concerning advances in science and technology.	CCIT
	Construction of School-Level Knowledge		T.D.: The teacher as the sole constructor of school-level knowledge. Reinforcement of the ideas presented by the teacher.	TSOL TREI
			P.D.: Attention to the students' interests.	PINT
			C.D.: Negotiation with the students on aspects of the curriculum.	CNEG

Reflection Analysis Categories System (R.A.C.S.) (Continued)

<table>
<tr><th>FRAMES</th><th>STRUCTURES</th><th>SUBSTRUC-TURES</th><th>CATEGORIES</th><th>CODES</th></tr>
<tr><td rowspan="27">Curricular</td><td rowspan="3">Types of Problems</td><td rowspan="3"></td><td>T.D.: Use of closed-answer problems.</td><td>TCAP</td></tr>
<tr><td>P.D.: Use of open problems.</td><td>POPR</td></tr>
<tr><td>C.D.: Use of research problems into the social and natural environment.</td><td>CRSP</td></tr>
<tr><td rowspan="3">Sequence of Activities</td><td rowspan="3"></td><td>T.D.: Rigid activities.</td><td>TRIG</td></tr>
<tr><td>P.D.: Flexibility in the sequences of teaching.</td><td>PFLE</td></tr>
<tr><td>C.D.: Sequences of flexible diversified activities taking the students' different rates of learning into account.</td><td>CDIV</td></tr>
<tr><td rowspan="6">Curricular Materials or Information Sources</td><td rowspan="3">For the Teacher</td><td>T.D.: Use of the textbook as principal source of information.</td><td>TTEX</td></tr>
<tr><td>P.D.: Use of different sources of information.</td><td>PSOU</td></tr>
<tr><td>C.D.: Socially important problems.</td><td>CSIM</td></tr>
<tr><td rowspan="3">For</td><td>T.D.: Use of the textbook as principal source of information.</td><td>TTEX</td></tr>
<tr><td>P.D.: Use of different sources of information.</td><td>PSOU</td></tr>
<tr><td>C.D.: Socially important problems.</td><td>CSIM</td></tr>
<tr><td rowspan="12">Evaluation</td><td rowspan="3">Objective (what to evaluate)</td><td>T.D.: Objectivity of evaluation.</td><td>TOBJ</td></tr>
<tr><td>P.D.: Subjectivity of evaluation.
Evolution of the students' ideas.</td><td>PSUB
PIDE</td></tr>
<tr><td>C.D.: Acquisition of student-centred skills.</td><td>CCEN</td></tr>
<tr><td rowspan="3">Purpose of Evaluation</td><td>T.D.: Santioned evaluation.</td><td>TSAN</td></tr>
<tr><td>P.D.: Evaluation as summary and overview of the process.</td><td>PSUM</td></tr>
<tr><td>C.D.: Evaluation as centred on the development of the individual as a social entity.</td><td>CDEV</td></tr>
<tr><td rowspan="3">Participation in Evaluation</td><td>T.D.: The teacher as sole guarantor of the evaluation process.</td><td>TGUA</td></tr>
<tr><td>P.D.: Participation of the student in the evaluation process.</td><td>PSTU</td></tr>
<tr><td>C.D.: Co-evaluation carried out by teachers and students.</td><td>CCOE</td></tr>
<tr><td rowspan="3">Instruments of Evaluation</td><td>T.D.: Use of the final examination.</td><td>TEXA</td></tr>
<tr><td>P.D.: Diversity of sources for the evaluation.</td><td>PDSE</td></tr>
<tr><td>C.D.: The students' productions based on metacognition and self-evaluation.</td><td>CMET</td></tr>
</table>

References

Baird, J.R., Fensham, P.J., Gunstone, R.F. & White, R. (1991). The importance of reflection in improving science teaching and learning. *Journal of Research in Science Teaching*, **28** (2), 163-182.

Bartolomé, M. (1984). La Pedagogía experimental. In A. Sanvincens (Ed.) *Introducción a la pedagogía experimental*. Barcelona: Barcanova.

Bell, B. & Gilbert, J. (1994). Teacher development as professional, personal and social development. *Teaching & Teacher Education*, **5**, 483-497.

Bell, B. (1998). Teacher development in science education. In B. J. Fraser & K. G. Tobin (Ed.): *International Handbook of Science Education* (pp. 681-693). Dordrecht: Kluwer.

Bonil, J. & Pujol, R. M. (2005). La aventura de integrar la complejidad en la educación científica de la ciudadanía. *Actas del VII Congreso Internacional sobre Investigación en la Didáctica de las Ciencias. Número Extra. Enseñanza de las Ciencias*. Barcelona.

Bryan, L. & Atwater, M.M. (2002). Teacher beliefs and cultural models: A challenge for science teacher preparation programs. *Science Education*, **86**(6), 821-839.

Cachapuz, A. (1995). Da investigação sobre e para professores à investigação com e pelos professores de Ciências. In L.J. Blanco & V. Mellado (eds): *La Formación del Profesorado de Ciencias y Matemáticas en España y Portugal* (pp. 243-254). Badajoz (Spain): Diputación Provincial.

Carr, W. & Kemmis, S. (1988). *Teoría crítica de la enseñanza*. Barcelona: Martínez Roca.

Cochran-Smith, M. & Lytle, S.L. (2003). Más allá de la certidumbre: adoptar una actitud indagadora sobre la práctica. In A. Lieberman & L. Miller (Ed.): *La indagación como base de la formación del profesorado y la mejora de la educación* (pp. 65-79). Barcelona: Octaedro.

Colás, M. P. & Buendía, L. (1998). *Investigación Educativa*. 3rd edition. Sevilla: Alfar.

Confrey, J. (2000). Improving research and systemic reform towards equity and quality. In Anthony E. Kelly & Richard A. Lesh (Eds.): *Handbook of Research Design in Mathematics and Science Education* (87-106). Mahwah, New Jersey: LEA.

Cook T. D. & Reichrdt, CH. S. (1997). *Métodos cualitativos y cuantitativos en investigación educativa.* 3rd edition. Madrid: Morata.

da-Silva, C., Mellado, V., Ruiz, C. & Porlán, R. (2007). Evolution of the conceptions of a secondary education biology teacher: Longitudinal analysis using cognitive maps. *Science Education*, **91**(3), 461-491.

De La Orden, A. (1985). *La investigación educativa. Diccionario Ciencias Educación.* Madrid: Anaya.

Elliot, J. (1993). Conocimiento, poder y evaluación del profesor. In W. Carr (Ed.): *Calidad de la enseñanza e Investigación-Acción* (pp. 155-174). Sevilla: Díada.

Elliot, J. (1999). La relación entre comprender y desarrollar el pensamiento docente. In A. Pérez-Gómez, J. Barquín & J. F. Angulo (Ed.): *Desarrollo profesional del docente: Política, investigación y práctica* (pp. 364-378). Madrid: Akal.

Feldman, A. (1994). Erzberger's dilemma: Validity in action research and science teacher's need to know. *Science Education*, **78**(1), 83-101.

Feldman, A. (2002). Multiple perspectives for the study of teaching: Knowledge, reason, understanding, and being. *Journal of Research in Science Teaching*, **39** (10), 1032-1055.

Fourez, G. (1994). *La construcción del conocimiento científico*. Madrid: Narcea.

Fox, D. (1981). *El proceso de investigación en educación*. Pamplona: EUNSA.

García Díaz, J. E. (2005). Complejidad y construcción del conocimiento. *Actas del VII Congreso Internacional sobre Investigación en la Didáctica de las Ciencias. Número Extra. Enseñanza de las Ciencias*. Barcelona.

Gárritz, A. & Trinidad-Velasco, R. (2004). El conocimiento pedagógico del contenido. *Educación Química*, **15**(2), 1-6.

Gess-Newsome, J. & Lederman, N.G. (1993). Preservice biology teachers' knowledge structures as a function of professional teacher education: A year-long assessment. *Science Education*, **77**(1), 25-45.

Gess-Newsome, J. & Lederman, N.G. (1999). *Examining Pedagogical Content Knowledge.* Dordrecht: Kluwer A.P.

Goodman, J. (1987). Reflexión y formación inicial del profesor: Estudio de casos y análisis teórico. *Revista de Educación*, **284**, 223-244.

Gunstone, R., Slattery, M., Baird, J R. & Northfield, J.R. (1993). A case study exploration of development in preservice science teachers. *Science Education*, **77** (1), 47-73.

Gunstone, R. & Northfield, J. (1994). Metacognition and learning to teach. *International Journal of Science Education*, **16** (5), 523-537.

Gunstone, R. (1999). Content knowledge, reflection, and their intertwining: a response to the paper set. *Science Education*, **19** (3), 394-396.

Hargreaves, A. (1996). *Profesorado, cultura y modernidad.* Madrid: Morata.

Herrán, A. de la, Hashimoto, E. & Machado, E. (2005). *Investigar en educación. Fundamentos, aplicación y nuevas perspectivas.* Madrid: Dilex S.L.

Hewson, P. W., Tabachnick, B. R., Zeichner, K.M & Lemberger, J. (1999). Educating prospective teachers of biology: Findings, limitations, and recommendations. *Science Education*, **19** (3), 375-384.

Hierrezuelo, J. et. al. (1998). *Ciencias de la Naturaleza, Física y Química, 3ºESO.* Málaga: Elzevir.

Howe, A.C. & Stubbs, H. (2003). From science teacher to teacher leader: Leadership development as meaning making in a community of practice. *Science Education*, **87**(2), 281-297.

Hoyle, E. (1974). Professionality, professionalism and control in teaching. *London Educational Review*, **3**, 13-19.

Jiménez, R., Pérez, J. & Rodríguez, C. (1999). Los condicionantes de la formación inicial del profesorado en España: Presupuesto para un nuevo modelo de formación. In A. Pérez Gómez, J. Barquín Ruiz & J.F. Angulo Rasco (Ed.): *Desarrollo profesional del docente: Política, investigación y práctica* (pp. 146-180). Madrid: Akal.

Jiménez, R. & Wamba, A. M. (2003). ¿Es posible el cambio en los modelos didácticos personales? Obstáculos en profesores de Ciencias Naturales de Educación Secundaria. *Revista Interunivesitaria de Formación del Profesorado*, **17**(1), 113-131.

Keeves, J. P. (1998). Methods and processes in research in science education. In B. J. Fraser & K. G. Tobin (Ed.): *International Handbook of Science Education,* 1127-1155. Dordrecht: Kluwer.

Kemmis, S. (1999). La investigación-acción y la política de reflexión. In A. Pérez-Gómez, J. Barquín & J.F. Angulo (Ed.): *Desarrollo profesional del docente: Política, investigación y práctica* (pp. 95-118). Madrid: Akal.

Kepler, K. (1999). Investigación en la enseñanza: implicaciones para los programas de formación del profesorado. In A. Pérez Gómez, J. Barquín & J. F. Angulo (Ed.): *Desarrollo profesional del docente: Política, investigación y práctica* (pp. 339-363). Madrid: Akal.

Kuhn, T. (1962). *The Structure of Scientific Revolutions.* Chicago: University of Chicago Press.

Lakatos, I. (1983). *La metodología de los programas de investigación científica.* Madrid: Alianza Editorial.

Latorre, A. (2003). *La investigación-acción. Conocer y cambiar la práctica educativa.* Barcelona: Graó.

Lesh, R. & Lovitts, B. (2000) Research agendas: Identifying priority problems and developing useful theoretical perspectives. In A. E. Kelly & Richard A. Lesh (Eds.): *Handbook of Research Design in Mathematics and Science Education* (pp. 45-71). Mahwah, New Jersey: LEA.

Louden, W. (1991). *Understanding Teaching: Continuity and Change in Teachers' Knowledge*. London: Casell / New York: Teachers' College Press, Columbia University.

Loughran, J.J. (1996). *Developing reflective practice: Learning about teaching and learning through modelling*. Washington, D.C.: Falmer Press.

Luft, J.A. (1999). Teachers' salient beliefs about a problem-solving demonstration classroom in-service program. *Journal of Research in Science Teaching*, **36** (2), 141-158.

Luft, J.A. (2001). Changing inquiry practices and beliefs: The impact of an inquiry-based professional development programme on beginning and experienced secondary science teachers. *International Journal of Science Education*, **23** (5), 517-534.

Lyons, L.L., Freitag, P.K. & Hewson P.W. (1997). Dichotomy in thinking, dilemma in actions: Researcher and teacher perspectives on a chemistry teaching practice. *Journal of Research in Science Teaching*, **34** (3), 239-254.

McRobbie, C. & Tobin, K. (1995). Restraints to reform: The congruence of teacher and students actions in a chemistry classroom. *Journal of Research in Science Teaching*, **32**(4), 373-385.

Marx, R.W., Freeman, J., Krajcik, J. & Blumenfed, P. (1998). Professional development of science education. In B. J. Fraser & K. G. Tobin (Ed.): *International Handbook of Science Education* (pp. 667-681). Dordrecht: Kluwer.

Mellado, V. (1998). The classroom practice of preservice teachers and their conceptions of teaching and learning science. *Science Education*, **82**(2), 197-214.

Mellado, V. (2003). Cambio didáctico del profesorado de ciencias experimentales y filosofía de la ciencia. *Enseñanza de las Ciencias*, **21** (3), 343-358.

Mellado, V., Ruiz, C., Bermejo, M. L. & Jiménez, R. (2006). Contributions from the philosophy of science to the education of science teachers. *Science & Education*, **15**(5), 419-445.

Mestre, J. P. (2000). Progress in research: The interplay among theory, research questions, and measurement techniques. In A. E. Kelly & Richard A. Lesh (Eds.): *Handbook of Research Design in Mathematics and Science Education* (pp. 151-168). Mahwah, New Jersey: LEA.

Milicic, B., Utges, G., Salinas, B & Sanjosé, V. (2004). Creencias, concepciones y enseñanza en la Universidad: Un estudio de caso de desarrollo profesional colaborativo centrado en un profesor de física. *Revista Española de Pedagogía*, **229**, 377-389.

Morin, E. (1995). *Introducción al pensamiento complejo*. Barcelona: Gedisa.

Nelly, A. E. & lesh, R. (2000) Trends and Shifts in Research Methods. In A. E. Kelly & R. A. Lesh (Eds.) *Handbook of Research Design in Mathematics and Science Education* (pp. 35-44). Mahwah, New Jersey: LEA.

Niaz, M. (2002). Understanding nature of science as progressive transitions in heuristic principles. *Science Education*, **86** (6), 684-690.

Oliva, J.M. (2005). Sobre el estado actual de la revista Enseñanza de las Ciencias y algunas propuestas de futuro. *Enseñanza de las Ciencias*, **23** (1), 123-132.

Parrilla, A. (1992). Análisis de procesos de clase. Una perspectiva ecológica. In C. Marcelo (comp.), *La investigación sobre la formación del profesorado. Métodos de investigación y análisis de datos.* (pp. 227-257). Madrid: Cincel.

Porlán, R. & Rivero, A. (1998). *El conocimiento de los profesores. Una propuesta formativa en el área de ciencias.* Sevilla: Díada.

Pozo, J. I. (1999). Más allá del cambio conceptual: el aprendizaje de la ciencia como cambio representacional. *Enseñanza de las Ciencias*, **17**(3), 479-492.

Reyes, L., Salcedo, L. E. & Perafán,G. (2001). *Acciones y Creencias IV: Análisis e interpretación de creencias de docentes de Biología y Ciencias Naturales*, Universidad Pedagógica Nacional, Bogotá.

Richardson, K. (2001). *Modelos de desarrollo cognitivo.* Madrid: Alianza Editorial.

Romberg, T. A. & Collins, A. (2000). The impact of standards-based reform on methods of research in schools. In A. E. Kelly & R. A. Lesh (Eds.): *Handbook of Research Design in Mathematics and Science Education* (pp. 73-85). Mahwah, New Jersey: LEA.

Roth, W.M. (1998). Science teaching as knowledgability: a case study of knowing and learning during coteaching. *Science Education*, **39**(3), 253-282.

Sánchez, G. & Varcálcel, M.V. (2000). ¿Qué tienen en cuenta los profesores cuando seleccionan el contenido de enseñanza? Cambios y dificultades tras un programa de formación. *Enseñanza de las Ciencias*, **18**(3), 423-437.

Sandín, M.P. (2003). *Investigación cualitativa en educación. Fundamentos y tradiciones.* Madrid: McGraw-Hill.

Schön, D. (1983). *The Reflective Practitioner. How Professionals Think in Action.* London: Temple Smith.

Schön, D. (1998). *El profesional reflexivo.* Barcelona: Paidós.

Shwartz, Y., Ben-Zvi, R. & Hofstein, A. (2005). The importance of involving high-school chemistry teachers in the process of defining the operational meaning of 'chemical literacy'. *International Journal of Science Education*, **27**(3), 323-344.

Shulman, L. S. (1986). Paradigms and research programs in the study of teaching: A contemporary perspective. In M. Wittrock (Ed): *Handbook of Research on Teaching* (pp. 3-36). New York: McMillan.

Stake, R. E. (1998). *Investigación con estudio de casos.* Madrid: Morata.

Stenhouse, L. (1987). *Investigación y desarrollo del currículo.* Madrid: Morata.

Tobin, K., Tippins, D.J. & Gallard, A.J. (1994). Research on instructional strategies for teaching science. In D.L. Gabel (ed.), *Handbook of Research on Science Teaching and Learning* (pp. 45-93), New York: McMillan P.C.

Tobin, K. & McRobbie, C. (1996). Cultural myths as constraints to the enacted science curriculum. *Science Education*, **80**(2), 223-241.

Tobin, K., McRobbie, C. & Andersos, D. (1997). Dialectical constraints to the discursive practices of high school physics community. *Journal of Research in Science Teaching*, **34** (5), 491-507.

Tobin, K. (1998). Issues and trends in the teaching of science. In B.J. Fraser & K. Tobin (eds.): *International Handbook of Science Education* (pp. 129-151). Dordrecht: Kluwer A. P.

Tom, A. (1984). *Teaching as a moral craft.* New York: Longman.

Van Manen, M. (1977). Linking ways of knowing with ways of being practical. *Curriculum Inquiry*, **6**, 205-228.

Vázquez-Bernal, B. (2006). La interacción entre la reflexión y la práctica en el desarrollo profesional de profesores de ciencias experimentales de enseñanza secundaria. Estudio de casos. [The interaction between reflection and practice in the professional development of secondary education science teachers: Case study.] *Proquest Information and Learning*: Michigan (USA).

Vázquez, B., Jiménez, R., & Mellado, V. (2007). La reflexión en profesoras de ciencias experimentales de enseñanza secundaria. Estudio de casos. *Enseñanza de las Ciencias*, **25**(1), 73-90.

Vázquez, B., Jiménez, R., Mellado, V. Martos, M. & Taboada, C. (2006a). Evolución de la reflexión y práctica de aula en la resolución de problemas. El caso de dos profesoras de ciencias de secundaria. *Campo Abierto*, **25**(1), 133-152.

Vázquez, B., Jiménez, R., Mellado, V. & Taboada, C. (2006b). El análisis de la epistemología en el medio escolar. Estudio de un caso. *Revista Mexicana de Investigación Educativa*, v. 11, nº 31, 1259-1286.

Vázquez, B., Jiménez, R., Mellado, V. & Wamba, A. (submitted for publication). The interaction between reflection and practice in the professional development of a secondary education science teacher. *Journal of Research in Science Teaching*.

Verjovsky, J. & Waldegg, G. (2005). Analyzing beliefs and practices of a Mexican high school biology teacher. *Journal of Research in Science Teaching*, **42**(4), 465-491.

Wamba, A.M. (2001). *Modelos didácticos y obstáculos para el desarrollo profesional: Estudios de caso con profesores de Ciencias Experimentales en Educación Secundaria*. Proquest Information and Learning: Michigan (USA).

White, R. T. & Arzi, H.J. (2005). Longitudinal studies: Designs, validity, practicality, and value. *Research in Science Education*, **35**(1), 137-149.

Zee, E. H. Van & Roberts, D. (2002). Using pedagogical inquiries as a basis for learning to teach: Prospective teachers' reflections upon positive science learning experiences. *Science Education*, **85**(6), 733-757.

Zeichner, K. (1993). El maestro como profesional reflexivo. *Cuadernos de Pedagogía*, **220**, 44-45.

Zeichner, K.M. (1995). Los profesores como profesionales reflexivos y la democratización de la reforma escolar. In *Volver a pensar la educación,* Vol. II, 385-398. Madrid: Morata-Paideia.

Zimpher, K. L. & Howey, K.R. (1987). Adapting supervisory practices to different orientations of teaching competence. *Journal of Curriculum and Supervision*, **2**(2), 102-112.

In: Science Education in the 21st Century
Editor: Ingrid V. Eriksson, pp. 165-182
ISBN: 1-60021-951-9

Chapter 5

CONCEPTUAL DEPTH AND CONCEPTUAL USEFULNESS IN CHEMISTRY: ISSUES AND CHALLENGES FOR CHEMISTRY EDUCATORS

Kevin C. de Berg
Avondale College, PO Box 19, Cooranbong NSW 2265, Australia

Abstract

In the 1990's two movements began to challenge the practice of science education from an historical and philosophical perspective. The International History and Philosophy of Science and Science Teaching Group (IHPST) maintains that not enough attention is being given to the intellectual heritage of many of our concepts in science with the result that, while students might be able to solve some quantitative problems in an algorithmic fashion, they have little understanding of the meaning or significance of the concepts they are using. The International Society for the Philosophy of Chemistry (ISPC) focuses its attention on the discipline of chemistry and has challenged practising chemists and chemistry educators to address, amongst other things, such fundamental questions as the meaning of the terms *atom*, *element*, *law*, and *theory* as they are used in chemistry. Is it possible to seek and accomplish conceptual depth, a desire of both international groups, without compromising conceptual usefulness, or must one be compromised in the pursuit of the other? In this paper this question is addressed in relation to the thermodynamic concepts of *energy*, *heat*, and *work*; the foundational concepts of *element* and the *mole*; and *quantum mechanical* concepts in chemistry. A textbook analysis of the definitions and use of these concepts in a tertiary level general chemistry course is given as is the implications of the study for chemistry education.

Introduction

At the time when the International History and Philosophy of Science and Science Teaching Group (IHPST) was being formed, journals like "Interchange", "Synthese", "Educational Philosophy and Theory", and "Studies in Philosophy and Education" dedicated a whole volume to the question of the place of history in science teaching. The topic was not a new one of course and, although past attempts at incorporating history into science teaching had met with only limited success, a new wave of optimism descended on the science education community in the late 1980's with the result that IHPST was formed, conferences were convened in North America or Europe every two years, and a dedicated journal, "Science & Education", was organised with Kluwer Academic Press and now Springer Press. Part of the reason for this new wave of optimism was the need to address falling enrolments in science classes with new approaches to science curricula which were more human-oriented. Ian Winchester's editorial for the dedicated "Interchange" volume in 1989 signified the urgency of the matter in these words: "When it (science) is introduced......two distinct tendencies are present. The first is to treat science (and mathematics) as something with results but without a history. And the second is to treat science as something which can be nearly completely captured by working examples from a textbook. Both of these activities, namely, dwelling on agreed upon scientific results and standard examples, are important and part of the initiation into the complex of activities involved in a scientific discipline. But they fail to convey to the student.....anything of the excitement of scientific discovery, of the conceptual problems involved in forging a new realm of scientific inquiry or of advancing in an old one, of the myriad difficulties of interpretation, theory construction and abandonment or the moral and aesthetic problems faced in the process" (Winchester 1989, p.(i)). It is argued by members of the history in science teaching movement that providing students with the historical background of a scientific concept gives a dynamic character to it and illustrates the progressive rather than static nature of science. Those more skeptical of the movement argue that conceptual clarity is compromised when students are introduced to the conceptual errors of the past.

As far as chemistry is concerned it is true that history of chemistry has traditionally received more attention from educators than has philosophy of chemistry (Scerri 1997). This lack of philosophical influence in chemistry teaching and learning has led to the establishment of the International Society for the Philosophy of Chemistry (ISPC). It is argued that chemistry education will benefit from discussions that detail the nature of chemical knowledge. As examples of this approach Erduran and Scerri (2002) address the problem of 'reduction' and how this might clarify the relationship between chemical composition, bonding, and molecular structure; they address the nature of 'explanation' in chemistry and claim that "there is a distinction to be made about the explanatory status of electronic orbitals in chemistry and their ontological status in quantum mechanics. Reflective classroom discussions based on such distinctions are likely to promote deeper understanding of chemical explanations among university students"(Erduran & Scerri 2002, p.15); they address the question of the nature of 'law' in chemistry and in relation to the periodic law conclude that, "The crucial feature which distinguishes this form of 'law' from those found in physics is that chemical periodicity is approximate. For example, the elements sodium and potassium represent a repetition of the element lithium, which lies at the head of the Group I of the periodic table, but these three elements are in no sense identical"(Erduran & Scerri 2002, p.16); and they briefly address the case of 'supervenience' in chemistry in terms of the maxim, "macroscopic identity need not imply microscopic identity"(Erduran & Scerri 2002, p.17).

What is the relationship, then, between conceptual depth, conceptual usefulness, and history and philosophy of chemistry? In this paper, conceptual depth refers to the multilayered structure of a concept and hence its meaning. Conceptual usefulness refers to the capacity of a concept to assist in calculation and problem-solving, that is, to arrive at answers to important questions that do not necessarily enrich the meaning of the concept. One could say that conceptual depth focuses on a concept as a signature of meaning while conceptual usefulness focuses on a concept as a problem-solving tool. However, depth and usefulness may not be necessarily mutually exclusive. The possibility of a concept contributing as both a tool and a signature of meaning will be explored in this paper. History and philosophy by nature contribute to meaning in slightly different ways. History provides the story behind the concept and philosophy clarifies the nature of the concept.

The use of concepts as problem-solving tools has been the focus of the research comparing the prevalence of qualitative problems with quantitative problems in chemistry (Reif 1983; Gabel et al. 1984; de Berg 1989; Bodner & Herron 2002). Algorithmic problem-solving exercises using quantitative data have been the main orientation of chemistry curricula. That is, the emphasis has been on the use of concepts in their quantitative dress as problem-solving tools. By comparison, only a small number of problem-solving qualitative exercises are typically used in textbooks (de Berg 1989). Reif (1983, p.952) compares these two problem-solving approaches as follows. "Students tend to place a great emphasis on remembering and using facts and mathematical formulas without trying to embed these in a rich framework of qualitative knowledge. Accordingly, students may be able to answer some quantitative questions by merely manipulating some mathematical formulas, but they may be quite unable to answer simple qualitative questions of a similar kind. Furthermore, students seldom use qualitative knowledge to plan solutions or to check whether results obtained by them make any sense". Stenhouse (1985, p.21) is critical of the algorithmic mathematical problem-solving approach and argues that, "What is inculcated in a great deal of science education is, all too often, not understanding, but rather a sort of recipe book acquaintance with a number of mathematical techniques which tend often to be used on an ad-hoc basis without any proper understanding either of the mathematical theory and assumptions on which the techniques are based, or of the subject matter to which they are applied". In this paper, problem-solving exercises, whether quantitative or qualitative, are classified in the category of 'conceptual usefulness' because of the use of the concept as a tool. It is exercises that enrich the meaning of a concept that are classified in the category of 'conceptual depth'.

Deepening one's understanding of a concept occurs during the process of *abstraction*. Peter Atkins (2003, p.110) gives an illustration of this process in relation to the steam engine as follows. "When we strip away the iron to leave the *abstraction* of a steam engine, we obtain a representation of the spring of all change. That is, provided we look at the *essence* of a steam engine, its abstract heart, and ignore the details of its realization – the steam, the leaky pipes, the drips of oil and grease, the rattles, bangs, and rivets – we find a concept that applies across the range of all events. Science is like that: science distils from reality its essence, its great ideas, then finds the same phantom spirit elsewhere in nature". Abstraction and conceptual depth depend so much on the provision of adequate time for students to experience such orientations in science teaching and learning. Students and teachers tend to find that an orientation towards the use of concepts as tools is more readily accessible given the practical problem of time constraints. Winchester (1989, p.(vi)) proposes the following solution. "As regards the question of whether or not one has time to teach everything one would like to teach in science, the answer seems to be that since we want scientists who are not only capable of solving the puzzles in normal times but also of advancing in revolutionary ones, we must try for a more liberally educated student. It would seem to be better not to teach too much in science (to echo Whitehead) but to teach what one does teach thoroughly, in context and in depth. And here the depth would include the history of what is taught in both its scientific and social context as well as the initiation into the technical mastery of the subject". But it is now twenty years since this suggestion was made and there appears to have been little progress in addressing the imbalance between teaching for understanding and a teaching orientation towards the solving of a set of standard problems. This is an issue and a challenge to chemistry educators which demands some attention. This paper now looks at some of the dilemmas chemistry students and teachers face when it comes to the teaching and learning of topics in thermodynamics, particularly *energy*, *heat*, and *work*; in the foundational concepts of *element* and the *mole;* and in some *quantum mechanical* concepts in chemistry. An exploration of these ideas may help to shed some light on the difficulties we face in teaching for understanding.

Energy

The term, "energy", is very much a part of the language of society in general in addition to its special significance in science. This has become particularly so in an era of the attempted political control of a country's greenhouse gas emissions and the consequent search for alternative energy sources. Society in general identifies the term 'energy' with its sources rather than as an abstract definable term. So people will think of the sun, of electricity, of motor vehicles, of power stations and such like. Scientists, on the other hand, attempt to be more precise with the term and try to distinguish it from its sources. Richard Feynman (1963, p.4-2), however, claims that, as far as the scientific concept of energy is concerned, "It is important to realise that in physics today, we have no knowledge of what energy is. We do not have a picture that energy comes in blobs of a definite amount. It is not that way. However, there are formulas for calculating some numerical quantity, and when we add it all together it gives....always the same number. It is an abstract thing in that it does not tell us the mechanism or the reason for the various formulas". This immediately presents a problem for chemistry educators. If 'we have no knowledge of what energy is', but are confined to using a mathematical principle that shows that, whatever energy is, it is always conserved, then we are forced to admit that 'energy' is a tool rather than a signature of meaning. But do we need to be as bland as this?

Joan Solomon (1992, p.107) takes issue with those purists who maintain that energy can only be thought of as a type of mathematical accountancy. In her opinion, "Adult physicists may possibly end up with the bare mathematical formulation, although it seems somewhat doubtful, but pupils' learning of such an important and difficult concept as energy must begin somewhere. Getting to understand energy is a long and gradual task during which the pupils' minds play with the idea in progressively more sophisticated ways". Even Feynman used the analogy of a child's twenty-eight playing blocks to illustrate conservation of energy while drawing our attention to the fact that as far as energy is concerned there are no playing blocks. Thus thinking of energy initially as some sort of 'stuff' that can be transferred from one object to another, and a little later as 'the capacity to do work' or 'the capacity to transform', and finally as a 'kind of accountancy', are important stages in a person's cognitive growth in relation to energy. Domenech et al.(2007, p.50) agree with this view and claim that initial qualitative considerations about energy such as those described above "should not be judged in the light of the body of knowledge accepted today (after much development) but in light of their capacity to facilitate the construction process, during which the initial ideas will evolve to more elaborated and fruitful conceptions". This multilayered structure for a concept, while containing some inadequate initial conceptions, helps to display the dynamic character of concept development and ways of understanding the concept even though we cannot actually say what energy is intrinsically. Thus while one may primarily regard 'energy' as a tool to account for changes in a chemical system for example, there is a sense in which its meaning is enriched as it is used as a tool.

First-year General Chemistry textbooks do not usually begin a discussion of energy as a purely mathematical accounting principle but instead focus on the different forms of energy such as solar energy, electrical energy, chemical energy, mechanical energy, and nuclear energy. While analysis into these different forms of energy is important, can a process of abstraction also yield fundamental energies common to all forms of energy listed here? And can these fundamental energies inform us about what is really meant by the term 'chemical energy'? Peter Atkins (2003, p.96) answers these questions in the affirmative as follows. "At this stage, we see that there are just two forms of energy (as far as energy stored in matter is concerned) – kinetic energy (the capacity to do work by virtue of motion) and potential energy (the capacity to do work by virtue of position). Although you will often encounter terms like 'electrical energy', 'chemical energy', and 'nuclear energy' there is really no such thing: these terms are just handy shorthand terms for special and particular combinations of kinetic and potential energy". For example, a battery stores electrical energy in the form of potential energy and when allowed to push electrons (voltage) around a circuit results in the conversion of potential energy into kinetic energy (current).

Table 1. Explanations of energy changes during molecular collisions across five General Chemistry textbooks

PE = potential energy; KE = kinetic energy

Textbook	Is rise in PE explained?	Is thermal energy released related to temperature difference?	Is temperature related to KE on product side?	On what basis is a reaction exothermic?
Kotz et al. 1999, pp. 710-715	Yes – twisting molecules	No	No	Energy of products less than energy of reactants
Moore et al. 2002, pp. 587-593	Yes – bonds stretched, twisted	No	No	Energy of products less than energy of reactants
Silberberg 2006, pp.697-700	Yes – electron clouds repel	No	No	Exothermic so reactants higher in energy than products
Spencer et al. 2006, pp.640-642.	No	No	No	E_a(forward) is less than E_a(reverse)
Whitten et al. 2000, pp.676-680	No	No	No	E_a(forward) is less than E_a(reverse)

The interrelationship between potential and kinetic energies has been shown to be very useful in detailing the energy changes which occur during a chemical reaction. Using the analogy of the pendulum as a potential energy (PE) – kinetic energy (KE) converter, de Berg (2006) explains why exothermic reactions result in a temperature rise and endothermic reactions result in a temperature fall. In this study it was found that very few textbooks detail this process with the result that the term 'energy' remains rather elusive. The responses of five textbooks are shown in Table 1.

Consider the energy changes shown in Figure 1 during the course of a reaction.

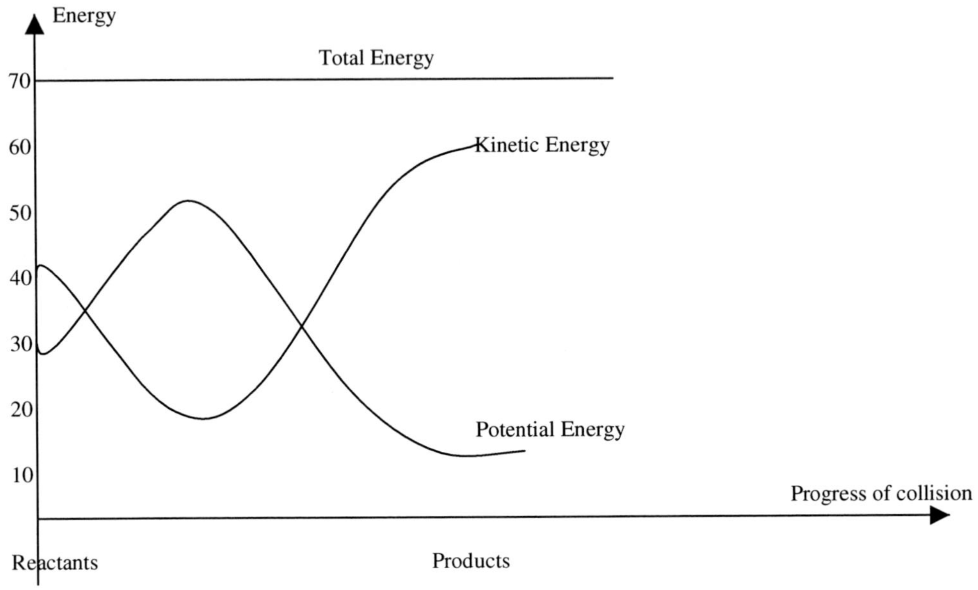

Figure 1. Energy exchange during a chemical reaction (exothermic case).

Some textbooks never specify that the vertical axis represents PE (in the case where only PE is plotted) and that PE rises as the molecules approach for collision. The PE – KE relationship can be described this way. As molecules begin the process of collision their electron clouds start to merge resulting in repulsion due to like charges and this raises the PE and therefore lowers the KE. One must remember that attraction of opposite charges lowers PE but repulsion raises PE. The KE of the molecules before collision is used to break chemical bonds by increasing the PE of the system. This is analagous to a spring or pendulum system where the bob is raised an increasing distance above the centre point of motion (pendulum) or stretched further from the point of support (spring). As new bonds form the PE starts to decrease and therefore the KE starts to increase again. If the products' PE lies above the reactants' PE, then the KE of the products is less than that of the reactants and since KE is proportional to absolute temperature, the temperature must fall. Conversely, if the PE of the products lies below that of the reactants, the temperature must rise since the KE of the products is greater than that of the reactants. None of the textbooks shown in Table 1 use KE differences to explain why a reaction is exothermic or endothermic. The textbooks talk about an energy difference between reactants and products but they do not specify what kind of energy this is. Figure 1 shows that (PE + KE) is conserved which is what might be deduced from a smoothly operating pendulum.

The discussion in the previous paragraph showing how PE and KE interrelate during the breaking and making of chemical bonds in a chemical reaction did not help us gain any insight into the intrinsic nature of 'energy'. But it did provide us with the tools for understanding the changes that occur during a chemical reaction. Thus while the concept of energy may be an abstract mathematical one at its heart, visual representations of this mathematical principle through the use of analogies like the pendulum or vibrating spring certainly enhance the usefulness of the energy concept and, to some extent, enhance the abstract nature of the energy concept.

Heat and Work

The concepts of heat and work are intimately tied to the concept of energy although they had independent histories till about the middle of the 19th century (de Berg 1997). Both the caloric theory of heat and the connection between the concept of work and the development of efficient engines for industry gave rise to the 'quantity' nature of heat and work. So, for example, in the field of calorimetry, one speaks about 'the amount of heat gained by the water' or 'the amount of heat lost by the metal'; and in the field of thermodynamics one speaks about 'how much work was done by the engine'. The difficulty for chemistry educators arises because heat and work are no longer regarded as 'quantities' but as 'processes'. Some textbooks use the 'quantity' understanding in problem-solving while stipulating in the definition of heat and work that they are not quantities but processes by which energy is transferred. Examples of this are given in Table 2. The textbook by Kotz and Treichel (1999) defines *q* as the quantity of energy

transferred as heat but in the end-of-chapter exercises questions such as, 'How much heat is required to heat a 168 g piece of metal from -15.2°C to +23.6°C?', is indicative of the style of questions across the textbooks with the possible exception of the textbook by Moore et al. (2002). Their standard questions are worded like this: 'How much thermal energy is required to heat 500 g of Al from 25°C to 250°C?'. This is the only textbook out of the five listed that takes this linguistic care in the wording of exercises to reflect the subtle differences exhibited in the definitions.

Atkins (2003, pp.102-104) clarifies the current understanding of heat and work by defining heat as the "transfer of energy that stimulates random motion of atoms in the surroundings", and, "heat is the transfer of energy as a result of a temperature difference"; while work is defined as "the transfer of energy that stimulates uniform motion of atoms in the surroundings". Heat is not the "energy possessed by anything". Heat and work, then, are ways of transferring energy, not the energy itself. Atkins (2003, p.102) gives a historical perspective here when he says, "The full understanding of heat and work, and how they were manifestations of energy, had to await the development of a molecular understanding of the distinction. As so often in science, with the understanding came the realization that they did not exist; there is no such *thing* as heat, and there is no such *thing* as work!". The problem, of course, is that it is very helpful, indeed useful, to think of heat and work as things or quantities when doing thermodynamic problems. In fact, the dilemma here is between emphasizing conceptual depth or conceptual usefulness. The clarity of the definition is lost when one focuses on the 'quantity' aspect in problem-solving while the usefulness of the concept tends to be lost if one focuses on the 'depth and clarity' aspect in the definition.

Table 2. The concepts of heat and work across five textbooks

Text-book	Definition of q	Definition of w	Heat exercise	Work exercise	Radiant energy
Kotz et al. 1999	Quantity of energy transferred as heat	Quantity of energy transferred by work	How much heat is required to heat a 168 g piece of metal from -15.2°C to 23.6°C?	-	Is mentioned as an exception to the rule that all energy transferred must be accounted for as heat and work
Whitten et al. 2000	Amount of heat absorbed by the system	Amount of work done on the system	How much heat is liberated when 0.0662 mole of sodium reacts with excess water?	Calculate the amount of work done in the oxidation of 1 mole of HCl (g) at 200°C.	Radiant energy not mentioned as a means of transferring energy
Moore et al. 2002	Quantity of energy transferred by heating the system	Quantity of energy transferred by doing work on the system	How much thermal energy is required to heat 500 g of Al from 25°C to 250°C?	If a system does 75.4 J of work on its surroundings and simultaneously there is 25.7 cal of heat transfer from the surroundings to the system, what is ΔE for the system?	Is mentioned as an exception to the rule that all energy transferred must be accounted for as heat and work
Spencer et al. 2006	Heat is energy in transit	Raising a weight requires work	The same quantity of heat is added to equal moles of CCl_4 and H_2O. The temperature increases more in CCl_4 than in H_2O. Which has the highest heat capacity?	What happens to the energy of a system when it does work on its surroundings?	Radiant energy not mentioned as a means of transferring energy
Silberberg 2006	Heat (or thermal energy) is the energy transferred between a system and its surroundings as a result of a difference in their temperatures only	Work is energy transferred when an object is moved by a force	A system receives 425 J of heat and delivers 425 J of work to its surroundings. What is ΔE?	What is ΔE of a system that releases 675 J of thermal energy to its surroundings and has 525 cal of work done on it?	Radiant energy not mentioned as a means of transferring energy

One must sympathize with textbook authors and the observations in Table 2 because it is difficult to be consistent in the use of the language because the 'quantity' aspect of heat and work have deep historical and cultural roots which are difficult to leave behind. The very framework of the laws of thermodynamics use heat and work as quantities. For example, when Holton and Brush (2001, p.245) discuss the first law in the form, $\Delta E = H\text{-}W$, H is defined as the "heat added to the body", and W is defined as the "work done by (the body) on its surroundings". Strictly speaking, according to current understanding, one should define H as the thermal energy transferred by heat to the body, and W as the mechanical energy transferred by work to the surroundings. But the historical relics of heat and work as quantities are so embedded in our culture that even though one might present the current definitions, when it comes to problem-solving and stating the laws one easily reverts to the old framework.

Textbook authors interpret the current understanding in two ways according to the analysis shown in Table 2. Moore et al. (2002) define q and w as Atkins (2003) has defined them, that is, as the energy transferred by heating and doing work respectively, and follow this up with a fairly consistent usage of the term 'thermal energy' instead of 'amount of heat' in the set exercises. Others, like Spencer et al. (2006), associate the term 'heat' with transfer of energy but use the term 'heat' to focus on the amount of energy transferred rather than on the process of transferring the energy. This is evident in the questions shown in Table 2. Spencer et al. (2006) consequently have no use for the term 'thermal energy' but simply use terms like 'amount of heat added'. Some textbook authors like Silberberg (2006) are inconsistent in that the term 'thermal energy' is used sometimes and 'amount of heat' is used at other times. This is a symptom of what happens as concepts receive a more refined definition. The challenge to chemistry educators is the question of how refined one should be in the use of definitions and the issue of whether qualitative and quantitative problem-solving should take precedence over conceptual clarity and depth. When Atkins (2003, p.104) was introducing readers to the modern interpretation of 'heat' and 'work', he added this footnote: "I am being pedantic, of course. I should really admit that all nouns – cats, dogs, heat, thermal energy, chemical energy – are but convenient ways of referring to things. But I want to purge and purify your thoughts".

The current definitions of *heat* and *work* have deep historical and philosophical roots which add great depth to one's understanding of the concepts. This leads to another dilemma for chemistry educators. While professional groups like IHPST and ISPC rightly advocate more emphasis on conceptual depth than currently happens in our science classrooms by exposing students to more historical and philosophical background, there is a danger that students will get lost in the quagmire of history and philosophy and lose the point of the strength of the concept as a problem-solving tool. Like many education reforms it is again a question of balance. But how to achieve this is a challenge.

Finally, it has been noted that the first law of thermodynamics in the form,

$$\Delta E = q + w,$$

is not always applicable. Domenech et al. (2007, p.56) comment that, "Energy variations may take place, not only through work and heat, but also by means of other processes, such as radiation exchange.....It cannot be forgotten that most of the energy resources available on earth (including fossil fuels) have their origin in solar radiation; neither can it be ignored that gaining a grasp of the greenhouse effect is not possible without reference to exchanges of energy by radiation". Atkins (2003, p.96) also notes that, "The exception to the universality of the terms kinetic and potential energy is the energy of electromagnetic radiation (for instance, the energy of light, such as that carried from the Sun to the Earth and used to warm us or drive photosynthesis and the production of food). As far as energy stored in matter is concerned, it is entirely composed of kinetic and potential energy". Two of the five textbooks in Table 2 recognize this fact by noting this exception to the $(q + w)$ contribution to ΔE.

Element

It is the philosophers of chemistry who have been recently drawing our attention to a renewed consideration of what a chemical element is. In a recently published book on the Periodic Table, Eric Scerri (2007) discusses two important ways of thinking about an element; the element as a *basic substance* and the element as a *simple substance.* The chances are that in our own chemistry education we have only been exposed to that view of an element as a simple substance. This view is commonly traced back to the late 18th century to the time of Lavoisier who established the idea that an element cannot be broken down into simpler substances by ordinary chemical means. This view and the one that emerged in 19th century atomic theory that saw an element as consisting of only one type of atom are those expressed in all our General Chemistry textbooks and summarized in Table 3. This is the view of an element as a simple substance.

Table 3. Definition of *element* and comments on elements of Group VII across five General Chemistry textbooks

Textbook	Definition of *element*	Comments on Group VII elements
Kotz et al. 1999	Substances like hydrogen and oxygen that are composed of only one type of atom are classified as elements.	Elements are quite similar but completely different from elements in other groups. All exist as diatomic molecules and combine violently with alkali metals to form salts.
Whitten et al. 2000	An element is a substance that cannot be decomposed into simpler substances by chemical changes.	Chemical properties of the halogens resemble one another closely but their physical properties differ significantly.
Moore et al. 2002	Substances that cannot be changed by chemical reactions into two or more new substances are called chemical elements (or just elements).	Diatomic molecules and highly reactive and form salts by vigorous reaction with alkali metals and with other metals as well. A picture of the different physical states of Cl_2, Br_2, and I_2 is shown but no comment made.
Spencer et al. 2006	Elements are substances that contain only one kind of atom.	All diatomic molecules and react with alkali metals to form ionic compounds or salts.
Silberberg 2006	An element is the simplest type of matter with unique physical and chemical properties. It consists of only one kind of atom and cannot be broken down into a simpler type of matter by any physical or chemical methods.	Regular trends in physical properties such as melting point increasing as one goes down the group. Exist as diatomic molecules that react with most metals and nonmetals to give many ionic and covalent compounds. A picture of the different states of Cl_2, Br_2, and I_2 given.

According to Scerri (2007, p.113), the view of an element as a *basic substance* can be traced back to the philosophy of Aristotle where "the elements themselves were to be regarded as abstract even though they gave rise to all the physical variety that is observed. The four elements (fire, earth, water, air) were considered as property bearers, responsible for the tangible features of substances although they were themselves unobservable. The elements were immaterial qualities impressed on an otherwise undifferentiated primordial matter and were present in all substances".

With the development of atomic theory at the beginning of the 19th century and with the expansion of the number of substances that appeared to be elements from the simple substance point of view, chemists were left with the question of what happens to the elements when they react to form a compound. Common salt, for example, neither contains the shiny metal element, sodium, nor the noxious yellow-green gas, chlorine, but mass was conserved when sodium reacted with chlorine to produce common salt. There was a sense (on the basis of mass) in which the elements remained and a sense (on the basis of physical properties) in which the elements disappeared. From the early 19th century this dual view of an element was established in order to explain this phenomenon. The dual view was that of an element as a *simple substance* and, drawing upon Aristotelian views, that of an element as a *basic substance.*

Simple substances did not survive in a compound but abstract elements did survive. The 19th century view of an abstract element differed from that of Aristotle in that an abstract element has a material ingredient as opposed to an immaterial ingredient due to the fact of the law of conservation of mass in chemical change discovered by Lavoisier. Scerri (2007, p.114) concludes that, "There are thus three important concepts regarding elements carried over into the 19th century: abstract element as property bearer owing its heritage to Aristotle; indestructible material ingredient of substances behaving according to Lavoisier's law of the conservation of matter; abstract element as unobservable whereas simple substances can be observed. In contemporary chemistry only the notion of simple substance has been retained in the concept of element". This latter conclusion is confirmed by the data in Table 3.

Mendeleev (1891, p.23) had this dual view of an element as evidenced by the following statement. "It is useful in this sense to make a clear distinction between the conception of an element as a separate homogeneous substance, and as a material but invisible part of a compound. Mercury oxide does not contain two simple bodies, a gas and a metal, but two elements, mercury and oxygen, which, when free, are a gas and a metal. Neither mercury as a metal nor oxygen as a gas is contained in mercury oxide; it only contains the substance of the elements, just as steam only contains the substance of ice, but not ice itself, or as corn contains the substance of the seed but not the seed itself". Scerri (2007, p.115) observes that this is probably the reason for Mendeleev's preoccupation with atomic weight in relation to the Periodic Table. "Mendeleev's genius now lay in recognizing that just as it was the 'basic element' that survived intact in the course of compound formation, so atomic weight was the only quantity that survived in terms of measureable attributes".

Can we expect chemistry educators to hone the definition of *element* to this level of precision particularly since it is not a topic of study in current textbooks? Scerri (2007, p.16) answers this question in the affirmative by giving the following example. "If a novice chemist were asked to group the halogen elements according to their appearances, it is inconceivable that he or she would consider classifying together fluorine, chlorine, bromine, and iodine. This is one instance where the subtle distinction between the observable and the abstract sense of the concept of element can be helpful. The similarity between them lies primarily in the nature of the abstract elements and not the elements as substances that can be isolated and observed". In Table 3, textbook authors draw attention to the similar chemical properties over and above the physical properties of the halogens and, in some cases, while showing the different physical states of the elements, do not present the difficulty either with respect to classification or with respect to the question of the sense in which elements are retained in compound formation.

The Mole Concept

The mole concept is central to understanding stoichiometric relationships in chemistry and is primarily a useful tool in determining amounts of reactants and products involved in chemical reactions. It is the means by which the number of entities involved in a chemical process can be obtained by weighing the appropriate amount of matter. All definitions shown in Table 4 focus on the *number* of entities involved and it is this emphasis on *number* which makes it such a useful concept in chemical calculations. However, that part of the definition that refers to *amount of substance* is not that commonly referred to because it is not explored to any depth in normal quantitative calculations. The amount of matter in 1 mol of Mg is not the same as the amount of matter in 1 mol of Zn, although they contain the same number of atoms, and this fact is usually not the focus of a calculation. *Number of moles* rather than *amount of substance* is the focus of the exercises shown in Table 4.

Table 4. Definitions of 'mole' and 'mass' across five General Chemistry textbooks

Textbook	Mole Definition	Mass Definition	Nature of Exercises
Kotz et al. 1999	The mole is the amount of substance that contains as many elementary entities (atoms, molecules, or other particles) as there are atoms in exactly 12 g of the carbon-12 isotope.	The mass of a body is the fundamental measure of the amount of matter in that body.	Calculate the number of moles represented by 0.876 g of arsenic.
Whitten et al. 2000	The mole is the amount of substance that contains as many entities (atoms, molecules, or other particles) as there are atoms in exactly 0.012 kg of the carbon-12 isotope.	Mass is the measure of the quantity of matter a body contains.	How many moles of substance are contained in 15.6 g of calcium carbonate?
Moore et al. 2002	The mole is the amount of substance that contains as many atoms, molecules, ions, or other nanoscale units as there are atoms in exactly 12 g of carbon-12.	The mass of an object is a fundamental measure of the quantity of matter in that object.	Calculate the number of moles represented by 0.0034 g of platinum.
Spencer et al. 2006	A mole of any substance contains the same number of particles as the number of atoms in exactly 12 g of the ^{12}C isotope of carbon.	-	Calculate the number of moles of NO_2 that could be prepared from 0.35 mol of NO and 0.25 mol of oxygen.
Silberberg 2006	The mole is the amount of substance that contains the same number of entities as there are atoms in exactly 12 g of the carbon-12.	The mass of an object refers to the quantity of matter it contains.	How many moles of CaO can be produced from 4.2 g Ca and 2.8 g O_2?

This might appear to be a rather pedantic point, but, is something of the concept of *amount of substance* lost if one focuses on the number of units of amount (number of moles) rather than on the concept of amount itself? An analagous situation arises when one asks for the number of grams of a substance rather than its mass. Consider one of the exercises given in Table 4: "Calculate the number of moles represented by 0.876 g of arsenic". To focus on what moles is a measure of, one could rephrase this question as: "Determine the amount of arsenic in 0.876 g of arsenic". One, of course, has to be clear here that *amount* is measured in units *mol*. While it is true that the majority of stoichiometric exercises in General Chemistry textbooks could be solved without insisting on this level of conceptual clarity, it is worthwhile pondering how one might answer the following question related to the concept, *amount of substance* in Figure 2.

Question: Are the following statements true or false? Give an explanation for your response.

1. The amount of matter in 1 mol Mg is less than in 1 mol Zn.
2. The number of atoms in 1 mol Mg is the same as in 1 mol Zn.
3. The number of atoms in 1 mol CO_2 is the same as in 1 mol C_3H_8.
4. The amount of matter in 1 mol CO_2 is the same as in 1 mol C_3H_8.
5. The amount of matter in 0.5 mol O_2 is less than in 1 mol CH_4.

Figure 2. A question relating to the concept, amount of substance.

All the textbooks in Table 4 that define *mass* use the idea of the fundamental measure of *amount* or *quantity of matter* and this idea is reflected in the formal definition of the mole when referring to the amount of carbon-12 as 12 g or 0.012 kg. So the *amount of a substance* that has the same number of separate entities as that in 0.012 kg of carbon-12 is referring in a fundamental sense to the mass of the substance. But this is a special mass that has the same number of separate entities as that contained in 12 g of carbon-12 and is given therefore the special name, *amount of substance*, with the unit, *mol*. It was in this context that Hudson (1981) questioned the definition of mole as 'the chemists' dozen', commonly used by secondary school chemistry textbooks of the 1970's and 1980's (de Berg 1986). Such a definition identified the mole with Avogadro's number of particles rather than with the amount of substance contained in Avogadro's number of particles. While it is true, as Hudson (1981) observes, that the use of the concept can precede its understanding, it does not necessarily follow that exercises involving the various calculations with a complex concept such as the mole can provide an understanding of it. This is a dilemma we all face in chemistry education.

The Uncertainty Principle

Peter Atkins (2003, p.201) commences his chapter on "Quanta" by quoting from Richard Feynman as follows. "If anyone claims to know what the quantum theory is all about, they haven't understood it". Jim Baggot (1992, pp.(ix)-(x)) assesses quantum theory as "unnerving. Not only is the theory mathematically complex, it is also conceptually challenging……Students are taught about a theory which they have to accept and which they have to learn how to apply, but for which they cannot expect to be told its meaning….Most undergraduate courses on quantum theory never touch on the theory's profound conceptual problems. This is because the theory brings us right back to some of the central questions of *philosophy* and, as we know, there is no room for philosophy in a modern science degree. I find this an absurd situation". Baggot touches here on the central issue of this paper: the challenge for chemistry educators to meet the demands for conceptual depth and meaning as well as the demand to demonstrate a concept's usefulness in calculation.

The Heisenberg Uncertainty Principle (HUP) is intimately connected to the particle-wave properties of the electron and so both these ideas appear in the textbook analysis in Table 5. Recent textbook authors like those appearing in Table 5 do a fairly stirling job of what even the experts agree is a complex topic. Three of the five textbooks not only state the HUP but give an explanation of it. In order to locate the position of an electron, radiation of a very small wavelength is required because of the very small size of the electron. All textbooks in Table 5 except one introduce the wave-particle properties of radiation in the form of de Broglie's, $\lambda=h/mv$, which illustrates that radiation of very small wavelength (λ) possesses photons (particles of light) of very high momentum (mv) and energy and this is imparted to the electron when the radiation is used to locate its position. Therefore, when locating the electron, its speed cannot be accurately determined because in the process of measurement momentum is imparted to it. Conversely, if one increases the wavelength in order to reduce the momentum imparted to the electron, one loses track of its position since a very small wavelength is required to track it. The textbook by Moore et al. (2002) uses an analogy of a high speed and low speed shutter to track a racing car; the high speed shutter giving a reasonably clear picture of the car and the low speed shutter giving a blurred picture. One must, of course, be aware of the limitations of such an analogy because the HUP, while applying to all objects, becomes recognizable only in the world of atomic dimensions. So, in our macroscopic world, the process of measurement does not impart sufficient momentum to the observed object to lead to a blurring of its position or momentum so both can be measured simultaneously.

Table 5. The Heisenberg Uncertainty Principle (HUP) across five General Chemistry textbooks

Textbook	Explanation for HUP	Wave-particle Relationship
Kotz et al. 1999	HUP is stated qualitatively and mathematically but no explanation given	Evidence is given: Photoelectric effect and the Davisson and Germer experiment - de Broglie's $\lambda=h/mv$ given
Whitten et al. 2000	HUP stated qualitatively-no mathematical statement-is a short explanation	Davisson and Germer experiment - de Broglie's $\lambda=h/mv$ given
Moore et al. 2002	HUP stated qualitatively- no mathematical statement – a fairly detailed explanation given – a photographic analogy given	Davisson and Germer experiment - de Broglie's $\lambda=h/mv$ given
Spencer et al. 2006	HUP stated qualitatively-no mathematical statement-is a very short explanation	Discusses properties of waves and particles –de Broglie's $\lambda=h/mv$ *not* given
Silberberg, 2006	HUP is stated qualitatively and mathematically but no explanation given	Davisson and Germer experiment - de Broglie's $\lambda=h/mv$ given

The textbook by Silberberg (2006) does give a hint of some of the philosophical dimensions of this topic by way of a short historical note to the effect that some scientists of the early 20th century regarded uncertainty as unacceptable in science. The argument went that because an accurate value of the momentum of a localized particle cannot be defined by measurement, it has no place in the theory. Feynman (1963, pp.38-8, 38-9), reflecting on this question, commented that, "Just because we cannot measure position and momentum precisely does not *a priori* mean that we cannot talk about them. It only means that we need not talk about them. The situation in the sciences is this: A concept or an idea which cannot be measured or cannot be referred directly to experiment may or may not be useful.....It is not true that we can pursue science completely by using only those concepts which are directly subject to experiment".

Another philosophical argument revolves around the interpretation of the HUP. According to Baggott (1992, p.33) Heisenberg and Bohr had a different understanding of the implications of the Uncertainty Principle (UP). To Heisenberg the UP taught that there are limits on what is measureable. The concepts of locality and momentum as far as the electron is concerned are legitimate concepts but they cannot be measured simultaneously. Bohr argued that it might be meaningless to talk about an electron's position and momentum since these are classical concepts. They only become of interest when we attempt to measure them. Baggott (1992, p.33) summarises Bohr's position as follows. "For him (Bohr) it was wave-particle duality that lay at the heart of quantum mechanics. All the rest – including the UP – were the physical and mathematical consequences of using two diametrically opposed classical concepts, waves and particles, to describe something that was fundamentally non-classical. According to Bohr, quantum theory tells us not what is measureable, but what is knowable". Fritz Rohrlich (1987, p.143) has a similar view of the implications of the wave-particle idea. "The classical physicist was confronted with the puzzle, 'how can an electron be both a particle and a wave?' With this question he had in mind the classical notions of particle and wave. The Quantum-Mechanical answer is: 'The electron is neither; it is neither a classical particle nor a classical wave. It is a quantum particle and as such it has some properties that resemble certain classical particle properties and others that resemble those of a classical wave".

To those of us involved in the teaching of chemistry at the tertiary level there is a real dilemma here. Yes, there are important historical and philosophical issues present in this topic. But a discussion of such topics only makes sense if students are also familiar with using the concepts and much time is required for getting to this stage. Peter Atkins (2003, p.225) reaches a stage in his discussion of quanta that illustrates the dilemma we face. "Quantum Mechanics gets philosophically interesting when we start to ask what it all means. That's the subject of the rest of this chapter. If you stop here, you will know the main principles of quantum mechanics and, in principle, can use it to do any calculation; if you go on, your ability to use it won't be helped, but you will know why people find it so deeply puzzling".

Conclusion

Chemistry educators as a professional group recognize the importance of teaching for understanding rather than just the skill of calculation with a set of algorithmic procedures. The IHPST and ISPC groups are all about assisting teachers in this quest by providing historical and philosophical analyses of the concepts we have to teach and learn. However, we have made very little progress in incorporating historical and philosophical elements for understanding in our curricula although some brave attempts are being made. Wandersee and Griffard (2002, p.31) note this deficiency but encourage us to be courageous in this matter. "Much of the research that has been peer-reviewed and published is prescriptive of approach, rather than being descriptive of implementation and bulwarked by evidence of effectiveness. We may be currently limited by this imbalance, but we hope that chemists and chemical educators will increasingly document their experiences in publication form".

There are a number of difficulties we face in teaching for understanding, for conceptual depth and meaning. The first difficulty is that most of us have never experienced this type of education ourselves. Most of the curricula we have studied have been loaded with information and standard exercises to be learned for a final examination. Consequently, we have naturally tended to perpetuate our own experience with our students. Wandersee and Griffard (2002, p.33) believe that this "rush to abstraction and the final form science.....evident in chemical education today actually short-circuits deep understanding by cutting off all human and epistemological contexts...We think rigour is comprised less of breadth than it is of depth and that both take much more time to develop than current chemistry curricula allow".

A second difficulty lies with the complex nature of historical and philosophical study. Students generally have enough difficulty handling the skills associated with calculation and problem-solving without having to deal with questions of historical and philosophical significance. IHPST and ISPC are making courageous efforts to help teachers and students in this regard and such efforts are to be encouraged and applauded. The fact that an international effort is being made to solve this problem augers well for the future.

A third difficulty is associated with the nature of the historical and philosophical elements that should be included in a chemistry course. Our accreditors rightly insist that we are responsible for teaching chemistry, not history and philosophy of science, which means that we must choose elements that *enhance* an understanding of the chemistry topics at hand and not elements that, although important, do not address the content being studied.

A fourth difficulty is related to how to find space in a content-rich curriculum for issues of historical and philosophical relevance. It is agreed that less breadth and more depth is required. The challenge is how to accomplish this.

My own initial response to these difficulties was to write a series of historical/philosophical case studies drawn from my own research and that of others and to use these as part of the assignment load each semester of my chemistry course. The idea was to use one of these case studies each semester so there was no great impingement on the course content covered. The case study would particularly relate to a chemistry topic for that semester. This approach met with mixed success. I overestimated the students' capacity to link historical ideas with the current ideas even though I tried to make these explicit. Using genuine historical data proved a stumbling block because of the units used for quantities like pressure and volume for example. The data had to be reworked for student use. This is currently an ongoing project. My immediate revised response has been to be more modest in my attempts and to include an extra question or exercise on a tutorial sheet and a laboratory report that impinges on the topic but which attempts to deepen a student's understanding of the topic. This extra question or exercise often depends for its content on historical/philosophical data. On some occasions a decision has been made to reduce the breadth of topics or laboratory exercises to allow time for more reflection. A sample exercise, pertinent to the concepts discussed in this paper, is shown in Figure 3. Ultimately one has to move beyond the evidence of personal experience as far as demonstrating the effectiveness or ineffectiveness of including historical/philosophical elements in chemistry teaching is concerned to a more encompassing research agenda, but at least it is a starting point. It is not clear at the moment what the nature of such a research agenda might be and this remains a major challenge to chemistry education researchers.

In Australia, as elsewhere, there has been great debate about the nature of the modern university and the political pressures put by government to enhance the role of the modern university in a nation's economic progress. Professor Alan Gilbert (2003), Vice Chancellor of the University of Melbourne at the time, in response to such pressures on the university of the 21st century asked, "But what of the other enduring characteristics of universities: the valuing of knowledge and inquiry for their own sake, and the civilizing mission of sustaining well-founded civil societies? When, arguably, the world needs powerful civilizing institutions more than ever, universities seem to be losing the capacity and even the will, to tackle the great philosophical and moral questions through which humankind seeks meaning and guidance, and through which humane, sustaining civil societies are built". Though the task of enhancing the soul of chemistry education, the depth of meaning, and clarity of understanding is a stupendous one as we have outlined, it is ultimately, as Alan Gilbert infers above, a deeply civilizing one.

Identify what you think is wrong or incomplete about these statements and rewrite the statement in a way you think corrects the errors or incompleteness.

1. When 50 J of heat is added to the system, the internal energy rises by 50 J.
2. When 20 J of work is done by the system, the internal energy falls by 20 J.
3. Common salt contains sodium and chlorine.
4. $6x10^{23}$ molecules of O_2 is one mole of O_2.
5. An electron is both a wave and a particle.

Figure 3. A sample exercise designed to assist understanding of concepts.

References

Atkins, P. (2003). *Galileo's Finger-The Ten Great Ideas of Science.* Oxford: Oxford University Press.

Baggott, J. (1992). *The Meaning of Quantum Theory.* Oxford: Oxford University Press.

Bodner, G.M. & Herron, J.D. (2002). Problem-Solving in Chemistry. In Gilbert, J.K., De Jong, O., Justi, R., Treagust, D.F., & Van Driel, J.H. (Eds.). *Chemical Education: Towards Research-based Practice* (pp.235-266). Dordrecht: Kluwer Academic Press.

de Berg, K.C. (1986). Textbook Analysis of the Mole and Its Underlying Concepts: A Teaching-Learning Perspective. *The Australian Science Teachers Journal* **31**(4), pp.33-43.

de Berg, K.C. (1989). The Emergence of Quantification in the Pressure-Volume Relationship for gases: A Textbook Analysis. *Science Education* **73**, pp.115-134.

de Berg, K.C. (1997). The Development of the Concept of Work: A Case where History can inform Pedagogy. *Science & Education* **6**, pp.511-527.

de Berg, K.C. (2006). Chemistry and the Pendulum – What have they to do with each other? *Science & Education* **15**, pp.619-641.

Domenach, J.L., Gil-Perez, D., Gras-Marti, A., Guisasola, J., Martinez-Torregrosa, J., Salinas, J., Trumper, R., Valdes, P., & Vilches, A. (2007). Teaching of Energy Issues: A Debate Proposal for a Global Reorientation. *Science & Education* **16**(1), pp.43-64.

Erduran, S., & Scerri, E. (2002). The Nature of Chemical Knowledge and Chemical Education. In Gilbert, J.K., De Jong, O., Justi, R., Treagust, D.F., & Van Driel, J.H. (Eds.). *Chemical Education: Towards Research-based Practice* (pp.7-27). Dordrecht: Kluwer Academic Press,.

Feynman, R.P., Leighton, R.B., & Sands, M. (1963). *The Feynman Lectures on Physics.* Reading, Massachusetts: Addison-Wesley.

Gabel, D.L., Sherwood, R.D., & Enochs, L. (1984). Problem-Solving Skills of High School Chemistry Students. *Journal of Research in Science Teaching* **21**, pp.221- 233.

Gilbert, A. (2003). Barbarians at the Gate. In *The Australian Higher Education Section,* October 15, p.29.

Holton, G., & Brush, S.G. (2001). *Physics, the Human Adventure.* New Brunswick: Rutgers University Press.

Hudson, M.J. (1981). A Tested Approach for Introducing the Mole. *Source Book for Chemistry Teachers.* Sixth International Conference on Chemical Education, p.113.

Kotz, J.C., & Treichel, P. (1999). *Chemistry and Chemical Reactivity* (4th ed.). Fort Worth: Saunders.

Mendeleev, D.J. (1891). *The Principles of Chemistry* (5th Russian ed.)(Vol.1)(1st English Trans. by G. Kemensky). New York: Collier.

Moore, J.W., Stanitski, C.L., & Jurs, P.C. (2002). *Chemistry – The Molecular Science.* Fort Worth: Harcourt.

Reif, F. (1983). How can Chemists Teach Problem-Solving? *Journal of Chemical Education* **60**, pp.948-953.

Rohrlich, F. (1987). *From Paradox to Reality.* Cambridge: Cambridge University Press.

Scerri, E.R. (1997). Are Chemistry and Philosophy miscible? *Chemical Intelligencer* **3**, pp.44-46.

Scerri, E.R. (2007). *The Periodic Table – Its Story and its Significance.* Oxford: Oxford University Press.

Silberberg, M.S. (2006). *Chemistry – The Molecular Nature of Matter and Change* (4th ed.). Boston: McGraw Hill.

Solomon, J. (1992). *Getting to know about ENERGY in School and Society.* London: The Falmer Press.

Spencer, J.N., Bodner, G.M., & Rickard, L.H. (2006). *Chemistry – Structure and Dynamics* (3rd ed.). New Jersey: John Wiley & Sons.

Stenhouse, D. (1985). *Active Philosophy in Education and Science.* London: Allen & Unwin.

Wandersee, J.H. & Griffard, P.B. (2002). The History of Chemistry – Potential and Actual Contributions to Chemical Education. In Gilbert, J.K., De Jong, O., Justi, R., Treagust, D.F., & Van Driel, J.H. (Eds.). *Chemical Education: Towards Research-based Practice*(pp.29-46). Dordrecht: Kluwer Academic Press.

Whitten, K.W., Davis, R.E., & Peck, M.L. (2000). *General Chemistry with Qualitative Analysis* (6th ed.). Fort Worth: Saunders.

Winchester, I. (1989). Editorial – History, Science, and Science Teaching. *Interchange* **20**(2), p.(i).

In: Science Education in the 21st Century
Editor: Ingrid V. Eriksson, pp. 183-204

ISBN: 1-60021-951-9

Chapter 6

ACCELERATION IN SCHOOL, IN EVERYDAY LIFE AND IN AMUSEMENT PARKS

Ann-Marie Pendrill*

Department of Physics, Göteborg University, SE-412 96 Göteborg, Sweden

Abstract

Newton's discovery of the laws of motion changed the way we view our universe. They are a standard part of any physicist's toolbox. Still, there is an overwhelming body of evidence that students often fail to master even the most basic aspects and everyday experience often appears to contradict the laws of motion.

This chapter focuses on acceleration, which, unlike velocity, can be detected by a body in motion and measured within the moving system. Acceleration is also more fundamental; it is absolute, whereas velocity is relative and depends on the frame of reference chosen. This makes it possible to build on, rather than reject, the experience by learners and start in situations which where everyday observations do not contradict Newtonian concepts. The vector character of acceleration does not need to be seen as an abstract concept, but is evident in our daily three-dimensional motions, without the need to invoke a formal mathematical treatment.

After an introduction presenting different aspects of acceleration, including different ways of measuring and experiencing acceleration in different types of motion, follows a presentation of an investigation of new physics students' understanding of acceleration. Consequences of introducing acceleration through the mathematical definition, $\mathbf{a}=d\mathbf{v}/dt$, or through Newton's second law, $\mathbf{a}=\mathbf{F}/m$ are discussed and related textbook presentations are analysed. Which aspects of acceleration are discerned and which are not? Can teaching be modified to make critical aspects of acceleration accessible to the students?

The last part of this chapter describes pilot projects with 10-11 year olds who were invited to guided physics experiments in an amusement park or playground. Is it possible and meaningful for them to investigate acceleration in three dimensions?

* E-mail address: Ann-Marie.Pendrill@physics.gu.se

1. Introduction

Noone teaching mechanics in school or in undergraduate physics can fail to notice how the concepts of force and acceleration cause problems for students. Physicists, who are used to studying the fundamental interactions in Nature, may be fascinated by students' common ways of discussing friction, normal force and centripetal force as fundamental forces in their own right.

Newton's discovery of the laws of motion changed the way we view our universe. The laws are a standard part of any physicist's toolbox. They are also among the most studied part of physics education and there is an overwhelming body of evidence that students often fail to master even the most basic aspects (see e.g. McDermott, 1998, McDermott and Redish, 1999, Redish, 2003). Can anything new be added to the research on learning and teaching mechanics? What if there is something fundamentally wrong in the approach to force and acceleration in school physics?

The laws of motion are traditionally introduced through non-motion, discussing forces on a system at rest. The study of forces then moves on to situations where all forces cancel. If the body is not at rest, it thus remains in uniform rectilinear motion and Newton's first law seems to contradict everyday experience, where a cyclist needs to pedal to keep going, and a car comes to rest unless the engine continues to provide energy.

Would it be possible to approach the teaching of the laws of motion by focusing instead on acceleration, related to force through Newton's second law? From a physics point of view, acceleration is more fundamental than velocity; it is absolute, whereas velocity is relative and depends on the frame of reference chosen. Acceleration, unlike velocity, can measured within the moving system. It can be detected by a body in motion and the "body" can be that of a learner. This makes it possible to build on, rather than reject, the experience by learners and start in situations which where everyday observations do not contradict Newtonian concepts. The vector character of acceleration does not need to be seen as an abstract concept, but is evident in our daily three-dimensional motions, without the need to invoke a formal mathematical treatment.

The investigations in this work are inspired by the theory of variation (Marton & Booth, 1996, 1997). The theory of variation builds on the phenomenographic tradition, where the researcher explores and cathegorizes different ways of experiencing a phenomenon. In the continuing development of phenomenography, the need for the *learner* to experience qualitative variation is emphasized as a key to learning and awareness. The variation of important aspects of the phenomenon of interest is essential for the learner to be able to develop meaning and a structure to observations. More recently work in that field has focused on what essential variation around critical aspects was offered to the students in a teaching situation: what constitutes the space of learning (e.g. Runesson, 2006, Ling et al, 2006).

The present work is restricted to analysis of the variation in students' understanding in connection to the aspects available from textbook presentations. Possible alternative strategies for teaching motion to younger learners are suggested, which would allow additional variation. Lessons have been implemented in a few pilot cases, but not yet evaluated systematically on a larger scale.

Section 2 gives a backgrund to the student group in focus, in terms of results on the common "Force Concept Inventory". A description of the use of amusement park activities for physics learning is followed by an summary of important aspects of acceleration and measurements. Section 3 presents student replies, collected through a quiz in class, to a questions concerning a relatively simple situation, with uniform circular motion in a vertical plane, used as a starting point for the investigation of their understanding. Group interviews are used to study and challenge the views of acceleration, starting from an open question. Interview excerpts are presented and discussed in Section 4. The student difficulties observed are then analysed in terms of the variation (or lack of it) which is available to the students from textbook presentations, as discussed in Section 5. Possibilities to expose younger learners to a wider variation of aspects of force and acceleration are explored in Section 6, including suggested experiments and interviews with 10-year olds who have tried them.

The main questions in this work are

- What difficulties are found in new students' understanding of acceleration?
- Which aspects of acceleration are discerned and which are not? Can teaching be modified to make critical aspects of acceleration accessible to students?
- Can it be possible and meaningful for young learners to investigate three-dimensional acceleration, by approaching the concept from the relation $\mathbf{a}=\mathbf{F}/m$ rather than from $\mathbf{a}=d\mathbf{v}/dt$.

2. Background

Motivation for this work arises from experiences with first-year engineering physics students. In spite of a competitive acceptance to the programme, and excellent scores on introductory diagnoses, including the Force Concept Inventory (section 2.1), many of the student have difficulties in the problem solving required for the mechanics examination. Different approaches have been used to probe deeper into the nature of the difficulties. Phenomenography and theory of variation inspired investigations of qualitatively different aspects of students' understanding of the concept of acceleration, and also to develop teaching strategies to make additional aspects available to students in the learning situation. Amusement park visits (section 2.2) are used as one way to allow students to experience and reflect on force and acceleration in contexts outside the classroom. The student presentations of the amusement park projects often reveal incomplete understanding of important concepts. Experiences from such presentations lay the basis for the choice of questions for interviews and diagnoses to probe deeper into their understanding.

2.1. The Force Concept Inventory

The Force Concept Inventory (FCI) is multiple-choice test with 30 questions designed to monitor students' conceptual understanding of force and related kinematics (Hestenes *et al.* 1992, Halloun *et al.* 1995). The answers in the test are based on common student replies to open-ended questions. Hestenes and collaborators conclude that an FCI score of 60% can be regarded as the 'entry threshold' to Newtonian physics and a score of 85% as being Newtonian 'mastery threshold'. Hake (1998) performed a large survey study, including data from 62 high-school, college and entering university groups. He introduced a normalized gain defined as (class post-test average - class pre-test average)/ (100% - class pre-test average).

and found that "conventional instruction" tends to give normalized gains around "1/4 of the possble gain, whereas the more interactive physics-education-research-based classes ... typically achieve twice as large fraction of the possible gain". The highest normalized gain in the survey was 0.69.

The engineering physics students regularly obtain an FCI score around 80 % as they start their education at Chalmers. This seems to be higher than any pre-test result in the study by Hake (1998). For other programmes where we have administered the test, including the university physics course at GU, the chemistry with physics and the chemistry/biology engineering programmes at Chalmers, the FCI scores have been much lower, around 55% at entrance.

In spite of the high entrance scores for engineering physics students, exam results have not matched expectations. A comparison between individual results indicates that a good FCI scores is a necessary, but not sufficient, condition for good exam results (Pendrill, 2005). During exam grading, it was obvious that mathematics skills also play an important role. The importance of mathematics also for conceptual understanding was emphasized further in an interview study (Adawi et al 2005). These results have been a strong incentive to analyse students' conceptual problems in more detail, and to invent ways to help students develop a deeper understanding of fundamental concepts.

2.2. Physics in the Amusement Park

An amusement park can be seen as a large physics laboratory, where the thrills and excitement are created using Newton's laws. Acceleration is experienced throughout the body during the rides - although the connection is not necessarily made by the rider. The use of Amusement Parks to study physics has a long tradition in the US (Bakken, 2007). Building on this tradition, we use the nearby Liseberg amusement park during the first weeks of the introductory course, as one way to help students develop their understanding of force and acceleration. Groups of 4-6 students are assigned one roller coaster segment and one or two other rides for closer analysis, and the results of their investigations are presented both in writing and orally in groups of around 30 students. Each group also has to read the report of another group and to ask questions after the presentation. The discussions in connection with reports and presentations indicate common areas of difficulty which would be worth to investigate further. In connection with the project work, students have been given diagnoses, and asked, e.g., to draw forces at various places in the classical Rainbow ride discussed in Section 3.

Experiments have also been designed for younger learners, with the intention to use the controlled motion in an amusement ride to illustrate fundamental physical principles, as discussed in Section 5, which also presents excerpts of group interviews after a visit. Section 5 also discusses observations from a physics lesson in a playground. More details about the amusement park project can be found in previous work (Bagge & Pendrill 2002, 2004, Bagge, 2003, Nilsson *et al.* 2004, Pendrill, 2005, Pendrill and Rödjegård, 2005) and at the project WWW-site (Pendrill, 2007).

2.3. What Is Acceleration and How Can It Be Measured?

Acceleration is closely linked to force through Newton's second law, but, in contrast to the understanding of force, seems to have received very little attention by Physics Education Research (see e.g. McDermott 1998, McDermott and Redish, 1999 and Duit, 2007). Acceleration is absolute - although, as Einstein (1916) pointed out, it can not be distinguished from a gravitational field. Acceleration can thus be detected and measured from within the accelerated coordinate frame, without the need to refer to another frame. The effects of acceleration are evident throughout the accelerated body, since every single atom must be exposed to a net force, which causes internal tension. A formal treatment of this aspect requires relatively advanced mathematical tools and introductory physics learners are typically sheltered from the complication. Learners are instead presented with simplified, idealised situations whose relevance for life are not necessarily apparent. Uniform or uniformly accelerated motion along a straight line is insufficient for life in three dimensions.

The experiences from real life are rarely from frictionless surfaces or without air resistance; the bicycle comes to a stop unless you pedal, the puck stops sliding even on very smooth ice and a feather falls to the ground slower than the stone. The Aristotelian world view of a force needed to maintain a constant velocity, and bodies falling with velocity depending on their mass, is a natural interpretation of years of continuous observations. It is given many years to develop before the physics teacher tries to interfere. Yet, acceleration, in its full vector capacity, is very much a part of everyday life. The child in the swing, on a slide or see-saw, or jumping down a from a climbing rack enjoys the sensation of changed acceleration and interplay of forces - without feeling the need to refer to Newton's laws.

Accelerometers can help develop insights for the understanding of force and acceleration. Acceleration as the time derivative of velocity can be obtained from a measurement of velocity or distance relative to another object, e.g using an ultrasound ranger. Such a measurement gives the difference in acceleration between the two systems. However, accelerometers measure acceleration without the need for an external reference, by building on Newton's second law in the form **a=F**/m. Simple mechanical devices can give a direct real-time visual measurement, reflecting the "g-force" experienced by the rider. For purely horizontal motion, acceleration can be obtained by measuring the angle to the vertical for a free-hanging pendulum - for young learners, this can be a soft toy on a string. Vertical acceleration can be measured through the extension of a spring. For illustration and qualitative measurement, this can be a small plastic "slinky".

Electronic accelerometers have the advantage of giving a data set that can be analysed in the classroom. They may use the changed distance between two condensator plates to obtain a voltage readout which is translated to acceleration. Acceleration in three dimensions can be obtained by combining three such devices Accelerometers, in spite of their name, do not measure pure acceleration, but rather components of the vector **a - g**, which can be seen as the non-gravitational force per unit mass. The values for the different components obviously depends on the orientation of the probe. A complete description of motion for a "motion tracker" or "inertial navigation system" requires also a way to measure rotation around the axes.

Which aspects of acceleration made are available to physics students in school teaching? What are the critical aspects and can a changed teaching strategy help students discern them?

3. Forces in a Circular Amusement Ride

First year engineering students were asked to draw forces in different parts of the classical Rainbow amusement ride (Figure 1) in connection with an amusement park project. They were told that the ride moves in a circle around a horizontal axis with essentially constant angular velocity and that the rider experiences near-weightlessness at the top.

Figure 1. The riders in the Rainbow amusement ride move in a circle around a horizontal axis.

For motion in a circle with constant angular velocity, the acceleration is always directed to the center and the resultant force, obtained by combining the force from the ride on the rider with the gravitational force, m**g**, should then always be directed toward the center. The magnitude of the centripetal acceleration does not change. In order for the rider to experience weightlessness at the top, the centripetal force must have a magnitude mg.

This task has been found to expose an incomplete understanding of forces. One aspect that is found to be problematic is that the force from the ride is neither purely horizontal or purely vertical (except at the top and bottom of the circular ride). On the sides, the total force from the ride is pointing 45° above the horizontal, towards the top of the circle.

Table 1 shows the results for three cohorts of engineering physics students. (In 2004 and 2005, the test was handed out only in subgroups of students.) A number of typical diagrams could be identified.The distributions of different types of replies is shown in Table 1. Even students mastering the apparent weightlessness at the top and the 2mg normal force at the bottom often fail to include all forces in the points to the left or right or in an arbitrary point. A number of students have drawn two additional forces at the sides, one upward force to counteract gravity and one inward force to provide the centripetal acceleration. These replies have been classifed as correct, but counted separately ("pair of forces"). Some students consistently assigned a normal force exactly cancelling the force of gravity. Other students included only a centripetal force of the same magnitude as the force of gravity. Not surprisingly, a centrifugal force was used in a few replies - but only in a total of two cases it was used in a consistent way. The Aristotelian concept of a force in the direction of motion ("velocity force") occurs in a few cases.

Table 1. Percentage of students drawing different types for force diagrams as discussed in more detail in the text. (Only students answering the question have been counted) . In 2004 and 2005, the test was administered just before the amusement park project presentations, whereas in 2006, it was given before the students had written their project reports. The amusement park project in 2004 was only loosely connected with the course.

Group	Correct[a] (pair of forces)	$\lvert N\rvert = m\lvert g\rvert$	m g and F_c	Centrifugal[b]	Correct, vertical	"Velocity force"	Misc.
F students (2004) (N=68)	19(6)	10	38	7(1)	6	7	12
F students (2005) (N=32)	36(15)	15	33	3	12	0	0
F students (2006) (N=89)	36(23)	0	26	7(1)	15	9	8

a) The first number is the total, and the second (in parenthesis) is the percentage of replies where the total force from the ride is written as two separate forces.

b) Of the students using centrifugal forces, only those in parenthesis drew a consistent force diagram

3.1. Discussion

Although students are usually comfortable with forces in circular motions in the horizontal plane, uniform circular motion in the vertical plane is found to raise problems for a large fraction of the students, except for the highest and lowest points, where all forces are vertical. This is consistent with observations from reading student reports on forces in roller coasters; most students are able to analyse forces at the top of a hill or bottom of a valley in a roller coaster but often run into problems if there is simultaneous sideways turn.

The force diagrams drawn for the Rainbow ride show that students do not automatically connect the experience of the body to the forces in Newton's second law. It also shows that many students treat the centripetal force as an interaction on its own, rather than as a resultant to real forces. In spite of high entrance FCI-scores, only about one third of the students drew consistent force diagrams in this relatively simple situation.

Comparison between the results for the different student groups show a significantly lower fraction of correct results for the F(2004) students. For this student group, the amusement park project was more loosely connected with the rest of the course. The importance of integration of informal learning activities with other course work, through preparation and follow-up, is known from other cases of informal learning (See e.g. Rennie and McClafferty, 1996). Discussion and follow-up, including challenges to students' descriptions, are important also to help students focus on important features of force and acceleration in connection with the rides.

4. Group Interviews on Acceleration

In order to gain a better grasp of students' understanding of acceleration, group interviews were performed with a few of the engineering physics students, during their first weeks in 2005. The groups were asked to discuss and reflect on the way they, themselves, consider the meaning of the concept acceleration, and how it can be measured and observed.

4.1. Group I

- I: What is acceleration?
- A: Change of velocity.
- I: Does anyone want to express it differently?
- B: In a given time.
- C: Yes, that sounds good.
- D: OK.
- I: Would time derivative be useful?
- A: Yes
- I: Are there different types of acceleration that you have worked with?
- B: Negative and positive acceleration.
- A: Yes, it's retarding or something like that. Yes it is ...
- B: When your break.
- C: Or retardation, or, yes, something like that, it was the opposite of acceleration.
- D: Yes it was.
- I: If I throw this thing up in the air, what is the acceleration at the highest point?
- C: Zero.
- D: Zero.
- B: No, 9.81.
- C: All the time.
- F: Obviously.

This group shows a basic understanding of acceleration in one dimension. They discuss positive and negative acceleration. A few of the students are tempted by the common assignment of a zero acceleration in the highest point, but when one student brings up the correct answer, the rest of the group immediately accepts it.

4.2. Group II

- I: If I throw this up in the air, what is its acceleration in the highest point?
- E: Zero.
- I: I guessed you would say that.
- F: The acceleration is g all the time.
- G: Is it???

- I: What is acceleration?
- G: We learned that it was zero momentarily at the turning point.
- E: Yes, I recognize that from school.

In this group, the thought of zero acceleration in the highest point shows more resistance to change. One student's assertion that it is g is met with serious doubts. Two of the members claim that this is what they learned at school.

Just preceding this discussion, the students had been discussing forces in a swing, and show the force diagram as the teacher enters.

- I: Here, the normal forces is as large as this component.
- G: That is how we did it in school.
- I: Then we get a component forcing the swing to go in that direction.
- G: Yes, that is what it looks like at the turning point.
- F: In the *turning point* it cant look like that? Can it? Then it must be gone?
- E: Yes
- G: That's how we have thought about it This is exactly at the turning point. It is free fall there.
- I: How would it move if it were free fall?
- F: (Hesitating) Straight down ?
- I: It can't.
- G: No, but just momentarily, We discussed this.
- F: You cannot take a point were it really looks like this. It is just in the moment of turning.

Here, the students' struggle with what happens momentarily in the turning point is more evident. Forces are discussed in the begining, since they were asked to draw forces. However, the students do not connect the possible motion for the swing to the acceleration available, nor to the implied restrictions of the resultant force.

4.3. Group III

- I: If we throw the ball into the air, what is the acceleration in the highest point?
- K: It depends on how you throw it.
- I: If I throw it straight up.
- (mumble...)
- L: Since the velocity is reduced, acceleration must be reduced?
- M: It should be ...
- K: It is the same as ... there the velocity decreases, but the acceleration is maximum
- L: It is just because... because acceleration is directed towards ... it is not in the direction of motion.

In this group the students bring in motion in two dimensions as a possibility. They show awareness of velocity and acceleration as vectors, but are unable to distinguish between them. When encouraged to keep in one dimension, they start to talk about reduction of velocity and acceleration, and the relation between acceleration and the direction of motion.

4.4. Group IV.

In this group, the students had a short time to discuss before the teacher enters, asking for a summary.

- N: The time derivative of velocity
- I: Does anyone want to express it differently?
- O: Gravity needs acceleration
- I: You were thinking about Einstein? Could you expand?
- O: That acceleration corresponds to curvature of space. That everything is relative in some way.

AM: Do you all go along with this?
P: No

- Q: It depends on what you mean with relative. It is something that depends on what reference system you use. In that way ... You get different values of acceleration depending on which coordinate system you use.
- N: Yes, but as long as the coordinate system does not accelerate, it is not relative.
- But if it accelerates in some way ...
- N: No it is not. You can distinguish between something that is at rest and something that accelerates away from it
- I: Can you say that what is at rest accelerates away from something that moves?
- O: Yes, I think you can
- N: No, I don't think you can. I think there is a difference. I don't know how.
- Q: Yes, but if you say that you use a coordinate system so that the other accelerates away from it. Then the other accelerates away from the coordinate system
- N: Yes, but now one feels a force. One does experience inertia.
- O: One can't have acceleration without a force
- N: It depends on what you require from your reference system. If you only talk about Galilean reference systems, where this law of inertia holds, that an object continues in uniform etc. In such a system, you can distinguish between what accelerates and what doesn't. Otherwise you are no longer in a Galilean coordinate system

In this group, the student N immediately gives a relatively complete answer. Student O tries to expand, bringing in concepts from general relativity. Student R takes up the thread of relativity and brings the discussion on to relative acceleration. Again, N is aware that there is a difference between accelerated and non-accelerated coordinate systems, and after some hesitation introduces force, and O agrees on the importance of force. N introduces inertial forces. Student P is mostly quiet, and probably somewhat bewildered by the discussion of quite difficult concepts.

4.5. Discussion of Group Interviews

From the group discussions above, we see a wide variety of conceptual awareness of acceleration. Acceleration as change in the direction of motion is sometimes brought up spontaneously, although not in the interviews presented here. In group IV, the concept of inertial forces is close. The relation between force and acceleration is clearly present only in group IV, although one of the members may have got somewhat lost in the discussion.

An important dichotomy is whether the description of acceleration refers to another object or involves the own body. In these interview excerpts, only one reference is made to the experience of the body. Is the human body an underutilized teaching tool in mechanics? The next section investigates to what extent are textbooks make use of it.

5. Textbooks and the Role of Mathematics, Statics and Dynamics

In view of student difficulties, the treatments of acceleration was studied for a number of textbooks for school, as well as for undergraduate physics courses. For end-of-chapter problems, a similar picture emerges with very few exceptions: For the case of horizontal one-dimensional motion, the most likely body to accelerate is a car - or possibly "a particle". In one of the introductory university textbooks (Halliday et al 1997), a car appears more than twice as often as other objects in the exercises for the first of the sections describing acceleration, followed by trains, rockets, motorcycles, elementary particles, unspecified objects and an aeroplane. (It does, however, also compare the acceleration of a car to that of rattlesnake head and the text of the chapter discusses the discomfort of an astronaut in a rapidly retarding rocket sled.)

One-dimensional vertical motion is usually restricted to free fall starting from rest. The accelerating objects are mainly balls, stones and unspecified objects, possibly a planet or a runaway elevator. Two-dimensional motion is usually restriced to projectile motion, uniform circular motion and possibly a pendulum.

In engineering textbooks on mechanics, an extensive treatment of statics often precedes the study of motion. In the "dynamics part", Newton's second law could be perceived as a giving correction terms to Newton's first law, where the presence of acceleration is dealt with by including inertial forces in the accelerated systems.

University physics textbooks are more likely to focus the Copernican revolution, with motion of planets playing an important role in the "clockwork universe". Galilean invariance is given a more prominent role and Newton's first law, with absence of forces, is seen as a special case. Independent of approach, the "body" is rarely human.

Below, we consider in more detail the way undergraduate textbooks deals with the aspects of motion that can be measured from within moving systems, i.e. acceleration and rotation, and that are needed for a complete "motion tracking".

5.1. The Equivalence Principle and g-Forces

The concept of "g-forces", commonly used in connection with amusement rides, is closely related to the equivalence principle. Most of the university textbooks introduce this principle as an introduction to the general theory of relativity, Fishbane et al (2005, p 1101) also quote the principle, formulated by Einstein in 1911 as:

Provided that the observations take place in a small region of space and time, it is not possile by experiment to distinguish between an accelerating frame and an intertial frame in a suitably chosen gravitational potential.

Halliday et al (1997, p205) quotes Einstein's statement "I was in the patent office at Bern when all of a sudden a thought occurred to me: 'If a person falls freely, he will not feel his own weight. This simple thought made a deep impression on me. It impelled me toward a theory of gravitation'."

In the older, less colorful but more mathematical, treatment in Alonso and Finn (1992), the equivalence principle is introduced already in connection with the treatment of gravitation, with reference to Galileo.

Following the lead provided by Einsteins own textbook (1916), the most common example after the introduction of the equivalence principle is to consider elevators; elevators on the ground, free-falling elevators and elevators accelerating in deep space. Free-falling astronauts in orbits are also common.

Textbooks sometimes bring up the possibility to simulate gravity in a rotating space station. Hewitt (2005) is relatively unusual in including a discussion about the inhomogeneity in the apparent gravitational field if the space station is too small (p160). He discusses the fictive "centrifugal force" (p147) and emphasizing that it is as real as gravity to the person in a rotating frame, spending about 3 pages. Fishbane et al (2005) very briefly mentions centrifugal force, in less than half a page (p141). In most textbooks considered, the term "centrifugal force" is avoided. Of the engineering dynamics books studied (Meriam and Craige, 2003, Hibbeler, 2004, Bedford and Fowler, 2005) only Meriam and Craige (2003) mention the centrifugal force, and none introduces the equivalence principle.

How do textbooks treat "g-forces"? The only explicit example found is Halliday et al (1997, p66) who introduces "g-induced loss of consciousness", discussing the exposure of fighter pilots in loops and other maneuvers. They also refer to the experiment where Colonel John P Stapp was rapidly accelerated in a rocket sled and then braked to stop, being exposed to many g. The text emphasizes that it is not the speed causing discomfort. They also mention that acceleration is sometimes expressed in terms of "g units" (p17). In a later edition, Colonel Stapp is still included, but the fighter pilot example has been removed.

5.2. Rotation, the Coriolis Effect and the Foucault Pendulum

Rotation, like acceleration, is absolute, and can be measured from within the moving system. Rotation measurements are critical in inertial navigation systems - only in one dimensional motion are accelerometer data sufficient to determine motion. Three-dimensional motion involves 6 degrees of freedom; rotation around three axes, in addition to the acceleration in three dimensions

Although the fictive centrfugal force is mentioned in many of the books, the accompanying fictive force for moving bodies, the "Coriolis force" most often evades treatment in the college books. Searching the index for "Coriolis" brings one end-of-chapter problem in Fishbane et al. (2005). Giancoli (2005) gives an appendix discussing fictive forces, including a discussion of a ball rolling on the surface of a carousel. Serway and Beichner (2000) includes, in the context of pendulum motion, a photo of a science center Foucault pendulum, with a brief description of the history. Alonso and Finn (1992) spends about 3 pages on the Coriolis effect, including the the formula, a discussion of the Foucault pendulum and reference to weather systems.

All "Engineering Mechanics: Dynamics" books studied discuss the Coriolis effect in connection with moving machine parts, which is also what led Coriolis to consider it. Meriam and Craige (2003) includes an end-of-chapter problem involving the effect on a train moving north. Bedford and Fowler (2005) spend three pages discussing the effects in different parts of the world and the consequences for weather systems. None of them mentions the Foucault pendulum.

5.3. Discussion of Textbook Presentations

To a physicist it is obvious that the body in Newton's laws neither excludes, nor implies, a human body. To a learner studying textbooks this understanding is less obvious. In the textbook examples studied, accelerating humans are very rare. If they exist, they are never observing their own acceleration. In the examples that include people in the discussions about forces, the role of the person is usually to exert force on an object - although, of course, according to Newton's third law, this also means that a force is exerted on the person. This impersonal view of the laws of motions is reflected in the in most texts at all levels.

It can also be noted that commonly used tests, including the Force Concept Inventory discussed in 2.1, include no reference to the experience of the body.

In universities, mechanics is sometimes treated as part of mathematics, and mathematics is, indeed, an extremely useful tool to describe and find motions and forces. Mechanics may also be classified as "mathematical physics", and this tradition seems to penetrate school book presentations. An concern is that a mathematical approach may hide the connection to physical reality - which systems accelerates or rotates for relative motion? In addition, the purely mathematical approach clearly is not suitable for young learners. Young learners know of "g-forces", which connect directly to the experience of the body. However, they are very unlikely to be able to read about g-forces in a physics textbook.

The common initial focus on one-dimensional motion, which is a natural consequence of the mathematical approach, obscures important aspects of the concepts. Acceleration as a change of direction is not an obvious possibility in a simplified treatment where motion is studied only in one dimension. However, already in one dimension, failure to distinguish between speed and velocity leads to confusion: Many are the teachers who have been frustrated by the large number of students who state that the acceleration is zero in the highest point for a ball thrown up in the air - and are sometimes relucant to give up their view. Indeed, if retardation is described as different from acceleration, neither applies to the situation where the body is at rest. Defining retardation as "negative acceleration" implies a definition of acceleration as derivative of speed, i.e. of $|\mathbf{v}|$. In classroom discussions it is obvious that this concept is very well rooted among students, who also often incorrectly insist that the acceleration as a swing pass the lowest point must be zero, since the speed has a maximum, and forget all about the centripetal acceleration.

One undergraduate textbook states that "accelerometers" can be used as a navigation device because the measurement of acceleration can be turned into a measurement of position". This statement is true only in one dimension. In three dimensions, also rotation around three axes must be considered to obtain a "motion tracker" (as described e.g. by Rödjegård and Pendrill, 2005). Nevertheless, in discussions with students and colleagues, I often encounter the view that a three-dimensional accelerometer measurement would give a complete description of the motion, e.g. in a roller coaster. In an acceleration-based mechanics education, the focus on what can can be measured is natural, and opens for comparison with familiar situations.

Figure 2. Miniature "Foucault" pendulum experiment in a playground carousel.

Figure 3. Two fourth graders throwing ball in a playground carousel which rotates counter-clock-wise. In the first attempts, the ball arrives behind the person on the other side.

Foucault's classical pendulum experiment in Pantheon in 1851, demonstrating the rotation of the earth, risks to goes unnoticed in most first-year physics textbooks. The replies of new students to the question "How do we know that the earth rotates around its axis?" are consistent: Most students happily give a reply involving day and night, not reflecting that day and night on earth are much older than the insight of a spinning earth. The omission of Foucault's pendulum can thus be seen as a missed opportunity to illustrate physics as an experimental science, and to train students to reflect on "How do we know?".

6. Amusement Park and Playground Physics Experiments for Young Learners

Would it be possible to introduce concepts known to cause difficulty for university students at a much earlier state? Driver *et al* (1994) note: "There is considerable support for allowing pupils to develop their own dynamics - to clarify and label their own ideas. This is seen as a process which could begin early and which should precede any attempts to teach formal physics concepts, and which is better with 11-year olds than with 14-year olds!" The variation in the ways to learn mechanics can be expected to lead to a deeper understanding and more transferrable skills (Marton and Booth, 1997).

In planning lessons on acceleration for young learners, it is natural to build on strongly visual experiments and, if possible, also on the experience of the body. What can be learned by fourth-graders (10-11-year olds) from a lesson in a playground or in an amusement park? Earlier group interviews in connection with such activities have been analysed e.g. by Bagge and Pendrill (2002), Nilsson et al (2004), and Nilsson (2005).

A playground has the advantage that it can be visited frequently and experiments can be repeated if additional questions arise during follow-up discussions after the visit. Recently, a class of forth-graders were invited to experiment at a large playground in Göteborg. Experiments were also performed in swings and slides (Pendrill and Williams, 2005) and in the climbing rack. At the end of the visit, the whole class sat down and discussed the experience, and were asked what experiments had surprised them most.

In 2001 a class of 10-year olds were invited to experiment at Liseberg. All children were well familiar with the park from previous visits with their families. They had prepared the visit by writing letters with questions to a ride of their choice. This approach was suggested by the teacher, as a way to enable teaching to be more closely based on childrens thoughts. Below, we discuss their experiences in connection with two rides with vertical acceleration.

6.1. Experiments in Rotating Coordinate Systems

Figure 2 shows a fourth grader in a playground carousel letting a cuddly animal on a string swing back and forth as a pendulum. As the carousel moves around, the pendulum maintains its direction of motion, in a miniature version of the classical "Foucault pendulum". The rotation of the "Earth carousel" was discussed in connection with an amusement park version of this experiment, discussed by Bagge and Pendrill (2003), who found that it left a deep impression. In interviews three months after the visit, this experiment was recalled by many of the 10-year olds, e.g.:

- In the Pony Carousel, the cuddly toy on the strings started to move like this. I think it was to prove that the Earth is rotating.
- I learned that when going in the Pony Carousel, the cuddly toy kept going in the same direction, while I was going around

The larger freedom in a playground environment also allows for experiments unsuitable in an amusement park, such as attempting to throw a ball to the friend on the other side of the carousel (Figure 3). This experiment is best performed with assisting friends bringing back the ball from the ground; the first few attempts to reach the person on the other side are likely to fail.

Figure 4. Visual g-force measurements in the Frog Hopper, discussed in 6.2.

At the end of the playground visit, which included also forces in swings and bottle races down a slide (Pendrill and Williams, 2005), the class gathered and was asked what experiment they had found most surprising. The carousel was a clear choice, and they could discuss how they had to change the way to throw the ball in order to reach the person on the other side.

6.2. G-forces in Vertical Acceleration

Two of the rides used during an amusement park visit with a the class of 10-year olds featured vertical acceleration: The Frog Hopper, with one-dimensional bounces up and down, and the Circus Express, which is a children's roller coaster. In both these rides, the children took along a soft, small plastic spiral, a "slinky", which gives a visual measurement of acceleration, or rather of the g-force, that can be connected directly to the experience of the body:

- When you used the slinky in the Frog Hopper, you got to see how much you weighed.
- When you went in the Frog Hopper, at some places you were lighter and you were heavy at other places because the slinkies were pulled together when you are light and stretched out when you become heavy.

The children then tried to relate the observations to the different parts of the motion:

- If you went in the Frog Hopper with a slinky, the slinky was pulled together when you went down.
- When you went up, it was stretched out because there was so much speed.
- The slinky went up and down. When you were at the highest, the slinky was the shortest, which means that you were the lightest.

Only the last of these three quotes gives a correct description: It is not the motion or speed *per se* that causes the change in length, but rather acceleration, i.e. the change in velocity so the slinky is shortest at the top (and longest at the bottom).

For the Circus Express roller coaster we meet again children who associate long slinkys with high speed, and other children who are more precise in their observations, e.g.:

- When you went down a slope and you had speed downwards, then you felt heavy.
- When your went upwards and then went down, you became light. And when you went downwards and then went upwards, you became heavy.

The first quotes quotes for both rides illustrate the common need for a dialogue to find out what children mean by "went down" and "went up" in their descriptions. "Went down" can sometimes be used to describe the situation when the ride turns to go down, i.e. at the top. The comment about "so much speed" may indicate an Aristotelian expectation of a force needed to maintain speed, but could also mean that the ride had "so much speed" that needed to change. In the recorded interviews, these points were not pushed by the interviewer.

If possible, a dialogue to help children give more precise descriptions should take place directly after the ride, when memories are fresh and where children still have a chance to repeat the experiment. In the park, they can also observe the variation of the length of a slinky held by a friend in the Frog Hopper, making possible a joint discussion on the ground for the rest of the group.

6.3. Discussion of Physics Experiments for Young Learners

Many of children's questions in their initial letters to amusement rides were found to have significant physics content. The letter-writing approach has later been tried by collaborating teachers for various age groups. We have found that the character of the questions is essentially independent of the age of the learner and many questions involve height, speed and g-forces. Thoughts on safety and "what would happen if ..." are also high on the agenda.

The carousel experiments described above give first-hand experience with the fictive "Coriolis force" acting on objects moving in a rotating coordinate frame. The Coriolis force is certainly not part of the school curriculum. In college textbooks, it is in general not mentioned, except possibly in one of the end-of- chapter problems. When students encounter the mathematical treatment at university, in general, they have no direct experience of it, and find it a very abstract phenomenon, which little relevance for life nor technology. Still, experiments in slowly rotating system have been found to fascinate 10-year old learners (as well as older students), even if it did not relate directly to any question in their initial letters. The interest was created by the learning situation prepared in the park.

The carousel was also the experiment which made the strongest impression for the class doing the playground visit. They were absorbed by watching the little pendulum swing. They had also figured out that they needed to throw the ball a little bit to the front of the friend on the other side, since he would have moved a bit before the ball arrived.

The two amusement rides with vertical acceleration discussed above involve motion in one and three dimensions. In these rides children had the possibilities to measure g-forces for themselves. The replies are relatively similar for both rides, and also consistent with the experience from discussing with classes in the park, at these rides and also e.g. in the classical Rainbow ride discussed in Section 3. We see no evidence that the added spatial dimensions of the ride complicate observations. In fact, the accompanying horizontal movement makes it possible to use a small drawing of track of the roller coaster to ask children to be more precise about the points where they feel heaviest or lightest.

The use of a real time visual measurement makes it easier to discuss the experiences with a group of learners. This is particularly useful where observers can view the whole ride, or at least a significant part of it. In this way, the experience benefits also children (and sometimes teachers) who are reluctant to go on the rides. Children seem to have no difficulty associating their apparent weight with the length of the slinky, and thus make an initial connection between force and acceleration.

A dialogue is important for sharpening observations and conclusions and to clarify distinctions between different parts of the motion. This is no argument against 10-year olds performing the experiments. Replies similar to those quoted here are given by physics learners throughout school.

7. Discussion of Approaches to Acceleration

An essential part of learning a concept is to develop different ways of experiencing it. A deep understanding requires many different representations of the concept and links to different representations, as well as to other contexts. In everyday life, acceleration means increase of speed. Car advertisements express acceleration in terms of the time needed to reach 100 km/h. Only in the later school years and university, if at all, is the student likely to encounter acceleration as the time derivative of velocity, being a vector with both direction and magnitude.

New students' description of acceleration, as it shows in group interviews, indicates in most cases a very limited relation to the concept, which is unlikely to enable students to use it to understand new situations.

7.1. Time Derivative of Velocity

To appreciate the formal definition of acceleration as the time derivative of velocity requires an understanding both derivatives and of velocity, including its vector character. Both aspects have been found to be problematic also for entering university students. This approach directly excludes introduction to young learners. In analysing students' difficulties, we find that some simplifications introduced in school physics, may obscure, rather than illustrate the more generalised concept.

Students often forget the vector character of velocity and acceleration. Acceleration as a change of direction of the velocity is not an obvious possibility in a simplified treatment where motion is studied only in one dimension. However, already in one dimension, failure to distinguish between speed and velocity leads to confusion: Many are the teachers who have been struck by the large number of students who state that the acceleration is zero in the highest point for a ball thrown up in the air. Indeed, if retardation is described as different from acceleration, neither applies to the situation where $v=0$ and the concepts can not be generalised. If acceleration were derivative of speed, i.e. of $|\mathbf{v}(t)|$, then retardation would imply negative acceleration and the derivative would be undefined for $v=0$.

Thus, the restriction of discussions of acceleration to one-dimensional motion and to situations where the body starts from rest, not only excludes most everyday experience of acceleration, but may also lead to views that get in the way of an understanding of the more general concept. Although a mathematical description of 2 or 3 dimensional motion should not be attempted before the 1-dimensional situations are understood, there are alternative ways of viewing and introducing acceleration that makes the concept accessible also to younger learners.

7.2. Acceleration as a=F/m

Acceleration, in its full vector capacity, is very much a part of our life. The child in the swing, on a slide, trampoline or see-saw, or jumping down a from a climbing rack enjoys the sensation of changed acceleration and interplay of forces. Children do not feel the need to refer to Newton's laws, but may be helped to connect the experience of the body with a change in velocity - both as a change of speed or as a change of direction.

The formulation of Newtons second law as $\mathbf{a}=\mathbf{F}/m$ shifts the focus from force to acceleration. Whereas the forces acting on mother and child e.g. in an amusement park ride are very different, they may both share the same acceleration. They also experience the same "g-force".

8. Conclusion

Students' difficulties in understanding force and acceleration are well documented in the literature. The first investigation in this chapter focused on a group of students with good understanding as measured by very high scores on the commonly used "force concept inventory". It was found that many of them had difficulty in applying Newton's laws for circular motion in a vertical plane. Group interviews with some of these students about the concept of acceleration indicated a very weak connection to force. An analysis of textbook treatments showed that they are traditionally based in a mathematical tradition, and providing limited variation in the ways offered for experiencing the phenomenon of acceleration.

The common school focus on one-dimensional motion alienates school physics from everyday life experiences. This focus is also found to obscure important properties and to encourage ways of thinking that can not be applied to more general cases. Some misunderstandings that may be traced to the textbook treatments have been found to be very persistent. Focusing on acceleration, rather than on uniform, rectilinear motion makes it possible to build on the experiences of the body. Simple experiments are presented, which have been tried out in pilot projects in a few classes. These experiments relate to important physical principles and introduces a distinction between relative and absolute motion.

Experiments during a couple of hours in a park or on a playground can only be first steps toward an understanding of acceleration. Focusing on acceleration, rather than on uniform, rectilinear motion, makes it possible to build on the experiences of the body, offering qualitatively different ways of understanding the concept of acceleration. Observations and interviews show that physics experiments in a playground or amusement park can be both enjoyable and educational also for young learners. We hope that inviting children to make a distinction between acceleration, velocity and speed at an early age, can help them to interprete future observations in a more Newtonian way. The analysis in this work shows that the experience of the body is a severely underused resource for teaching about force and motion.

Acknowledgments

My first contact with educational science research was during inspiring discussions with Ference Marton and Inger Wistedt in connection with the development of the educational programme "Problem Solving in Natural Science" at Göteborg University. The investigations in this work have been carried out in creative collaborations with many colleagues, including Sara Bagge, Pernilla Nilsson and Åke Ingerman, and with helpful and generous assistance from the Liseberg amusement park. Financial support has been provided by the Swedish Research Council (VR), the Council for Renewal of Higher Education (RHU) and Chalmers Strategic Effort in Learning and Teaching (CSELT).

References

Adawi, T., Ingerman, Å. and Pendrill, A.-M. (2005). *How Mathematical is Conceptual Understanding?*, Contribution to Physics Teaching in Engineering Education (PTEE) 2005, Brno.

Alonso, M. and Finn, E. J. (1992). Physics, Addison Wesley Publishing

Bagge, S. and Pendrill, A.-M. (2002) Classical Physics Experiments in the Amusement Park, *Physics Education* **37**, 507-511.

Bagge, S. (2003). *Learning Physics by Experiment - An investigation of extramural learning* (licentiate thesis, Göteborg University)

Bagge, S. and Pendrill, A.-M.(2004). Extramural Learning at Liseberg. In E.K. Henriksen and M. Ødegaard (Ed.), Naturfagenes didaktik - en displin i forandring? - *Proc. 7th nordic research symposium about science in school*, Høgskolen i Agder.

Bakken, C. (2007). Physics/Science/Math Days @ Great America, physicsday.org/

Bedford A and Fowler W, *Engineering Mechanics: Dynamics*, Pearson Prentice Hall

Duit, R. (2007). Bibliography - STCSE Students' and Teachers' *Conceptions and Science Education,* www.ipn.uni-kiel.de/aktuell/stcse/stcse.html

Driver, R., Squires A., Rushworth, P. and Wood-Robinson, V. (1994). *Making sense of secondary science - research into children's ideas*, (Routledge)

Hake, R. R. (1998). Interactive-engagement versus traditional methods: A six-thousand-student survey of mechanics test data for introductory physics courses, *American Journal of Physics* **66**, 64-74.

Halloun, I., Hake, R., Mosca, E. and Hestenes, D. (1995). *Force Concept Inventory*, The test is available, password protected, at modeling.la.asu.edu/R&E/Research.html

Hestenes, D., Wells, M. and Swackhamer, G. (1992). Force Concept Inventory, *Physics Teacher* **30**(3), 141-158.

Hibbeler, R C (2004) *Engineering Mechanics: Dynamics*, Pearson Prentice Hall

Einstein, A (1916).Über die specielle und allgemeine Relativitätsteorie (Relativity: The Special and General Theory. New York: Henry Holt, 1920; Bartleby.com, 2000, www.bartleby.com/173.)

Fishbane, P M, Gasiorowicz, S G, and Thornton, S T, (2005). Physics for Scientists and Engineers with Modern Physics, Pearson, Upper Saddle River, NJ, 3rd ed.

Giancoli, D C (2005).Physics, International Edition (Pearson, Prentice Hall, Upper Saddle River, NJ) 6th ed.

Halliday, D, Resnick, R and Walker, J (2005). *Fundamentals of Physics*, Wiley, 7th ed.

Halliday, D, Resnick, R and Walker, J (1997). *Fundamentals of Physics*, Wiley, 5th ed.

Hewitt, P G, *Conceptual Physics* (2006). Pearson International Edition, (Pearson Upper Saddle River, NJ) 10th ed.

Ling, L. M., Chik, P. and Pang, M. F. (2006). Patterns of variation in teaching the colour of light to Primary 3 students. *Instructional Science*, **34**, 1-19.

Marton, F. and Booth, S. (1996). The learner's Experience of Learning. In D. R. Olson & N. Torrance (Eds.), *The Handbook of Education and Human Development* (pp. 534-564). Oxford, UK: Blackwell.

Marton, F., and Booth, S. (1997). *Learning and awareness*. Mahwah, New Jersey: Lawrence Erlbaum Associates, Inc.

McDermott, L. C. (1998). Students' conceptions and problem solving in mechanics, in A. Tiberghien, E.L. Jossem, and J. Barojas (Ed:s), *Connecting Research in Physics Education with Teacher Education*, ICPE

McDermott, L. C. & Redish E. F. (1999). Resource Letter: PER-1: Physics Education Research, *Am. J. Phys*. **67**, 755

Meriam, J L, and Kraige L G (2003). *Engineering Mechanics: Dynamics*, SI Version, Wiley, 5h Ed

Nilsson, P., Pendrill, A.-M. and Pettersson, H. (2004). *Learning Physics with the Body*, IOSTE IX.

Nilsson, P. (2005). Children's communication and learning in physics through practical experiments (in Swedish), *NorDiNa* **1**, 58-69

Pendrill, A.-M. (2005). University and Engineering Physics Students' Understanding of Force and Acceleration - Can Amusement Park Physics Help? *Proc. 8th Nordic research symposium about science in school,* Aalborg, Maj 2005,

Pendrill, A.-M. and Rödjegård, H. (2005). A Roller Coaster Viewed Through Motion Tracker Data, *Physics Education*, **40**, 522-526

Pendrill, A.-M. and Williams, G. (2005). *Swings and slides, Physics Education*, **40**, 527-533

Pendrill, A-M. (2007). *Project Impact*, http://physics.gu.se/LISEBERG/

Redish, E. (2003). Teaching Physics with the Physics Suite, Wiley Text Books

Rennie, L. R. and McClafferty, T.P. (1996). Science Centers and Science Learning, *Studies in Science Education*, **27**, 53

Runesson, U. (2006). What is it possible to learn? On variation as a necessary condition for learning. *Scandinavian Journal of Educational Research*, **50**, 397-410

Serway, R A and Beichner R J (2000) *Physics for Scientists and Engineers with Modern Physics,* Harcourt, Orlando, FL, 5th ed.

In: Science Education in the 21st Century
Editor: Ingrid V. Eriksson, pp. 205-240

ISBN: 1-60021-951-9

Chapter 7

ACTION RESEARCH AS PROFESSIONAL DEVELOPMENT OF SCIENCE TEACHERS

Pradeep Dass[1a], Avi Hofstein[2b], Rachel Mamlok[2c], Karen Dawkins[3d] and John Penick[4e]

[1]Department of Biology, Appalachian State University, Boone, NC 28608-2027 USA
[2]Department of Science Teaching, The Weizmann Institute of Science, Rehovot 76100 Israel
[3]Center for Science, Mathematics, and Technology Education, East Carolina University, Greenville, NC 27858 USA
[4]Department of Mathematics, Science, and Technology Education, North Carolina State University, Raleigh, NC 27695-7801 USA

Abstract

Achieving scientific literacy for all citizens has become a national goal in many countries. Although admirable, it represents a challenge for science teachers and for those responsible for professional development. To achieve this goal, there must be a reform in the way science is taught in schools and in the strategies used to professionalize science teachers in general as well as the development of leadership among science teachers in particular.

One of the more effective methods to attain the goals of reform and its proper implementation is to actively involve the teachers in the curricular processes. Action research is only one of many effective strategies (Loucks-Horsley, Love, Stiles, Mundry, & Hewson; 2003) that could be used in order to effectively prepare science teachers to cope with new content and pedagogical standards. Having a long and varied history, action research was first introduced by Kurt Lewin in the 1940s. However, only within the last 15 years have we discovered the potential of action research to bring about lasting change in the science classroom.

[a]E-mail address: dasspm@appstate.edu
[b] E-mail address: Avi.Hofstein@weizmann.ac.il
[c] E-mail address: Rachel.Mamlok@weizmann.ac.il
[d] E-mail address: DAWKINSK@ecu.edu
[e] E-mail address: john_penick@ncsu.edu

It has been suggested that long-term professional development activities could create a clearing-house for pedagogical ideas derived from the results of action research. This chapter presents four case studies, two from the USA and two from Israel, in which action research was used as a key strategy for professional development. For each case study, both the action research components and the professional development components are described. The chapter concludes with insights from these cases on issues and challenges involved in using action research as a professional development strategy.

Introduction

Achieving scientific literacy for all citizens has become a national goal in many countries. Although admirable, it represents a challenge for science teachers and for those responsible for professional development. To achieve this goal, there must be a reform in the way science is taught in schools and in the strategies used to professionalize science teachers in general as well as the development of leadership among science teachers in particular. One of the more effective methods to attain the goals of reform and its proper implementation is to actively involve the teachers in the curricular processes. Action research is one of many effective strategies (Loucks-Horsley, Love, Stiles, Mundry, & Hewson; 2003) that could be used in order to effectively prepare science teachers to cope with new content and pedagogical standards and become actively involved in the curricular processes. Action research was first introduced by Kurt Lewin in the 1940s. However, after going through a long and varied history, it's only within the last 15 years that the potential of action research to bring about lasting change in the science classroom has been recognized.

Long-term professional development activities have the potential to create a clearing-house for pedagogical insights derived from the results of action research projects conducted by participants in professional development programs. Thus, action research can serve as a significant vehicle for professional development of science teachers for the purpose of achieving the goal of scientific literacy among students. This chapter presents four case studies, two from the USA and two from Israel, in which action research was used as a key strategy in professional development.

Theoretical Framework

Teaching science in a 21st century classroom requires much more than a straightforward implementation of a specified curriculum. It involves explicit attention to classroom discourse, to overt and covert messages conveyed when asking students questions and conducting discussions, and to participation in mutual learning activities. Consequently, in addition to content and curriculum knowledge, the teacher needs a fairly sophisticated knowledge of science teaching and learning theories. The content, pedagogy, assessment, and professional development standards published in the USA (National Research Council, 1996) toward the end of the last century specify the scientific content that all students should know and be able to utilize. The teaching and professional development standards within the same document define the knowledge, skills, and experience that teachers must have if they are to engage effectively in teaching science. With regard to professional development standards, Bybee (1995) wrote,

> *The Professional Development Standards describe the criteria for the preparation and ongoing opportunities that teachers must have to gain and maintain the knowledge base and skills, and to provide the learning opportunities for students that the teaching standards define. (p. 32)*

The key question in professional development planning is to decide which strategy to adopt in order to attain the goals of professional development promoted by the National Science Education Standards (NSES; National Research Council, 1996). Effective professional development plans should consist of a combination of strategies to form a unique design suited to meet the local needs of participants (Loucks-Horsley et al., 2003).

One such strategy for professional development is action research. Action research means that teachers systematically examine their own practice of teaching. It is an inquiry into their teaching and their students' learning in their own classroom (Feldman & Minstrel, 2000). According to Feldman (1996), the primary goal of action research is not to generate new knowledge, whether more local or universal but rather, to improve practice. Lewis and Munn (1987) indicated three main reasons for conducting teacher-based research:

- To try to find out what is actually going on;
- To monitor and thus formatively influence the direction of new developments; and
- To evaluate what is already taking place.

According to Feldman & Minstrel (2000), the main reason for action research to become an integral part of the implementation process of curricula is that it reduces the time lag between the generation of new knowledge and its application in the classroom.

In their book *Designing Professional Development for Teachers of Science and Mathematics, 1st Ed.*, Loucks-Horsley et al. (1998) contend that

> *Action research has evolved in the education community into an ongoing process of systematic study in which teachers examine their own teaching and students' learning through descriptive reporting, purposeful conversation, collegial sharing and critical reflection for the purpose of improving classroom practice. (p.95)*

With regard to the teacher as a curriculum developer, Ben-Peretz (1990) claimed that teachers may function as creators of their own curricula whether through involvement in the process called "teacher-based curriculum development" or as "participants in centralized curriculum efforts." This is fondly called 'bottom-up' as opposed to the 'top-down' process. According to Ben-Peretz, teachers come up with creative ideas about teaching in general and pedagogy in particular. They are also familiar with their students' needs and abilities. Thus, with proper training, teachers could serve as school-based curriculum developers and researchers (Feldman & Minstrel, 2000; Hofstein, Mamlok & Carmeli, 1997).

The strength of action research as a professional development strategy is that teachers either define the research questions or contribute to defining them in a meaningful way (Loucks-Horsley et al., 2003). In order for action research to be an effective means for helping teachers to reflect on their practice, they must be provided with opportunities to engage in life-long professional development. These opportunities will provide them with an environment of support, collegiality, and collaboration with professional researchers and other teachers who teach the same or related subjects, an environment that encourages reflection on their classroom practice and the results of their research efforts. Typically, teachers who are inexperienced in action research need support and training regarding its methodology, procedures and activities. This includes designing tools, collecting data, analyzing and interpreting the results, and finally applying the findings in the science classroom. Engaging in professional development provides the teachers with an opportunity to share the results of their classroom research and related pedagogical activities with their fellow teachers, who will later provide them with feedback and other ideas.

Each case study described below focuses on both the action research components and the professional development components. The chapter concludes with insights from each of these cases on issues and challenges involved in using action research as a professional development strategy.

CASE 1. The Collier Chautauqua Program: Action Research Leading to Authentic Assessment

Making science relevant to the lives of students requires, among a variety of other factors, a classroom environment in which students can be actively involved in making meaning of the information within a relevant context. Teachers need to learn to create an environment and employ strategies that encourage active questioning and identification of issues and answers by students. They need to be able to encourage students to challenge the information presented and discuss its personal relevance. These abilities cannot be developed through brief, 'one-shot' in-service sessions that have been traditionally dubbed professional development. They require carefully designed, sustained, professional development opportunities that involve teachers actively in the learning process. One key component of involving teachers actively in the learning process is to get them to reflectively analyze their instructional practices, identify or develop ways of collecting information that can inform their reflective analyses, and employ the results of these analyses to improve their instructional practices. In other words, teachers should become involved in what has come to be known as 'action research' in order to become active learners.

This case describes one specific result of action research undertaken by 24 elementary and middle school teachers in Collier County (Florida, USA) while they participated in a year long professional development program promoting the teaching and learning of science through Science-Technology-Society (STS) approaches (the teaching and learning of science in the context of human experiences, as described by the National Science Teachers Association, NSTA, 1992-93, pp. 168-169). The Collier Chautauqua Program (CCP) was born in Collier County as a result of the dissemination efforts of the Iowa Chautauqua Program (ICP, a year-long professional development program for K-12 in-service science teachers promoting STS approaches to science instruction) through the National Diffusion Network (NDN) of the U.S. Department of Education. Primary leadership for the initial planning of CCP was provided by the district science coordinator of Collier County Public Schools. Emulating the ICP model, the CCP evolved into a district-wide professional development program. During the first two years of implementation, the program focused on professional development of elementary and middle grades teachers with regard to science instruction.

Due to its focus on elementary and middle school teachers, the CCP fostered an integrative approach to teaching and learning using the content of science. The activities of the CCP included the following:

1. **Summer Leadership Institute**: A four-day institute for lead teachers designed to prepare them for leadership roles for the summer and academic year series of workshops.
2. **Summer Training Institute**: A three-week institute for participating teachers during which they experience the STS approaches as students and subsequently develop instructional modules based on these approaches for use in their own classrooms.
3. **Academic Year Series of Workshops**: Three-day workshops during Fall and Spring semesters to evaluate teaching trials of the modules designed during the summer institute, make further plans and refinements, and develop appropriate assessment schemes.
4. **Interim Communication**: Monthly meetings, electronic communication, and site-based meetings to continually share information, assess progress, and provide support and encouragement to peers.
5. **Interim Teaching Projects**: Teaching trials of modules developed during the summer institute and incorporation of new teaching strategies into the entire curriculum.

During the first year of implementation, 24 teachers participated in the CCP. Brief demography of these teachers is presented in Table 1.

Table 1. Participant Demography, CCP Year 1

Name of School	# of Teachers Involved	Grade Levels affected	Approximate # of Students affected
Avalon Elementary	6	K-5	120
Big Cypress Elementary	1	Kindergarten	25
Lake Park Elementary	1	5	17
Laurel Oaks Elementary	5	K-5	100
Pine Crest Elementary	1	2	26
Village Oaks Elementary	1	3	40
Vineyard Elementary	1	5	32
Gulf View Middle	1	6-8	100
Manatee Middle	1	7	90
Oak Ridge Middle	4	7-8	400
Pine Ridge Middle	2	6-8	393

Action Research in the CCP

A major focus of action research conducted by teachers participating in the CCP during its first year was to examine the teaching and learning in their classes as they used the STS modules and determine the extent to which their current classroom assessment approaches matched the instructional approaches inherent in the STS modules. This examination led to the realization that their current approaches were lacking in assessing student growth in several domains of science that were addressed in their modules. They also realized that they did not have appropriate ways of evaluating and documenting specific elements of instructional practices (such as the use of technology and the use of local human resources) promoted through the STS modules. In order to deal with these assessment issues, they initially tried using some of the assessment instruments that had been developed by the ICP and used by teachers in Iowa (Enger & Yager, 2001). However, as they used these instruments and shared their results with each other during early part of the fall semester, they discovered that many components of these instruments were not suitable for their particular setting. For instance, some items on student assessment instruments were not intelligible to students in Collier County because they were not familiar with the examples and illustrations used in those items. Majority of students in Collier County came from 'limited English proficiency' (LEP) background and had a difficult time understanding some of the items on these questionnaires. With regard to teacher assessment of their own instructional practices, the ICP questionnaires were found to be unsuitable in several respects. For instance, questions regarding the use of technology did not address some of the ways Collier County teachers were using technology as well as the types of technology available to Collier County teachers. Upon careful analysis of the results of administering these instruments (an action research project), CCP participants came to the conclusion during the fall workshop that they needed assessment instruments whose development takes local situations into account, thus providing items that can capture appropriate information within relevant local contexts.

As a result, CCP participants organized an Assessment Committee (comprised of a small group of participating teachers) that was charged with the responsibility of developing locally relevant assessment instruments. Using the ICP instruments as templates and input from the rest of the teachers in the program as well as the program evaluation specialist, the assessment committee completed the development of the following assessment instruments by the spring workshop.

- Student Attitudes about Science
- Classroom Environment Survey
- Teacher Self-Assessment of Instructional Processes
- Technology Use (by Teachers) Survey
- Teacher Self-Evaluation of Basic Computer Use
- Community Involvement Inventory

These instruments, which were used during the second and subsequent years of CCP, are presented in the Appendix.

The CCP case may initially not appear to fit the typical action research cycle, such as the one offered by Sagor (1992), which involves the following five steps: *Problem formulation, data collection, data analysis, reporting results, and action planning.* However, a close examination of this case will indicate that each of these steps are present in one form or another, albeit not necessarily in the same sequence. For instance, 'problem formulation' step took place twice in this case—first, when teachers initially realized the need for assessment instruments to measure specific elements of science instruction involving their STS modules and second, when they realized the unsuitability of the ICP instruments and the need for locally relevant instruments. Organization of the assessment committee was certainly representative of the 'action planning' step and their presentation of the new assessment instruments at the spring workshop represented a form of 'reporting results' step. In the process of developing locally relevant assessment instruments, the participants—particularly members of the assessment committee—gained a newfound confidence in their ability to do something they had never attempted before, namely development of authentic assessment instruments that provided useful information regarding their instructional practice and student achievements. This confidence and the resulting product (the assessment instruments) enhanced the degree of ownership teachers took in the process of assessing their teaching and their students' learning. In the spirit of learner centered pedagogy, it can be argued that the greater the ownership, the greater will be the usefulness of instructional tasks to the teacher (in this case assessment activities). Collaboratively developing locally relevant assessment instruments, presenting them to their peers at the spring workshop, and seeing them used during the next year of CCP provided a sense of leadership to the members of the assessment committee. The sense of leadership, together with increased confidence, contributes much more significantly to professional development of participants than the traditional in-service workshops in which teachers usually play the role of information recipients. It can safely be argued that in this case action research lead to the development of leadership, enhancement of confidence, and a sense of ownership (of assessment activities) within these CCP participants.

The CCP case can serve as a good example of the contribution action research can make in developing leadership among teachers to undertake tasks they have not undertaken before. It also reflects how action research can help teachers take ownership of the most critical component of their work—assessment of teaching and learning. Both the elements of leadership and ownership development are critical components of the type of professional development that leads to real educational reform (Loucks-Horsley et al., 2003). As such, the CCP case clearly attests to the potential of action research in professional development of science teachers in ways that contribute to translating reform tenets (American Association for the Advancement of Science, 1994; National Research Council, 1996) into classroom instructional practice.

CASE 2. High School Chemistry Pedagogy: Improvement through varying student assessment methods

A key element of effective science education is implementation of new and varied instructional techniques by teachers in their classroom. The rationale is that instructional techniques in science should be matched with individual learners' characteristics and needs, in order to maximize the effectiveness of the teaching and learning processes and enhance student motivation to learn science (Hofstein & Kempa, 1985; Hofstein & Walberg, 1995). The implementation of new instructional techniques in the science classroom implies a prerequisite of matching to each technique an assessment tool or method to measure student achievement and progress. According to Gitomar and Duschl (1998):

> *The most promising efforts in assessment reform are those that address directly the relationship of assessment and instruction, specifying precisely how assessment can be used to support improved instructional practice. (p. 803)*

Assessment has typically focused on only a small fraction of student learning and activities (Lazarowitz & Tamir, 1994). The final grades given to students are usually based on their performance on paper and pencil tests. Student achievement in other learning processes that they were involved in, such as inquiry in the science laboratory, group and personal projects, reading of scientific literature, and many other instructional activities were rarely assessed and included in students' final grades. This was usually due to lack of valid and reliable criteria as well as suitable assessment tools for these learning components. Additionally, teachers have been less confident in using alternative assessment methods. Consequently, there is a need for teacher professional development to enable them to implement both varied instructional and assessment approaches.

The Study

An experimental project was conducted in Israel in one high school, aimed at bringing about improvement in the way chemistry is taught and learned through new pedagogical and assessment standards. The study was initiated in 1995 as a part of a nation-wide research project, involving 22 schools in which an alternative assessment method was tried out. The idea was that these school-based assessment procedures, planned and implemented by the school teachers would replace the national matriculation ('*Bagruth*') examinations, that are created by the Israeli government (as part of the central education system that exists in the country) and administered at the end of the 12th grade. Because of its structure, the '*Bagruth'* covers only a limited scope of student cognitive abilities and learning experiences (mainly assessed by a one-time paper and pencil test). The theory behind this project was that the assessment should be valid and reliable, authentic (close to the students' way of learning), and comprehensive, i.e. should cover a wide range of abilities and ensure the acquisition of a wide range of learning goals.

The Chemistry Project

The chemistry project took place in an urban type high school, representing 100 students who studied chemistry between the 10th and 12th grades (The same cohort over a period of 3 years). These students opted to major in high school chemistry. In the 10th grade the students studied chemistry for 3 periods per week and in the 11th and 12th grades 8 periods per week. The course content was based on the regular syllabus used in Israeli high schools. A team of nine chemistry teachers from one high school was involved in the part of the project that is reported here. They were led by the school chemistry coordinator (head of department) and guided by an experienced teacher (tutor) who was appointed by the government to provide guidance in conducting the project. Together, they were involved in course development and design of relevant assessment tools. Among the assessment tools developed by the team were tests, quizzes, and assessments of inquiry type experiments and mini-projects. The teachers determined the weights given to each of the assessment criteria in order to increase the objectivity of the assessment. The role of the tutor and the coordinator was to enhance the team's confidence regarding the importance and effectiveness of the project. In fact, they acted as guides for what could be regarded as long-term, dynamic, school-based professional development of the chemistry teacher team. The team also served as an "open platform" for dealing with problems that emerged in the teachers' daily work. This platform supported the development of an environment of collegiality, in which it was possible to reflect on one's work and to critically evaluate different pedagogical ideas and alternative assessment methods. During the years of the project, teachers in the team developed the ability to cooperate and to communicate, which became a model for other schools.

Table 2. Assessment categories in chemistry

Learning objectives	Tests	Quiz	Work on articles	Inquiry type experiments	Mini-Projects
Knowledge, comprehension and application	✓	✓		✓	✓
Understanding of relationship between chemistry and the environment	✓		✓	✓	✓
Development of thinking skills in chemistry			✓		✓
Understanding of inquiry methods (and abilities) used in chemistry				✓	✓
Understanding of the industrial applications in chemistry	✓		✓		
Understanding of the nature and history of chemistry			✓		
Development of scientific thinking			✓	✓	✓
Development of practical (laboratory) abilities				✓	✓
Using sources of information					✓
Development of inquiry skills				✓	✓
Development of critical reading skills using scientific articles			✓		✓
Presentation of information, results, and data			✓		✓
Development of the ability to work individually					✓

This project could be regarded as action research, which is increasingly viewed as central to the restructuring of schools and as a strategy for teacher professional development. In this context, Loucks-Horsley et al. (2003) claimed that one of the forms of action research is when teachers are working together in inquiry and collaborative teams to attain certain pedagogical goals and to bring about change in classroom practice. The team was given time and support from the school administration to develop the new chemistry program. Access to research resources for the development was provided by the tutor from the 'National Center for Chemistry Teachers' at the Weizmann Institute of Science. Finally, the team was given opportunities to share the results of their project at chemistry teacher conferences that were held in various centers throughout the country. According to Loucks-Horsley et al. (2003) these three factors (time, resources and ability to share results) are vital for action research to be successful and beneficial. Table 2 describes the correspondence between teaching and learning objectives and the method of assessment that was used in the 'alternative chemistry assessment' project resulting from the team's collaborative discussions and decisions.

Students' and Teachers' Attitudes and Perceptions towards the Project

Following three years of the project, the team of teachers administered a questionnaire aimed at probing into students' attitudes and perceptions regarding the program in which they were involved.

In general, it was found that the students who participated in the project were satisfied with the way they learned chemistry. Clearly the project established a healthy learning environment in the chemistry classroom. These findings are especially important in an era in which creating an effective and a better classroom learning environment is becoming a major goal.

Regarding the teachers who participated in the program, the project significantly changed their perceptions of their role as chemistry teachers in an era of reform.

CASE 3. Transforming High School Chemistry Instruction through Action Research

Ten high-school science teachers participated in a two-year professional development program promoting innovations in the teaching and learning of chemistry in eight high schools in Israel. As an integral activity of the program, participants systematically analyzed the teaching and learning in their classes. The program involved monthly one-day meetings throughout the year at the Science Teaching Department of the Weizmann Institute of Science in Rehovot, Israel. During the workshop, the participants identified and discussed problems they had encountered with their students, which they wanted to focus on and investigate. The research questions, objectives, and methodology for each investigation were discussed and analyzed by the participants and workshop coordinator. All the investigations were developed in stages, and conducted stage-by-stage in each class. In each meeting the teachers reported on their progress and reflected on their work. In addition to promoting instructional innovations, the program also sought to meet the following objectives.

- To establish a lead teachers team who will perform action research investigations with other teachers at their respective schools.
- To provide support by the academic staff regarding the topic of "Action Research".
- To create an atmosphere of cooperation between the participants in the workshop at different stages of the investigation and the research.

Program Description

The workshop for high-school chemistry teachers was conducted during 2000 - 2002. The workshop participants consisted of ten chemistry teachers from eight high schools (grades 10-12) in central Israel. Each participant had more than 10 years of experience teaching high school chemistry and had previously participated in a few professional development workshops.

The participants met for a four-hour meeting each month, for a period of 30 weeks. In the first four meetings, the participants attended lectures and a workshop regarding the following:

- What does research in science education mean?
- What is the difference between quantitative research and qualitative research?
- What do we mean by "Action Research"?
- What is a good research question for "Action Research"?
- How do we design our study (outline, time line....)?
- How can we design tools for our study?

During the following meetings, the workshop coordinators supported the participants in the stages of their classroom-based action research: (1) identify a general problem; (2) get deeper insight into the situation and understand it; (3) focus on the problem and prepare an outline of the planned research; (4) perform the research and carry out the required changes; (5) evaluate the situation resulting from the changes they made; and (6) introduce changes in view of findings, and start all over again.

After each meeting, the participants met with their chemistry colleagues at school, and shared with them the topics and subjects discussed at the workshop. In this way, they involved the whole team of chemistry teachers at their school in the action research process. During the following six meetings, the participants reported on their work, discussed the difficulties that had arisen throughout their team work, and received advice, clarification, critical comments, and support from other participants and the workshop coordinators.

Examples of "Action Research" Questions Pursued by Participants

- Will changes in the teaching strategies motivate and improve non-science oriented students who study chemistry?
- Will changes in the teaching strategies motivate and improve 10th grade chemistry students?
- What misconceptions do students who study electrical conductivity of metals and ionic materials have and how can we cope with them?
- Can we change students' attitudes toward science by integrating relevant and everyday issues in their science curriculum?

In the following section, we elaborate on the question: *Will changes in the teaching strategies motivate and improve non-science oriented students who study chemistry?*

The team of teachers who chose this question consisted of three chemistry teachers from the same high school. In this school, all 10th graders (both science-oriented and non-science oriented) study the same curriculum in chemistry to enable all students to have the same opportunities to study science. However, the non-science oriented students could not cope with it, and it caused a drastic decrease in their achievement and deterioration in their classroom behavior. In the non-science oriented classes, the traditional curriculum became non-relevant and included very few issues that were relevant to the students. The teachers made great efforts to simplify it and to attract students, however, many students failed in chemistry at the end of the 10th grade. A high percentage of the students had to pass summer examinations in order to continue to the 11th grade. Thus, the students' motivation to study chemistry decreased. However, the same teachers taught the non-science oriented classes and the science oriented classes, where the students majored in chemistry and performed well. The teachers were aware of the problems in those classes, and discussed it in the team meetings. They made efforts to teach in those classes the traditional curriculum with slight changes, and indeed they made some progress but still the teaching in those classes was considered to be problematic. The teachers even considered using a new STS curriculum, but with very little success. The biology teachers and the physics teachers had the same problems. Based on these problems, the biology and physics teachers joined the workshop in order to change the curriculum and to upgrade their science-oriented students. They wanted to change their strategy of teaching and observe its impact on the motivation and behavior of the students.

The first step was to eliminate parts from the regular curriculum and to add relevant parts of the STS type. The study team planned the lessons together, taking into account the teaching strategies. The second step was to divide the students into two sections. The third step was the replacement of the traditional curriculum by an STS one.

Assessment of the Workshop

Research tools were developed with the goal in mind of assessing students' change and teachers' satisfaction from the workshop. They consisted of classroom observations, interviews of students at different stages of the study, and interviews of teachers.

Observations conducted in the chemistry classroom

A chemistry teacher, a colleague who belonged to the team of chemistry teachers functioned as an observer in each class. He observed four students during several lessons, with regard to: (1) their active participation in the classroom and their ability to express their ideas, (2) their active listening, (3) their way of preparing homework, and (3) their motivation to write notes in their journals. All of the teachers were asked to evaluate the development of students' writing skills, from the first assignment, at the beginning of the program, to the last one, toward its end. Each teacher chose four different students randomly. Additionally, the observer took notes of special events in the classroom.

Interviews

The teachers interviewed students who were selected randomly. The interviews were conducted twice: at the beginning of the study and at its end. The interviews were short, and consisted of the following three questions:

a) *How do you feel about the changes that your teacher made, regarding the curriculum and the class management?*
b) *What is the difference between the teaching and assessment methods of this program compared to regular tests?*
c) *Were your learning habits influenced due to the changes made by your teacher?*

The interviews were audio recorded and the content was analyzed for patterns in student responses. At the completion of the workshop, the workshop coordinators interviewed the participating teachers, in order to find out its effectiveness and contribution to the participants.

Findings

The students

After a series of observations, the high school management approved the replacement of the traditional curriculum by an STS one, and allowed the division of students into two sections. It seemed that teaching classes in the earlier format, with 40 students, could not be effective or successful. During the first classroom observation that was performed after the students were divided into two sections, the teachers reported that much better learning process took place. In addition, there was a meaningful change in the students' behavior and attitude toward science:

- The class was very active and about 60% of the students took notes in their journals and did not hurry to leave the classroom when the lesson was over.
- Students' interest in experiments and laboratory demonstrations increased.
- Students started to ask more questions, to perform experiments at home, and to look for enrichment materials. They said that their learning habits changed, from studying only before an examination to continuous study.

The results of the interviews that the teachers conducted with their students validated the observed findings. The students, with no exception, favored the new teaching strategies and the alternative assessment methods and emphasized the following points:

1. Satisfaction with the assessment methods. These methods were linked to the teaching methods and the various assignments they had completed. Some students also said that they had managed to achieve a final grade higher than the one they would have achieved, if they had merely sat and studied for the matriculation examination. They indicated that the possibility of correcting their work and the weight given to numerous assignments provided more opportunities to reflect and consider all possible answers.

> *"At the beginning, I thought that perhaps it would be better to do the matriculation exam, but on the other hand, if I had been given a low grade on the exam, I wouldn't have been able to improve it. But with the assignments, if I got a low grade on one, I had others with which to raise my grade, and even with the first paper, I had a 'second chance' to improve it."*

Students spoke of the assessment's reliability and noted that ongoing assessment provides a better reflection of their knowledge, diligence, and creativity.

> *"The assessment system is great. It's a correct system; you work right through the year. You show your ability and the grade is a true one, not just for one matriculation exam. Over the year you can improve your work, correct it...." Another student said, "In your papers you can express yourself in many ways, not only in dry writing but also by presenting a paper on the computer or in a story about radon. It's easier for me to express myself this way."*

2. Interest in the process of learning. The students said that studying through this program changed their learning habits, from studying only before an examination to continuous study:

> *"In the other matriculation exams I did this year I crammed for two weeks beforehand, while here it was a case of intensive study throughout the year." Or, "Throughout all my years at school I wouldn't sit down and study, but in this subject I sat myself down and made an effort, read and took an interest; it taught me to prepare my work, it opened new horizons for me. I didn't do it as a punishment; I knew that my papers were my grade. And [a great deal] remained from my work, it wasn't like studying for an examination."*

<u>3. Satisfaction from the student-teacher relationship.</u> Some students claimed that in the past they had never experienced such a serious attitude toward them on the part of the teachers. Typical student remarks were, *"Are we so important to you? You're devoting so much time to us."*

The teachers

The teachers enthusiastically claimed that the workshop contributed a lot to the learning environment in the classroom and to their ability to find solutions to their problems. Moreover, it strengthened their teamwork at school and encouraged the collaboration of the team of teachers and the researchers. We should not forget that it is not typical for teachers to take part in qualitative reflective research. They are usually involved in positivistic studies, in which there are usually experimental groups and control groups. However, the lectures that they attended in the workshop enabled them to undergo a conceptual change and realize that a reflective study has its own value and is in fact beneficial to their work.

Summary

This case-study illustrates the contribution of action research in developing leadership among teachers, more specifically, enabling them to undertake tasks they had not undertaken before and to take control of the most critical component of their work—the teacher-student environment. The interactions that developed between teachers and students in the classroom turned out to be an important factor in motivating students to learn. The students mentioned that in the past the teachers had never treated them so seriously. "Are we so important to you?" and "You're devoting so much time to us" were typical remarks made by overwhelmed students. In this study we found that involving teachers in an intensive program dealing with both instructional innovations and investigation of their impact on students is extremely effective in professional development of participants.

We believe that teachers who participate in an "Action Research" workshop learn how to undergo a change, get new insights regarding their teaching, improve it, and promote their professional work. Moreover, this case study indicates that teachers who are involved in such a process are satisfied with their work and their accomplishments. They develop more confidence for teaching, have a closer contact with academic institutions on a professional basis, and are recognized as highly professional experts. The active learning we strive for with our students in order to stimulate and motivate them also stimulates and motivates the teachers. We believe that reflective teachers should incorporate action research into their teaching as a means for both their professional growth as well as for strengthening their collaboration with their colleagues. Since one of the main goals of the workshop was to find a problem and to try to solve it, the changes that occurred due to their activities, were beneficial to the students as well as to the teachers.

CASE 4. The Earth-View Project: Professional Development through Action Research in High School Earth/Environmental Science Courses

Context

This descriptive study relates to an action research experiment that took place in the context of an NSF-funded project for 45 teachers from school districts throughout North Carolina. Focusing on content, pedagogy, and leadership, the *Earth-View* project approached these three strands from the perspective of teacher research. Summer courses provided instruction and experiences related to content, with implications for pedagogy and leadership. During the school years, teachers designed and implemented research projects and reported on them at the spring meeting with their peers. During Year I, the research projects took the form of scientific research; teachers designed research projects to be conducted in their home communities during the school year following their field experience in Summer I. Research assignment during Year II focused on pedagogy, and teachers were asked to design action research projects that examined a problem relevant to their teaching. Year III targeted leadership and the associated teacher assignments took the form of development and testing of materials to be used in professional development sessions with their peers. Table 3 summarizes the research components in the context of the three-year project. This case study description, however, focuses on the action research component of the project and its impact on teachers' professional development.

Table 3. Research Components of *Earth-View*

Content	Pedagogy	Leadership
Teachers will design and implement a research study in environmental science, focusing on a topic of relevance in their home communities. After completing the study, they will report to their peers in a professional meeting.	Teachers will design and implement an action research study in their own classrooms. They will report on their findings in a professional meeting with peers.	Teachers will use research findings in the areas of adult learning, professional development, and facilitation of change to prepare professional development modules to be used with other teachers and with administrators. Those materials will address earth/environmental science content as well as teaching.

Teachers worked on developing a plan for their action research during Summer II and the first two-day follow-up session in October 2001. Instructors provided information about the purpose of action research and several examples. A research proposal format was provided for the teachers, along with a sample proposal. Although teachers were given the freedom to make their own decisions, project leaders provided a variety of examples from which they might choose if they needed assistance. We describe the choices teachers made and, significantly, the options they chose not to consider.

What Teachers Chose To Do

At a follow-up meeting in the fall of 2001, teachers worked in their previously assigned research groups to begin the development of action research proposals. In some cases they chose to conduct similar studies in different school settings; in other cases they designed a plan specifically relevant to their own teaching situation. In general, the teachers' proposed projects addressed the following areas: student learning (as a result of various interventions), student attitudes, outside factors affecting student performance, student understanding of the nature of science, and correlation of achievement with socio-economic/ethnic factors. Table 4 provides more details of the action research projects.

Table 4. Teachers' Choices for Action Research Proposals

Ten teachers focused on **student learning**, examining the effect of various interventions. Examples of studies include effects on student learning of . . . • Use of "foldables" with students in alternative school setting • Inquiry setting vs. textbook-based setting • Use of CBLs in field trip setting • Tutoring sessions • Field trip vs. classroom book work • Use of customized learning activities tailored to students' needs • Use of alternative assessments
Fifteen teachers focused on **student understanding of the nature of science**, using a survey instrument provided by project instructors. These teachers collaborated on the project, determining results for their own students and for the entire group of students from all schools.
Four teachers focused on **student attitudes** in relation to factors such as . . . • Increased numbers of school yard field trips • Use of the internet • Various earth science topics
Three teachers collaborated on a project to examine relationships between **student achievement and socio-economic and ethnic backgrounds** of students in three different schools.
One teacher focused on correlations between **home communication and performance of exceptional students**.

Being allowed to design projects of their own choosing, teachers addressed areas of particular interest or concern to them. For those who were uncertain about their ability to design and conduct a study, the *Nature of Science Survey* became a default. The results from the surveys were certainly interesting to the group and especially to the university instructors, but they probably had minimal practical value for the teachers in their own practice, unless they decided to focus on student understanding of the nature of science as a major instructional goal.

What Teachers Chose NOT To Do

From the perspective of the instructors, the projects teachers chose to conduct were perhaps not as significant as the projects they did not consider—more specifically, the option to look at their own teaching. Although there were a number of teachers that looked at certain teaching choices and their effects on learning or attitudes, no teacher proposed to examine his/her teaching through the use of videotaped lessons. During the discussion of possibilities for action research, the instructors gave many examples, including the videotape option. In fact, project leaders encouraged teachers to consider analyzing their teaching through videotapes, emphasizing that no person would see the tapes except the teacher and/or another person they might choose to help them with the analysis (perhaps a peer). In previous sessions, teachers had read research reports and conducted discussions on such topics as teacher talk vs. student talk, questioning techniques, and wait time. They were challenged to reflect on those aspects of teaching in the context of their own practice. Despite the extensive discussions and the experience several teachers previously had with videotaping themselves for National Board Certification, the teachers avoided this option.

In the context of gathering data about the value of the experience, the instructors prepared a survey to determine the teachers' reasons for avoiding the use of classroom videos. A summary of results is found in Table 5 below. There were multiples responses from each teacher.

A disturbing aspect of the responses was a recurring reference to being judged. Although there was much discussion about the purpose of their action research projects, the specter of videotaping obviously triggered emotional responses related to evaluation of their teaching as seen in the statements below:

> *I fear that those analyzing you will find nothing good with your teaching style based on the one episode you teach—not take into account your teaching over time.*

> *Teachers are under so many evaluations and criticized on style, not content.*

> *The types of evaluations used in many schools are one-shot observations with a very rigid set of standards, generally not related to the kinds of teaching practices advocated by the National Science Education Standards and other reform documents.*

In spite of the teachers' hesitation to use videotaping, they expressed that there were benefits to this method of analyzing their own teaching. In fact, every teacher surveyed except for one conceded that there were advantages, including seeing themselves as their students see them, analyzing new strategies and students' responses to them, and looking for specific behaviors (such as questioning techniques). The one teacher who was an exception thought the practice would be helpful to others trying to critique his teaching, but not helpful to him in his own professional development as a practitioner. Although there was nearly unanimous agreement about the usefulness of the method, there were a few disclaimers:

> *If a new teacher undertakes this, they may be more successful in modifying their behavior/presentations since they are still learning how to teach.*

> *I think it would depend on how much a person would really want to improve their methods.*

> *I could learn more about my strengths and weaknesses, but changing is the problem.*

Table 5. Teachers' Reasons for Avoiding Videotaping

No. of Teachers (n = 27)	Reasons for choosing options other than videotaping	Example Teacher Quotes
6	Takes too much time	*I have enough problems getting ready for class and students without having to add another task to my responsibilities.*
12	Self-conscious about appearance and speech	*I am camera shy and not photogenic* *Like any other woman, I hate to see how fat I really am.* *I don't like the way my voice sounds on tape.* *I stutter in front of a camera.*
7	Difficult to face the reality of being an imperfect teacher	*Reality is a tough thing to swallow—no taping, please!* *I guess it would expose flaws that I don't want to see but need to see.*
10	Technical complications (equipment, etc.), getting permission forms signed	*It's a hassle to find the equipment and figure out how to use it.*
6	More relevant problems to investigate	*The research I chose to do did not involve taping.* *I was more interested in another project.*
7	Distractions in classroom	*This might be an intrusion that would impact the behavior of the students.*

Conclusions and Implications

The use of videotape analysis is considered by the education community as a productive means of improving the practice of teachers, and is widely used with pre-service teachers and teachers in graduate programs. In the U.S. it is a requirement for those teachers seeking certification from the National Board for Professional Teaching Standards. The researchers in this project were surprised by the unanimous avoidance of this procedure as an option for action research by the *Earth-View* teachers. Obviously, in the context of university courses and National Board Certification, videotape analysis may be a requirement; in the *Earth-View* project, it was an option. Given the option, teachers chose other research alternatives.

If professional developers are to encourage use of videotape analysis, they must reduce the factors that discourage teachers from using this method. This study suggests several actions that might be helpful:

(1) Provide equipment and help with setting it up.
(2) Emphasize that videotapes will be viewed only by the teachers and others of their choosing and that the teachers will decide on specific aspects of teaching to be examined—not every aspect.
(3) Provide suggestions for looking at positive teaching behaviors, not just negatives.
(4) Suggest ways of making the process less intrusive in the classroom.
(5) Be willing to videotape yourself and allow teachers to critique.
(6) Discuss the difference between using videotapes to improve teaching and observations used by administrators for the purposes of evaluation.

Action research as a component of professional development has great promise for the improvement of teaching practice. The responses of teachers in this study provide insights into how videotape analysis might be encouraged as one strategy among others that teachers might use in action research in their classrooms.

Conclusion

As evident in the cases presented above, action research by teachers contributes significantly to their professional development and enhances their instructional practice either directly or through improved assessment of the teaching and learning that occurs in their classrooms. As a result, both teachers and students are positively impacted. At the same time, these cases reveal that effective use of action research as a professional development strategy is not an easy task and requires motivational and support skills on the part of professional development providers. For instance, most members of the assessment committee in Collier Chautauqua Program had never been involved in a committee of that nature before and had never engaged in developing locally relevant assessment instruments. They were initially nervous about the task and were apprehensive about their ability to accomplish it. The external assessment specialist for their program (the first author of this chapter) met with the committee frequently and communicated with them regularly to encourage, motivate, and guide them in the process of using the data gathered regarding their program assessment to develop instruments that would lead to authentic assessment of the impact of their program.

Similarly, the role of mentors and program coordinators in the two high school chemistry cases described here cannot be overstated. In these cases too, frequent meetings, regular communication, and classroom support by program staff was found to be crucial in ensuring the progress of action research projects undertaken by participating teachers. While the same kind of support was also provided to the participants of the Earth-View project, they shied away from using a very effective tool of action research—video recording of their teaching. This implies that in certain situations, even a high level of support and encouragement may not be able to overcome specific barriers, which prevent teachers from using action research to its maximum potential toward their professional growth and development.

Another factor that enhances the use and effectiveness of action research in science instruction is tangible products. All the cases described here involve specific products or changes, which were the end products of action research. For instance, the Collier Chautauqua Program benefited from the production of locally relevant assessment instruments developed by their own assessment committee. This allowed for more authentic assessment of program implementation and impact while promoting leadership development in the assessment committee members. Similarly, in one of the two high school chemistry cases, new student assessment methodologies were developed, which made more authentic assessment of student learning possible. In the other chemistry case, new and more appropriate curriculum was developed and implemented for students who were not being well served by the old curriculum. In the case of the Earth-View project, participants were expected to develop professional development modules, which incorporated the findings of their action research projects, to be used with other teachers in their school districts. The expectation to translate the results of action research into tangible products that can be used with students or other teachers provides the impetus and motivation for teachers to engage in action research. These tangible products give the teachers a sense of ownership and expertise in the products they have developed. This sense of ownership and expertise amounts to growth, which is what professional development is all about.

Therefore, if a professional development program wants to use action research effectively to bring about teacher growth and student improvement, then it must ensure two components: 1) High level of ongoing mentorship support to participants and 2) Implementable tangible products expected of participants as culmination of the action research projects.

Appendix

Assessment Instruments Developed by the Assessment Committee of the Collier Chautauqua Program

Learning By Doing
COLLIER COUNTY
CHAUTAUQUA

To: Chautauqua Teachers
From: The Assessment team
Re: Student Program Surveys

Thank you for taking the time to read and administer this survey. As a participant in the Chautauqua program, your prompt administration and return of this survey is important. This survey will be used as a tool for program evaluation, and not for assessment of any individual student or class's growth from your participation in the project.

The survey should be administered within the first two weeks of school, prior to starting your module and implementing constructivist practices. The second survey will be administered midterm. Enclosed is a copy of the survey and an overhead. Choose an administration procedure appropriate for your class and return a scantron sheet for each student.

In order to document growth for our program evaluation, it is important that you indicate your grade level and the date of administration on each scantron sheet. Read your directions carefully and return the surveys to the Science Coordinator, as soon as possible. Your participation in this program is vital to its success.

Learning By Doing
COLLIER COUNTY CHAUTAUQUA

How I Feel About Science

Grades K-3

This survey is to measure students' attitudes about science. Teachers should explain to the students that this is to find out their likes and dislikes and that there are no right or wrong answers. It may also be necessary to describe some of the science activities that are appropriate for your grade level to help clarify for the students what science is all about. Please ask them to respond to the following questions either **YES, I DON'T KNOW**, or **NO**. Mark the scan sheet accordingly.

A: Yes **B: I Don't Know** **C: No**

---

1. It is fun to do science activities, such as study the weather, collect bugs, or grow some plants and flowers.

---

2. I do science activities at home.

---

3. Things I learn in science time help me understand things at home.

---

4. I can do some science activities by myself.

---

5. I have lots of questions about science that I want to ask my teacher.

---

6. I feel that it is O.K. to ask my teacher questions during science time.

---

7. I like to help my teacher find out answers to my questions.

---

8. I might want to do a science job when I grow up.

Learning By Doing
COLLIER COUNTY
CHAUTAUQUA

To: Chautauqua Teachers
From: The Assessment team
Re: Student Program Surveys

Thank you for taking the time to read and administer this survey. As a participant in the Chautauqua program, your prompt administration and return of this survey is important. This survey will be used as a tool for program evaluation, and not for assessment of any individual student or class's growth from your participation in the project.

The survey should be administered within the first two weeks of school, prior to starting your module and implementing constructivist practices. The second survey will be administered midterm. Enclosed is a copy of the survey and an overhead. Choose an administration procedure appropriate for your class and return a scantron sheet for each student.

In order to document growth for our program evaluation, it is important that you indicate your grade level and the date of administration on each scantron sheet. Read your directions carefully and return the surveys to the Science Coordinator, as soon as possible. Your participation in this program is vital to its success.

CLASSROOM ENVIRONMENTAL SURVEY

DIRECTIONS (1-23): Think about your class when you read each of the following sentences. Do you agree with the sentence? Do you disagree with it? Do you understand the sentence? Each sentence is followed by three choices. Decide which choice is the closest to the way you honestly feel about that sentence. There are no wrong answers. On the separate answer sheet, mark your answer in the row of circles for each sentence by filling in the circle that has the same letter as the answer you have chosen. Use a pencil to mark your answer sheet.

1. My teacher asks us if we like the things we do in class.
   A Yes
   B No
   C Don't understand the sentence

2. We learn that safety is important when we use materials in class.
   A Yes
   B No
   C Don't understand the sentence

3. Sometimes we go outside of the school building to study.
   A Yes
   B No
   C Don't understand the sentence

4. The things we read and do in class make me think about how to solve problems.
   A Yes
   B No
   C Don't understand the sentence

5. We learn about things that are new to me.
   A Yes
   B No
   C Don't understand the sentence

6. Our class has gone on field trips.
   A Yes
   B No
   C Don't understand the sentence

7. My teacher has assigned projects for us to do at home.
   A Yes
   B No
   C Don't understand the sentence

8. Sometimes we work in small groups.
   A Yes
   B No
   C Don't understand the sentence

9. There are enough textbooks, supplies, and equipment in our class so that I can work alone or with a small group of other students.
   A Yes
   B No
   C Don't understand the sentence

10. Learning is fun for me.
    A Yes
    B No
    C Don't understand the sentence

11. My teacher asks us questions in class that make me think.
    A Yes
    B No
    C Don't understand the sentence

12. My teacher gives us tests.
    A Yes
    B No
    C Don't understand the sentence

13. My teacher helps me if I do not understand what to do.
    A Yes
    B No
    C Don't understand the sentence

14. I am pleased with the marks on my report card.
    A Yes
    B No
    C Don't understand the sentence

15. Our classroom has materials with directions or ideas about things we can do on our own.
    A Yes
    B No
    C Don't understand the sentence

16. We sometimes use books other than our textbooks.
    A Yes
    B No
    C Don't understand the sentence

17. My teachers in my school sometimes help each other teach lessons.
    A Yes
    B No
    C Don't understand the sentence

18. Our classroom is organized so that when I need something, I know where to find it.
    A Yes
    B No
    C Don't understand the sentence

19. Our projects and workpapers are put out for people to see.
    A Yes
    B No
    C Don't understand the sentence

20. Sometimes people come to our class or school to talk about things we are studying.
    A Yes
    B No
    C Don't understand the sentence

21. My teacher uses things such as movies, tapes, filmstrips, or the computer to teach.
    A Yes
    B No
    C Don't understand the sentence

22. If I am given homework, I usually finish it on time.
    A Yes
    B No
    C Don't understand the sentence

23. If I have trouble with my homework assignments, someone helps me.
    A Yes
    B No
    C Don't understand the sentence

CHAUTAUQUA
TEACHER SELF-ASSESSMENT

Learning By Doing

COLLIER COUNTY CHAUTAUQUA

As a participant in the Chautauqua project, you will be asked to respond to this set of questions twice, once prior to the summer workshop and again in the fall, after having taken the summer course and having worked with your students, using "Constructivist" techniques discussed in the course. The survey will be used as a tool for program evaluation, and not for assessment of any individual's growth from participation in the project. Please mark all responses on the Scantron sheet.

If you wish to evaluate your personal growth in using "Constructivist" teaching techniques, mark your responses on the questionnaire as well, and retain for your records.

Section I
Directions: Using the following scale, indicate the frequency most closely matching the occurrence of each event in your classroom or in your teaching experiences.

A = Not applicable **B** = Never **C** = once per Semester
D = once per Month **E** = once per Week **F** = more Frequently than once per Week

---

___ 1. My students compare ideas with one another.

___ 2. I reward my students for expressing their ideas and for explaining their thinking about a topic.

___ 3. I connect areas of the curriculum to each other (i.e., math, social studies, science, language, spelling, reading).

___ 4. My students use resources other than the textbook.

___ 5. I use alternative assessment techniques (i.e., portfolios, journals, student demonstrations, presentations, projects, observations).

___ 6. I challenge students to apply learned concepts to new situations.

___ 7. I work with other teachers on classroom projects.

___ 8. I use resources from colleges/universities and/or business/community.

___ 9. Students use the textbook in my classroom as a primary resource.

___ 10. I deal with controversial subjects and/or current events in my classroom.

___ 11. I read professional journals and/or publications.

___ 12. I focus lessons on activities, without necessarily reaching an answer.

___ 13. I use class-time to investigate questions that arise unexpectedly.

___ 14. In my classroom, students use manipulatives appropriately.

___ 15. My students plan and carry out their own learning based upon their interests.

___ 16. I answer student questions with a question.

___ 17. I provide individual feedback to my students.

___ 18. I discuss teaching with other teachers.

___ 19. In my classroom I value cooperative work on problems and issues.

___ 20 My students use the textbook sequentially.

Section II

Directions: For the following statements, answer:

A if the statement is true or nearly true for you as a teacher,
B if the statement is false or nearly false for you as a teacher,
C if the statement is not applicable.

___ 21. I sometimes involve students with a topic or issue about which I have little knowledge.

___ 22. I am comfortable working with other teachers to improve my teaching.

___ 23. I help my students to apply their learning to real-life situations.

___ 24. I sometimes use self-evaluation to improve my teaching (i.e., video taping, audio taping, peer evaluations, reflection journal-writing).

___ 25 My classroom assessment includes student "self-assessment".

___ 26. I am comfortable when students ask questions that I cannot answer.

___ 27. I teach my students how to ask questions.

___ 28. I assess students for creativity.

___ 29. I voluntarily participate in workshops to improve my teaching.

___ 30. I enjoy discussing teaching with other teachers.

___ 31. I often create my own teaching materials.

___ 32. I involve administrators when I need support for trying something new in my classroom.

___ 33. My students see a need to affect change in themselves and others.

___ 34. As a teacher, I learn along with my students.

___ 35. Career awareness and exploration are integrated into my teaching.

___ 36. Application of concepts and principles is evident in my teaching;

___ 37. I utilize a variety of teaching styles/strategies in my teaching.

___ 38. My students value process, as well as outcome.

___ 39. Parent involvement is an important aspect of my classroom program.

___ 40. I feel uncomfortable when students ask questions I can't answer.

Chautauqua Technology Survey

Learning By Doing
COLLIER COUNTY CHAUTAUQUA

As a participant in the Chautauqua project, you will be asked to respond to this set of questions twice, once prior to the Summer workshop and again in the Fall, after having taken the Summer course and having worked with your students using "Constructivist" techniques discussed in the course. The survey will be used as a tool for program evaluation, and not for assessment of any individual's growth from participation in the project. Please make all responses on the Scantron sheet.

If you wish to evaluate your personal growth in using "Constructivist" teaching techniques, mark your responses on the questionaire as well and retain it for your records.

Directions: Using the following responses, please answer the following questions regarding technology at your school.

A=Yes **B=No** **C=Not Applicable** **D=The technology being addressed is not available at my school**

For example: If a teacher is at a school that has a CD Tower and the teacher makes use of it, they would respond Yes to Question 3. If a teacher is at a school that has a CD Tower and the teacher does not use it, they would respond No to Question 3. If a math teacher is at a school that has a CD Tower but there are no math CDs they would respond Not Applicable to Question 3. Finally, if a teacher is at a school that does not have any CD Towers they would respond that the technology being addressed is not available at my school.

___ **1.** The hardware in my room is adequate for my subject area.
___ **2.** The network software fits my curriculum.
___ **3.** CDs are provided for my subject area/grade level.
___ **4.** Laserdisks are provided for my subject area/grade level.
___ **5.** Telecommunications are important to my subject area.
___ **6.** More training on hardware is needed.
___ **7.** More training on software is needed.
___ **8.** More training on technology use in the classroom is needed.
___ **9.** My team has actively used an integrated curriculum.
___ **10.** My team has used an integrated curriculum with emphasis on technology.
___ **11.** Technology has become a major part of our integrated curriculum.
___ **12.** Technology has become a major part of my subject area/grade level.

____ **13.** More training is needed to integrate the curriculum.
____ **14.** More training is needed to integrate the curriculum with technology.
____ **15.** A full time person is needed to support the technology.
____ **16.** Technology has enhanced my teaching this year.
____ **17.** Technology has motivated my students.
____ **18.** Technology has assisted my students to become life long learners that can adjust to change.
____ **19.** Technology presents real life situations for use in the classroom.
____ **20.** Computers and display panels are used for presentation in my classroom.
____ **21.** There are enough computers in my classroom.
____ **22.** The subject area software is familiar to me so that I can use it with my students.
____ **23.** The level of software is appropriate for my students.
____ **24.** The software fits the needs of my students.
____ **25.** Students us CDs and Laserdisks at least 50% of the time in my class.
____ **26.** Students use computers 50% of the time in my class.
____ **27.** Students know more about technology than I do.
____ **28.** Technology is available to all on an equal basis.

Please answer the following questions regarding computer usage at your school.

____ **29.** The computer(s) in my room is(are) used:
A. mainly by me for management purposes.
B. mainly by students for subject area curriculum.
C. mainly by students to play games/rewards.
D. for the above three purposes about equally.
E. not available at my school.

____ **30.** Students use the network software for subject areas:
A. not at all.
B. once or twice a week.
C. three or four times a week.
D. every day of the week.
E. not available at my school.

____ **31.** Students go to the computer lab to work on subject area curriculum:
A. not at all
B. once or twice a week.
C. three or four times a week.
D. every day of the week.
E. not available at my school.

Learning By Doing
COLLIER COUNTY
CHAUTAUQUA

A Self-Evaluation for Basic Teacher Computer Use

As a participant in the Chautauqua project, you will be asked to respond to this set of questions twice, once prior to the summer workshop and again in the fall, after having taken the summer course and having worked with your students using "Constructivist" techniques discussed in the course. The survey will be used as a tool for program evaluation, and not for assessment of any individual's growth from participation in the project. Please make all responses on the Scantron sheet.

If you wish to evaluate your personal growth in using "Constructivist" teaching techniques, mark your responses on the questionaire as well and retain for your records.

Directions: Please indicate your level of achievement for each of the following competencies. Choose the letter that best reflects your current level of skill attainment. (Be honest, but be kind.)

___ **1. Basic computer operation:**

A. I do not use a computer.

B. I can use a computer to run a few specific, pre-loaded programs. It has little effect on either my work or home life. I am somewhat anxious I might damage the machine or its programs.

C. I can set up my computer and peripheral devices, load software, print and use most of the operating system tools like the scrapbook, clock, notepad, and command.

D. I can run two programs simultaneously, and have several windows open at the same time. I can customize the look and sound of my computer. I use techniques like ALT-TAB to work with multiple programs. I look for programs and techniques to maximize my operating system. I feel confident enough to teach others some basic operations.

___ **2. File management:**

A. I do not save any documents I create using the computer.

B. I save documents I've created, but I cannot choose where they are saved. I do not back-up my files.

C. I have a filing system for organizing my files, and can locate files quickly and reliably. I back-up my files to floppy disk on a regular basis.

D. I regularly run a disk-optimizer on my hard drive, and use a back-up program to make multiple copies of my files on a weekly basis. I have a system for archiving files which I do not need on a regular basis.

___ 3. Word processing:

A. I do not use a word processor, nor can I identify any uses or features it might have which would benefit the way I work.

B. I occasionally use the word processor for simple documents which I know I will modify and use again. I generally find it easier to handwrite or type most written work I do.

C. I use the word processor for nearly all my written professional work: memos, tests, worksheets, and home communication. I can edit, spell check, and change the format of a document. I feel my work looks professional.

D. I use the word processor not only for my work, but have used it with students for all of the stages of writing as a process.

___ 4. Spreadsheet use:

A. I do not use a spreadsheet, nor can I identify any uses or features it might have which would benefit the way I work.

B. I understand the use of a spreadsheet and can navigate within one. I can create a simple spreadsheet which adds a column of numbers.

C. I use a spreadsheet for several applications. These spreadsheets use labels, formulas, and cell references. I can change the format of the spreadsheets by changing column widths and text style. I can use the spreadsheet to make a simple graph or chart.

D. I use the spreadsheet not only for my work, but have used it with students to help them improve their own data keeping and analysis skills, showing them how to explore questions and the power of mathematical relationships.

___ 5. Database use:

A. I do not use a database, nor can I identify any uses or features it might have which would benefit the way I work.

B. I understand the use of a database and can locate information within one which has been pre-made. I can add or delete data in a database.

C. I use databases for personal applications. I can create a database from scratch: defining fields and creating layouts in order to support inquiry. I can sort and print the information in layouts which are useful to me.

D. I can use formulas with my database to create summations of numerical data. I can use database information to mail merge in a word processing document. I use the database not only for my work, but have used it with students to help them improve their own data keeping and analysis skills.

___ **6. Graphics use:**

A. I do not use graphics in my word processing or presentations, nor can I identify any-uses or features they might have which would benefit the way I work.

B. I can open and create simple pictures with the painting and drawing programs.

C. I use both pre-made clip-art and simple original graphics in my word processed documents. I can edit clip-art, change its size, and place it on a page. I can purposefully use most of the drawing tools, and can group and un-group objects. I can use the clipboard to take graphics from one application to use in another. The use of graphics in my work helps clarify or amplify my message.

D. I use graphics not only for my work, but have used them with students to improve their own communications.

___ **7. FIRN operation:**

A. I do not use FIRN, nor can I identify any uses or features it might have which would benefit the way I work.

B. I can log onto FIRN , but I spend little time doing so.

C. I am able to make profitable use of FIRN and its resources.

D. I use FIRN not only for my work, but have shown my students how to use it.

___ **8. Internet research:**

A. I do not use the Internet or World Wide Web menus, nor can I identify any uses or features it might have which would benefit the way I work.

B. I can connect to the Internet and use World Wide Web menus to find basic information on the Internet, but I spend little time doing so.

C. I am able to make profitable use of Web searching software as well as lists of Internet resources to explore educational resources.

D. I use the Internet not only for my work, but have shown my students how to mine the great information resources available on the Internet.

___ 9. Telecommunications use:

A. I do not use electronic mail, nor can I identify any uses or features they might have which would benefit the way I work.

B. I understand that there is a large amount of information available to me as a teacher which can be accessed with electronic mail. I send occasional requests for information and messages using e-mail - mostly to friends, family, and district colleagues.

C. I use e-mail to access professional information from listservers. I am an active participant in on-line discussions and check my e-mail account on a regular basis.

D. I involve my students in using e-mail to communicate with other students and various kinds of experts from other states and nations.

___ 10. Ethical use understanding:

A. I am not aware of any ethical issues surrounding computer use.

B. I know that some copyright restrictions apply to computer software.

C. I clearly understand the difference between freeware, shareware, and commercial software and the fees involved in the use of each. I know the programs for which the district or my building holds a site license. I understand the school board policy on the use of copyrighted materials. I demonstrate ethical usage of all software and let my students know my personal stand on the issue. I have a personal philosophy I can articulate regarding the use of technology in education.

D. I am aware of other ethical issues involving technology use including medical and equitable access ones. I can speak to a variety of technology issues at my professional association meeting, to parent groups, and to the general community.

___ 11. Technology integration:

A. I do not see the need to blend the use of new technologies into my classroom learning activities.

B. I would like to blend the use of new technologies into my classroom learning activities more often than I do, but there just isn't much time or enough access to equipment and I need more help understanding what strategies will work and how to do it.

C. From time to time I encourage my students to employ new technologies to support the communicating, data analysis and problem solving outlined in the district technology plan.

D. I frequently encourage my students to employ new technologies to support the communicating, data analysis and problem solving outlined in the district technology plan. We have moved dramatically toward a more student-centered, technology-supported kind of classroom.

COLLIER COUNTY PUBLIC SCHOOLS

Tracking Community Involvement

Name__________ Grade______ Subject__________ Year______ Chautauqua Participant yes____ no____

Contact Person:	Type of Resource						Method and Date of Communication:					I/C	F/T	N/A	Other	Time/Duration of the Visit	Fees & Costs	Materials Donated	Purposes/Objectives	Comments	Call Again Y/N
	B	P	S	E	O		E	T	F	M	P										
Name____ Address____ Phone____ FAX ____ E-Mail ____ Organization ____																					
	B	P	S	E	O		E	T	F	M	P										
Name____ Address____ Phone____ FAX ____ E-Mail ____ Organization ____																					
	B	P	S	E	O		E	T	F	M	P										
Name____ Address____ Phone____ FAX ____ E-Mail ____ Organization ____																					
	B	P	S	E	O		E	T	F	M	P										
Name____ Address____ Phone____ FAX ____ E-Mail ____ Organization ____																					

B:Business
P:Parent
S:Student
E:Educator
O:Other

E:Email
T:Telephone
P:In-person
M:Mail
F:Fax

I/C:In Class
F/T:Field Trip

Send to science coordinator when the sheet is full.

References

American Association for the Advancement of Science. (1994). *Science for all Americans.* New York, NY: Oxford University Press, Inc.

Ben-Peretz, M. (1990). Teachers as curriculum makers. In T. Husen & N.T. Postlethwaite (eds.), *The International Encyclopedia of Education.* Oxford: Pergamon.

Bybee, R, W. (1995). Science curriculum reform in the USA. In R. W. Bybee & J. D. McInereny (eds.), *Redesigning the Science Curriculum; A Report on the Implication of Standards and Benchmarks for Science Education.* Colorado Springs: BSCS.

Enger, S. K., & Yager, R. E. (2001). *Assessing student understanding in science: A standards-based K-12 handbook.* Thousand Oaks, CA: Corwin Press, Inc.

Feldman, A. (1996) Enhancing the practice of physics teachers: Mechanisms for generation and sharing knowledge and understanding in collaborative action research. *Journal of Research in Science Teaching,* **33**, 513-540.

Feldman, A. & Minstrel, J. (2000) Action research as a research methodology for study of teaching and learning science. In A. E. Kelly & R. A. Lesh (eds.), *Handbook of Research Design in Mathematics and Science Education,* 429-455.

Gitomar, D.H., & Duschl, R.A. (1998). Emerging issues and practices. In B. J. Fraser, & K. G. Tobin. (eds.) *International Handbook of Science Education.* Dorderecht: Kluwer Academic Publishers.

Hofstein, A., & Kempa, R. F. (1985). Motivating aspects in science education: an attempt at an analysis. *European Journal of Science Education,* **7**, 221-229.

Hofstein, A., Mamlok, R., & Carmeli, M. (1997). Science teachers as curriculum developers of science and technology for all. *Science Education International,* **8**, 26-29.

Hofstein, A., & Walberg, H. J. (1995). Instructional Strategies. In B.J. Fraser & H.J. Walberg (eds.), *Improving Science Education: an International Perspective. The NSSE Yearbook* (pp. 1-20). Chicago: The University of Chicago Press.

Lazarowitz, R., & Tamir, P. (1994). Research on using laboratory instruction in science, in D. L. Gabel. (ed.) *Handbook of Research on Science Teaching and Learning.* New York: Macmillan Press.

Lewis, I. & Munn, P. (1987). *So you want to do research! A guide for teachers on how to formulate research questions.* Edinburgh: Scottish Council for Research in Education.

Loucks-Horsley, S., Hewson, P. W., Love, N., & Stiles, K. E. (1998). *Designing Professional Development for Teachers of Science and Mathematics.* Thousand Oaks, CA: Corwin Press.

Loucks-Horsley, S., Love, N., Stiles, K. E., Mundry, S., & Hewson, P. W. (2003). *Designing Professional Development for Teachers of Science and Mathematics, Second Edition.* Thousand Oaks, CA: Corwin Press.

National Research Council (1996). *National Science Education Standards.* Washington, DC: National Academy Press.

National Science Teachers Association. (1992-93). *National science teachers association handbook.* Washington, DC: Author.

Sagor, R. (1992). *How to conduct collaborative action research.* Alexandria, VA: Association for Supervision and Curriculum Development.

In: Science Education in the 21st Century
Editor: Ingrid V. Eriksson, pp. 241-262

ISBN: 1-60021-951-9

Chapter 8

TESTING THE PER-METHOD TO TEACH STUDENTS TO EVALUATE AND RANK ENVIRONMENTAL THREATS TO THE ENVIRONMENT

K.-O. Waara* and S. Waara
Department of Public Technology, Mälardalen University, Box 883,
SE-721 23 Västerås, Sweden

Abstract

There is a great need for science-based methods to estimate and rank the strength of different threats to the environment and Man. This demand has increased during the last decade, driven principally by the challenge to society of embracing sustainable development. In the case of the university sector, the inclusion in programmes and courses of facts and tools of general and subject-specific relevance for sustainable development is an appropriate educational response..

In working with environmental management and protection, science-based methods for risk-assessment and the ranking of environmental threats are of the utmost importance. Therefore, in environmental science education, it is necessary to include in course syllabi, a presentation, use and evaluation of relevant, available methodologies. The Potential Ecological Risk method (PER) represents one practicable approach. Whilst the PER-method was initially developed for ranking chemical threats to Swedish water ecosystems, it was subsequently extended for use not only in terrestrial ecosystems but also to characterise biological and physical threats having a global perspective.

The PER-methodology has been installed as a course module (value 2 ECTS credits) in an Erasmus Mundus Masters programme entitled "Coastal and Water Management". To date, the course module has been taught twice to international students at the University of Algarve, Faro, Portugal and once at University of Bergen, Bergen, Norway. The module has also been integrated into a general Ecological Risk Assessment course that is regularly delivered at Mälardalen University, Västerås, Sweden. Additionally, it was included in the mandatory course "Ecological Risk Assessment" in the joint Masters programme entitled "Master of Inland Water Quality Assessment" which commenced in 2007.

This paper presents the methodology, educational results, including student responses, reflections and chosen fields of assignment in relation to PER. Student-feedback concerning

* E-mail address: karl-otto.waara@mdh.se. (Corresponding author)

the methodology, learning outcomes and applicability of the PER method for managing threats to the environment are detailed. Potential enhancement of the course-model for integration into education for sustainable development is discussed.

Introduction

Around the world, human activities interfere with the environment affecting and changing ecosystem structure and function and probably also the potential for evolution of new species. As a result of increased indirect and direct uses of the environment, most human activities also affect human health.

Environmental threats vary in severity and exposure to some threats may quickly reduce biodiversity and functionality in an ecosystem (see for example Begon *et al.* 2006; Cunningham *et al.* 2006; Walker and Salt 2006 and related course books for undergraduate students in Environmental Science and Ecology). The effects of other threats are variable and may be dependent upon resistance to change by an ecosystem. For example, acidification occurs faster in poorly-buffered lakes compared to those with a high buffering capacity. Once an ecosystem has been changed from its original state, an intrinsic ability to return to its original condition, i.e. resilience, also varies according to the ecosystem and the environmental threat. Thus, the fate of an ecosystem exposed to environmental stress depends upon space, time and the severity of the effect. Recognition of this fact and its implication for reaching sustainable development in the near-term (*circa* 50 years) needs to be acknowledged and accepted in natural science education.

The term "Sustainability" currently remains loosely defined. Wals and Jicklings [2002] outline the many uses and misuses of the term and their consequences for higher education. Presently, it is perceived as a delicate balance between the economic, environmental and social health of a community, nation and, of course, the earth [Fricker, 1998]. Public awareness is often a key element in social reform. The Brundtland Commission recognized that, in addition to legal and institutional changes, adjustments in public attitudes and awareness are necessary in order to integrate environmental and economic concerns [World Commission, 1987] for attaining a satisfactory standard of ecosystem and human health. Furthermore, Chapter 36 of Agenda 21 ("Promoting Education, Public Awareness and Training") [1992], notes that there is still a considerable lack of awareness of the interrelated nature of human activities on the environment, owing to insufficient information. Although the amount of overall information has increased over time, the situation has not yet significantly changed in favour of science-based knowledge. This is probably attributable to the fact that the knowledge gained is to some degree dependent on the number of stakeholders involved and the activities they generally represent. For example, it is very likely that economists and ecologists will perceive different aspects of reality when observing the state of interactions between human and natural systems [Orr, 1994]. By drawing on systems-thinking, knowledge of thermodynamics and a sense of place in the ecosystem, the ecologist will see an entirely different picture (including manifestly unsustainable resource-exploitation, climate change, fragile ecosystems and alienation between humans and nature) to that of the economist, who places unwavering faith in an unlimited potential for economic growth [Diduck, 1999].

In adopting an ecological perspective towards education for sustainability, this paper describes and discusses a science-based methodology for ranking environmental threats. It thus contributes to enhancing environmental assessment and education for sustainable development.

Risk assessment is an integral part – and often a major driver of sustainable development [Reinert et al, 2006]. Sustainability is made possible by the assessment of risks to human health and the environment. Ecological risk assessment is simply the determination of potential to cause harm, integrated with estimated actual exposure, and permitting both qualitative and quantitative comparisons to be made [US EPA, 1998; Suter II, 2007]. Since resources for environmental management in relation to sustainable development are usually limited, it is important to optimize management efforts and rank the severity of different environmental threats. In view of the key role played by risk assessment in supporting sustainable development programs, it follows that environmental impact analysis and risk assessment should form a prominent component of education for sustainable development.

There is an urgent need for science-based methods to estimate the risk caused by different environmental threats to the environment and humankind. From an environmental perspective, these are mainly focused on one or more similar threats towards biotic (species or population scale) or abiotic factors. Whilst this aspect makes them useful for evaluations of the hazard(s) from one or more environmental threats to parts of the environment, it does not facilitate comparisons between hazard levels for different kinds of environmental threats. The latter is more likely to be a management challenge for regional, national and/or international authorities. At present, environmental managers need a simple, scientifically-based methodology which does not require an expert knowledge of the natural sciences in relation to optimizing efforts towards sustainable development. The potential effect of different environmental threats varies with activity, including strength/magnitude, area distribution and persistence over time. Since different ecosystems show varying sensitivities to different kinds of disturbances, it is important to identify the most sensitive ecosystems and to include both spatial and time dimensions in estimating the effect of the threat. The analysis may be modified according to the regional scale of the hazard evaluations. Currently, there exists an initial analytical stage of risk assessment– that of comparative risk assessment. One description of the technique is that it ranks issues/problems according to urgency, cost and likelihood of success [Finkel & Golding, 1993]. This paper presents and discusses such a prior comparative risk assessment method in which urgency, costs and likelihood of success are environmentally valued as estimations of ecological risks caused by environmental "threats".

As this paper describes and assesses the suitability of a method for use in a science-based education for sustainable development, it focuses on a description of the student- and teacher-learning processes. Given that the complexity and change in resource and environmental management contribute to unpredictable situations, is highly desirable that the reasons and factors contributing to uncertainty are evaluated. An approach to education, with its emphasis on critical thinking, empowerment and problem solving would therefore be a commendable step. Education may be viewed as a social interaction that either empowers or domesticates the learner. In that context, a democratic dialogue that shifts the centre of the learning process from the facilitator (or teacher) to the learner is significant. The approach should be participatory, discursive and action-oriented. The education poses critical problems for inquiry relating to important features of learner experience. This allows the students to have their thoughts and language reflected in the course of study. The discursive approach invites learners to think critically not only about the subject matter and related doctrines, but also the learning process itself and society. Moreover, it challenges learners and teachers to empower themselves for social change and to advance democracy and equality [Freire 1970, Shore 1993].

This paper presents and discusses learning outcomes for an environmental risk assessment methodology originally devised and published by Professor Lars Håkansson [Håkansson 1999]. The methodology is called Potential Ecological Risk (PER) and it is based on estimations of an effect variable (E-value), an area variable (A-value) and a time variable (T-value). The method is science-based and gives simple general evaluations that facilitate the ranking of hazards by different environmental threats. The method could and should be criticized and further developed, as also the method developer argues for. In these presentations the author presents examples from Sweden. The method, which is separately outlined below, was developed for chemical threats to freshwater and coastal ecosystems in Sweden. The principal objective of this paper is to describe the learning-development among students who undertook a course module entitled “PER-Potential Ecological Risk”. The aim and goal of the module was to teach a method of estimating and ranking environmental threats. Because the methodology was specifically developed for chemical threats in Sweden, the main part of the course module focused on threats and examples from that country. However, the student assignments were neither limited to Swedish water ecosystems nor to chemical threats. There was a free choice of study region and environmental threats. As a result, students from outside Sweden were motivated to test the methodology for environmental threats in their home country or region. On the other hand, many students chose assignments that did not include chemical threats and/or water ecosystems.

Methodology

The PER-Methodology

The PER-methodology focuses on effects at the ecosystem level rather than on specific sites, communities or individual animals/plants, organs or cells. Since very large ecosystems, such as oceans or large lake, may be divided into different (sub)ecosystems, it is necessary to define the boundary of the ecosystem(s) before the risk is calculated. The question of boundary definition is well described in the course book [Håkansson 1999].

The potential ecological risk is estimated from 3 variables: effect (E), area (A) and time (T). All three are characterised by means of a numerical scale ranging from 1 to 10 (Table 1-3). The PER-level is obtained by multiplying the E-value by the A-value and the T-value. In the case of the E-value, the gradings from 1 to 10 are analogous to the choices in the well-known and usefully practical Richter scale. On the scale, an E-value 1 indicates a zero effect-threat whilst a threat that causes a total change of the ecosystem is given an E-value of 10 (Table 1). In situations where $E<10$, it is important to seek an operational guideline value for E for the determinations of A and T. The operational guideline value is a negative effect level on the ecosystem, here termed Ecrit, which occurs at a chosen magnitude of the threat. An operational guideline value might be a guideline concentration, as for example, the limit concentration for mercury in pike for human consumption.

Table 1. The scale used for evaluating effect (the E-value) in PER-analysis [from Håkanson 1999]

Score	Effect
1	no known or no likely ecosystem effects
2	unlikely ecosystem effects using statistical (i.e. probabilistic) methods
3	probable but small ecosystem effects using statistical methods
4	probable ecosystem effects using statistical methods
5	Small, real (=manifested) ecosystem effects
6	Clear, real ecosystem effects
7	Substantial, real ecosystem effects
8	Large, real ecosystem effects
9	very large, real ecosystem effects
10	total collapse of real ecosystems

Once the E-value has been selected, the A-value may be determined. This represents the geographical extent of the ecosystem defined earlier in the study that has this E-value. The area distribution is graded, having values from 1 (i.e. no ecosystem has the selected E-value) to 10, i.e. all ecosystems have the selected E-value (Table 2)

Table 2. The scale for the area variable (the A-value) used in PER-analysis [from Håkanson 1999]

Score	Area
1	no ecosystems with E = 10 or E = Ecrit
2	a few ecosystems (e.g. lakes or coastal areas in Sweden) with E = 10 or E = Ecrit (n<25)
3	several ecosystems with E = 10/Ecrit (25-100)
4	many ecosystems with E = 10/Ecrit (100-400)
5	most ecosystems in a region with E = 10/Ecrit
6	ecosystems in many regions with E = 10/Ecrit
7	more than 25% of ecosystems with E = 10/Ecrit
8	more than 50% of ecosystems with E = 10/Ecrit
9	more than 75% of ecosystems with E = 10/Ecrit
10	all ecosystems (in Sweden) with E = 10/Ecrit

The T-value is an estimation of the time predicted for the threatened environment to revert to "normal" conditions if the threat is stopped. A predictive, validated model is the preferred method for this estimation. The T-value varies from 1 (no effect) -10 (the effect is predicted to persist for more than 320 years) (Table 3).

Table 3. The scale used for the time variable (the T-value) used in PER-analysis. [from Håkanson 1999]

Score	Time
1	no effects
2	effects (E-value) for less than 1 month
3	more than 1 month
4	more than 1 year
5	more than 10 years
6	more than 20 years
7	more than 40 years
8	more than 80 years
9	more than 160 years
10	more than 320 years

According to the methodology, the PER-value, which is an estimation of the total risk, is given by multiplying E-, A- and T-values (E * A * T). Examples of PER-value estimations obtained in this way are presented in Table 4.

Table 4. Examples of PER-value estimations developed for Swedish aquatic ecosystems [from Håkanson 1999]

Swedish aquatic ecosystems						
Environmental threat	**Ecosystem**	**(E)**	**(A)**	**(T)**	**PER= E*A*T**	**PER ranking priority for action**
Acidification	Lake	10	7	10	700	1
Mercury contamination	Lake	3	8	10	240	4
Radiocesium contamination	Lake	2	6	6	72	6
Organic toxin contamination	Marine/coastal	6	6	6	216	5
Eutrophication, phosphorus	Lake	10	6	5	300	3
Eutrophication, phosphorus and nitrogen	Marine/coastal	10	8	7	560	2

The Educational Methodology

The PER-methodology has been presented as a course module, valued at 2 ECTS credits, in an Erasmus Mundus Masters programme entitled “Coastal and Water Management”. The course module has been twice delivered to international students at the University of Algarve, Faro, Portugal and once at University of Bergen, Bergen, Norway. The methodology has also been integrated into a general Ecological Risk Assessment (ERA) course that has been taught several times at Mälardalen University, Västerås, Sweden. Moreover, ERA comprised one part of the mandatory course entitled "Ecological risk assessment" that was included in the new joint European masters programme "Master of Inland Water Quality Assessment" that launched in 2007 [Waara et al. 2006].

The course module is composed of three parts: lectures, discussions and a student assignment. The lectures presented the PER-methodology including examples of PER-value estimations for diverse environmental threats (acidification, eutrophication, contamination by mercury, radiocaesium and organic toxins in Swedish ecosystems). All of the examples are described by Håkansson [1999] and his publication was also used as the principal course reference book. In order to further highlight and compare the PER-scores for different threats and ecosystems, supplementary facts and comments were sourced from the scientific literature. All lectures were open for questions and discussions concerning methodological results and various environmental threats. Discussions on the applicability of the method for other environmental threats and other parts of the world comprised the second division of the module. In order to further promote discussion, two other kinds of threats, namely water power plants (physical threat) and alien species introduction (biological diversity threat) were assessed from the perspective of the PER-methodology. The third part consisted of a mandatory assignment in which students undertook a PER-analysis of at least two environmental threats, followed by a written report of the findings. All students freely selected the environmental threats, ecosystems and a region in the world. In addition, they were offered advice on their selected topics and geographical regions for PER analysis, as well as constructive criticism of their draft assignment reports before submission for evaluation.

Results and Discussion

Part 1 – PER Methodology

The concept of environmental complexity -including the different chemical, biological and physical factors that may affect the functioning of the ecosystem appeared to pose a keen intellectual challenge for students A number questioned their ability to achieve the course goal of learning a method of estimating risk of changes to the ecosystem induced by various anthropogenic inputs. To be able to describe in detail the different ecosystems and their steering parameters and to measure the influence of any new variable on the "normal" processes seemed overwhelmingly difficult. Introduction of the PER-method further compounded student misgivings, as did discussions of the choice of target parameters, such as ecosystem and effect variables, and driving variables.

However, whilst students quickly accepted the importance of identifying steering factors in ecosystem structure and function, they readily acknowledged the poor level of understanding both of ecosystem processes and the effects of introduction of new compounds or increases in existing compounds on ecosystems. All concurred, nevertheless, on the use of statistics to enhance our level of understanding of ecosystem-effects.

A singular development of class reviews of the PER-methodological examples was a gradual change from hesitation to acknowledgment by students of the methodology. In the first instance, the A-value was accepted as an uncomplicated, easy-to-use factor.

In the case of T-value estimation, there was a consensus that only predictive, preferably validated, models could give appropriate and reliable values. Without the support of a robust model, the T-value would be significantly influenced by personal judgment. Clearly, the strength of the particular influence would positively correlate with the quality and quantity of scientific data for a particular threat. Thus, sound scientific data reduce personal influence and result in less variation in T-value estimations, whilst poor data encourage vague personal judgments that invariably lead to incorrect estimates of the T-value.

Typically, E-value selection provoked extensive discussion within class-groups since, in most instances, a selected E-value could prove difficult to explain and defend. There was agreement on the difficulty of distinguishing between different values on the scale because the levels appeared to be dictated by "grammatical differences" and, in addition, were not explained in detail and/or qualified with environmental examples. It was concluded by students that personal influence would especially bias the setting of value levels 2, 3, 4 and 6, 7, 8, 9. Although the extinction of key, functional organisms was immediately and uniformly acknowledged to warrant an E-value of 10, continuing discussions centred on a definition of key functional organisms, particularly in relation to the applicability of the method for other environmental "threats" such as introduction of alien species.

Part 2 – Applying the PER-Method

This part of the course module created awareness in students of potential changes in the environment caused by water power plants and introduction of alien species. Both issues were used as examples to focus student attention on environmental threats for PER-analysis apart from those presented in the course book. Follow-up discussion of estimations of PER-value for their identified "threats" consolidated the learning outcome of this part-module.
On the specific theme of water power plants and their environmental effects, there was a consensus among students that:

Water-damming of the river locally changed and/or destroyed the littoral zone.
Animal populations, especially fish, would be reduced by the turbines.
Above-ground electric cable-systems would also affect the terrestrial zone
Reduced water flow could further affect even downstream ecosystems

There was an extensive discussion concerning E-value selection for water power plants. It was concluded that the methodology should be tested for several types of water power plants and that general guidelines should be developed to indicate the choice of appropriate E-value. This conclusion stimulated an evaluation of a study by The Swedish National Protection Agency of the effect on threatened species of different environmental activities in freshwater in [Swedish Environmental Protection Board, 1994 The study found that the greatest threat to red-listed species was posed by water power plants. Water power plants were perceived to threaten twice as many red listed species as eutrophication, the second biggest threat. Arising from their evaluation of the Swedish finding, the students deduced that the present PER-methodology did not take threats to red listed species into account and, hence, failed to acknowledge the growing importance of biodiversity in sustainable development. In relation to the A-value for power plants, the key question concerned the distance upstream and downstream of the plant that effects of dam construction on the normal water level would be exerted. The view adopted by the students was that the entire, developed water basin should be regarded as affected. However, since it would be difficult to give a general guideline for the distance downstream that manifested an effect that was sufficient to merit its inclusion in the estimation of A-value, it was reasoned that this had to be evaluated on a case-by-case basis. Since the use of water power plant technology will persist well into the future, its environmental effect will be equally long-lived. Hence, a T-value of 10 is appropriate in this instance.

The introduction into Sweden of two alien species, the American signal crayfish (*Pacifastacus leniusculus*) and the mink (*Mustela vison*) generated robust discussion of the potential of the PER methodology to contend with biological threats. For example, whilst both crustaceans are mutual competitors, the American species carries the crayfish fungus (*Aphanomyces astaci*), which is lethal to its Swedish relative. The crayfish are preyed upon by the mink, which is moreover a fish predator, and also affects bird reproduction by eating bird eggs. These examples created intense student interest because they posed a challenge of estimating the potential effects of different biological interactions (e.g. competition, disease and predation,) on the ecosystem.

Part 3 -The Student Assignment

The students felt quite well prepared for assignments, since their course literature had given information on ecological risk assessment, albeit mainly focussed on chemical threats. The chief problem concerned the E-value and how to distinguish between levels 1-4 and 7-9 respectively. The group response favoured the precautionary principle, comprising arguments for and against a selected value. During the assignment writing-procedure, a number of queries concerned the choice of a suitable chemical threat as well as the difficulty of distinguishing the separate effects of different chemical present in a complex. Given the introductory nature of the ERA-module, and its specific learning outcomes, students were advised in the first instance to hypothesise an effect-level for their A- and T-value estimations. They were additionally recommended to undertake a literature search in support of their hypothetical effect-levels.

The results of the assignments - which were assessed with reference to the ECTS scale - ranged from good to excellent. Successful assignments were alphabetically graded from A to E, where grade A is considered to be excellent. In order to merit a Grade-A award, an assignment was required to demonstrate compliance with the following criteria:

- show a comprehensive understanding of the PER-methodology
- use science-based knowledge to evaluate the environmental threat. In this context, relevant scientific references should underpin the reasoning and evaluations.
- exclude a personalised view.

In order to receive a Pass-grade, the assignment should clearly show that the PER-analysis of the selected environmental threat is conducted according to the methodology.

Assignments that fail to do so or that otherwise do not reach a Pass standard are recorded as F or FX.

There was a wide geographical spread of the environments studied and a majority of evaluations were associated with chemical threats. The selected topics are shown in Table 1. Some students failed to distinguish between ecological and human risk-effects. Others, because of their education background, were more interested in human risk-assessment and accordingly sought to transform human risk-assessments into ecological risk assessments by putting forward their E-value level determined for human, not ecosystems/biota effect-levels. However, this error may be traceable in the literature to two examples that quoted Swedish recommended human, safe consumption-levels for E- value level. Notwithstanding the more detailed, subsequent clarification of this module-topic, the inaccuracy persisted in several assignments. The difficulty of differentiating between ecological and human risks may be related to the student's background. In general, those presenting assignments using human risk levels in their determination of E-value, had a background in ecotoxicology or medicine. Understandably, their fields of interest may well have disposed their evaluation towards human risk assessments.

Table 5. Examples of assignments comparing chemical threats

Geographical area	Environmental threat 1	PER-value (E*A*T=PER)	Environmental threat 2	PER-value (E*A*T=PER)
India	Union Carbide tragedy	10*5*6=300	Endosulfate	7*5*6=210
Bangladesh	POP´s	7*7*9=441	Arsenic	9*9*10=810
Thailand	Eutrophication	9*6*4=216	Lead contamination	6*3*9=162
Lake Victoria, Africa	Eutrophication	10*7*7=490	Mercury contamination	5*5*6=150
Ethiopia	Eutrophication	9*9*6=486	Fluoride contamination	3*8*10=240
Ukraine	Manganese contamination	4*2*3=24	Iron contamination	2*1*2=4
Bulgaria	Eutrophication	10*6*5=300	Salinization	10*2*5=100
Chesapeake Bay, USA	Eutrophication	10*8*7=560	Acidification	10*7*10=700
India	Coastal eutrophication	6*6*8=288	TBT contamination	10*6*8=480
Spain	Coastal eutrophication	7*4*7=196	Metal contamination	6*8*10=480
Dalarna, Sweden	Eutrophication	6*3*5=90	Acidification	6*4*8=192
Västernorrland, Sweden	Caesium contamination	3*6*9=162	Acidification	10*6*10=600
Sweden	Cadmium contamination	3*7*10=210	Lead contamination	3*1*10=30
Sweden	Bromated flame retardant	5*1*7=35	TBT contamination	5*2*3=30
Finland	Eutrophication	9*6*5=270	Acidification	8*6*8=384
Västmanland, Sweden	Eutrophication	10*7*6=420	Acidification	10*6*10=600
Baltic Sea, Sweden	Eutrophication	10*10*10=1000	Potential oil discharge	7*2*5=70
West India	Eutrophication	6*6*8=288	TBT contamination	10*6*8=480
Brazil	Metal contamination	4*3*3=36	Org compounds	5*3*7=105

Whilst the course book describes the methodology for chemical threats, a number of students felt motivated to apply the PER-analysis methodology to biological threats. This initiative generated a large number of assignments that featured biological threats in their PER-analysis (Table 6). The assignments specific to this topic had a wide geographical spread, were meritorious and scored high marks in all cases. The difficulty of separating ecological from human risks did appear in assignments in areas that were closely related to human risk assessments. In relation to the assessment of PER-value for non-chemical threats, two different kinds of difficulties were evident. One concerned the effects of microbial discharges: do bacteria and/or viruses introduced by sewage have an ecological risk effect and how is the risk-level affected by the outputs? The second concerned illness introduced by insects - in this case, malaria. Does malaria infestation have an ecological effect and how is its E-value determined? Both issues are generally considered as human risks and in our experience, are not studied from an ecological point of view and are extremely challenging to evaluate. Given that literature references are predominantly concerned with human effects, it would be desirable to extend the information base with environmental science research results.

Table 6. Examples of assignments comparing chemical versus biological effects and biological versus biological effects (* effect evaluation against the Swedish noble crayfish (*Astacus astacus*)

Geographical area	Environmental threat 1	PER-value (E*A*T=PER)	Environmental threat 2	PER-value (E*A*T=PER)
Philippines	Dam construction	10*8*8=640	Gold mining	3*9*10=270
Global	DDT	9*6*7=378	Malaria infestation	6*7*10=420
China	Eutrophication	9*8*7=504	Dam construction	6*7*8=336
Portugal	Eutrophication	10*7*7=490	Microbial pollution	6*6*6=216
USA	Over fishing	9*8*8=648	Eutrophication	8*8*7=448
India	Interlinking of rivers	10*9*10=900	Arsenic pollution	8*6*6=288
China	Dam construction	6*8*10=480	Introduction of alien species	7*2*5=70
Tanzania	Mercury pollution	3*7*10=210	Water hyacinth distribution	10*8*8=640
River Nile, Africa	Aswan Dam	9*8*8=648	Heavy Metals	7*6*8=336
Ukraine	Eutrophication	10*10*7=700	Introduction of alien species	10*10*6=600
Sweden	Eutrophication	10*9*7=630	Over fishing	8*9*8=576
Lake Victoria	Eutrophication	10*9*8=720	Introduction of alien species (*Nile perch*)	9*9*8=648
Africa	Dam construction	9*9*6=486	Desalinization	4*1*10=40
Jamaica	Eutrophication	7*9*7=441	Coral reef bleaching	8*3*10=240
Australia	Coral reef bleaching	10*8*9=720	Pollutant threats to the reefs	7*4*5=140
Brazil	Genetically modified bean	8*8*10=640	Water construction	10*10*10
India	Coastal erosion	9*10*5=450	Wetland losses	8*8*6=384
Sweden	Introduction of mink (*Mustela vison*)	6*6*10=360	Introduction of American signal crayfish (*Pacifastacus leniusculus*)	2*2*10=40 10*8*10=800*

The results presented in the assignments proved extremely interesting because of the ambitious evaluations of various kinds of physical and biological threats to the environment in many parts of the world. Significantly, the most popular chemical threat-analysis was eutrophication (19), followed by assorted metals (15) and disparate organic compounds (5). The topic of acidification attracted five assignments of which four dealt with the Nordic countries. Amongst physical and biological threats, most PER-analyses involved dam construction (5) and the introduction of alien species (5).

These assignments, with their wide-ranging environmental effects-analysis stimulated enthusiastic pedagogical discussions of environmental education, particularly environmental management for sustainable development. The following discourse on evaluations of environmental threats for sustainable development solely embrace results that involved two or more PER-analyses, with a focus on the determination of E-values. The results for A- and T-factors are very difficult to compare as, on one hand, estimations of A-values depend upon the dimension of the region studied, whilst on the other hand, the estimation of T-values is a still-tentative methodology. Thus, the discussion focusses on facts and uncertainties central to the educational process.

Across assignments, acidification received a high mean PER-value (495). Nevertheless, the geographical distribution of the PER-analysis was modest, comprising only 3 from Sweden and 1 each from Finland and USA. It might be reasonably concluded that the acidification threat was narrowly distributed geographically and/or that other environmental threats attracted greater interest. From a pedagogical point of view, we noted with interest that although two students assessed the effect-value at 6 and 8 respectively, Håkansson [1999] attributed an effect-value of 10 because an acidified state induces a total change of the ecosystem. However, the explanation was that, in their study, both students included some lakes that were either incompletely acidified or else intermittently acidified due to sporadic calcareous inputs. The students concluded that the process-state to be used for E-value determinations should be defined so that evaluators may reach similar determinations for similar data.

Since acidification acts synergistically on the effects of many environmental threats, the following questions are thus appropriate to the issue of ranking of threats:-

- are synergistic effects caused by acidification on other environmental threats included in estimates?
- should the synergistic effect have an influence on the PER-analysis value for acidification?

The most popular topic for the PER-analysis was eutrophication. The threat dealt with different regions of the world and received a high mean PER-value (476). On a global scale, eutrophication may be one of the most widely-distributed environmental threats and therefore is evaluated as one of the greatest threats to global sustainable development.

Evaluation of the E-value for acidification appeared to be influenced by a personal view of the state of Eutrophication in the lake region/coastal area being studied. This gave some E-values that were lower than 10, the value presented in the course book and by the lecturer using supporting scientific data.

Whilst the E-valuations determined for coastal eutrophication also appeared to be less than entirely objective, the dimensions of the coastal areas might also have influenced the students' estimations. Instead of estimating the total coastal area that was wholly changed to a eutrophic state (E=10) - giving, in turn, a lower A-value - students accorded an E-value less than 10 to eutrophication, thereby unintentionally enlarging the coastal area affected by that eutrophication level.

Concurring with that for acidification, an overall conclusion of this study-area pointed to a great need for a clear definition of the process-state to be used in the determination of E-values for environmental threats. In a newly-proposed E-value scale, shown in the Appendix (*q.v.*), this process development has been partially taken into account. The technique imparted in the course literature for determination of the T-value for eutrophication is questionable. Its author [Håkansson 1999] uses a T-value of 5-6 for eutrophication, based on a theoretical recovery time following a halt to nutrient inputs. However, since a different technique was used for estimating the acidification T-value, the course lecturer judged that this approach would give rise to a methodological conflict. Insofar as student estimations yielded T-values that were greater than 5 – the quoted T-value for lakes in the course book – the lecturers contention was undoubtedly influential.

Concerning metals there was a wide variation in obtained PER-values (min 4, max 810). As there mostly were single assessments the metal evaluations was clumped as metals resulting in a mean PER-value of 235, which was the lowest mean PER-value found. Variation in total PER-values is expected because varying toxicity amongst metals and metal form is presented in scientific literature resulting in variation in the E-values. Furthermore, some metals as Arsenic and Chromium are classified as genotoxic compounds while others are not. Also a variation in the spatial distribution of effects (the A-value) and the persistence of the effect (the T-value) is probable resulting in increasing variability in the PER-value. Concerning the E-value estimations you find a similar inbuilt uncertainty as for eutrophication and acidification. The uncertainty concerning this is also shown by the fact that in these assignments

1) Cadmium did not belong to the more toxic metals, receiving an E-value of 3 as for Gold and Lead and it was only evaluated as more toxic than Iron.
2) The two TBT values are very different, 10 and 3 respectively.

A special case is Arsenic for which the PER-analysis was conducted for areas in India and Bangladesh. The very high E-value determined by the students principally relied on the fact that Arsenic has caused severe human health problems in those area. In attempting to convert this result into an ecosystem-effect, the students were subjectively influenced by the hypothesis that the probability for a strong ecosystem-effect is high. Unfortunately, the hypothesis is not supported with field results from scientific studies. In order to reduce this uncertainty, the concentration levels used to establish the E-value (i.e. a link between the dose-response curve of a chemical and the potential ecological risk) need to be defined.

A further observation concerned the wide range of T-values for metals, which were in the range 2 –10, signifying that the estimated duration of the metal pollution varies from less than 1 month to more than 320 years. It is highly likely that the T-value (= 2) was obtained by the same estimation-methodology as that for eutrophication. According to that approach, if inputs of metals were halted, then a recovery to "normal" condition would take less than one month – an unlikely scenario in view of the fact that metals are not biodegradable. The T-value of 10 (for gold mining, amongst other threats) is probably influenced by pessimistic, personal views as the available quantities of extractable metals will unquestionably not last for 320 years or beyond.

For purposes of discussion, assignments involving organic persistent pollutants were, like those for metals, grouped into one set. A mean PER-value for the findings was 287. As noted by Hakansson [1999] and also emphasised in the course, there is a general scarcity of data on the environmental effects of individual organic compounds. The popular view presented in the scientific literature is that organic compounds, especially the polyaromatic and/or chlorinated types, are persistent and ecotoxic. However, most of the data have been obtained in laboratory-based, toxicological tests of single-species and/or transformed subsequently to human and ecological risk assessments (for example, see guideline values generated by RIVM (Dutch National Institute for Public Health and the Environment), USEPA (United State Environmental Protection Agency and CCME (Canadian Council of Ministers of the Environment). Oddly, some students failed to identify ecotoxicological endpoints for these environmental threats and, instead, used human values. Similarly, the methodology for caesium and mercury, described in the course book in which the Ecrit-values is estimated from metal concentration levels found as human risk guideline levels, was probably applied in the submitted PER-analysis. The results show a probable influence of the estimation-methodology described in the course literature and in lectures. Area and time dimensions are also difficult to establish thus contributing to further variance in PER-value estimations.

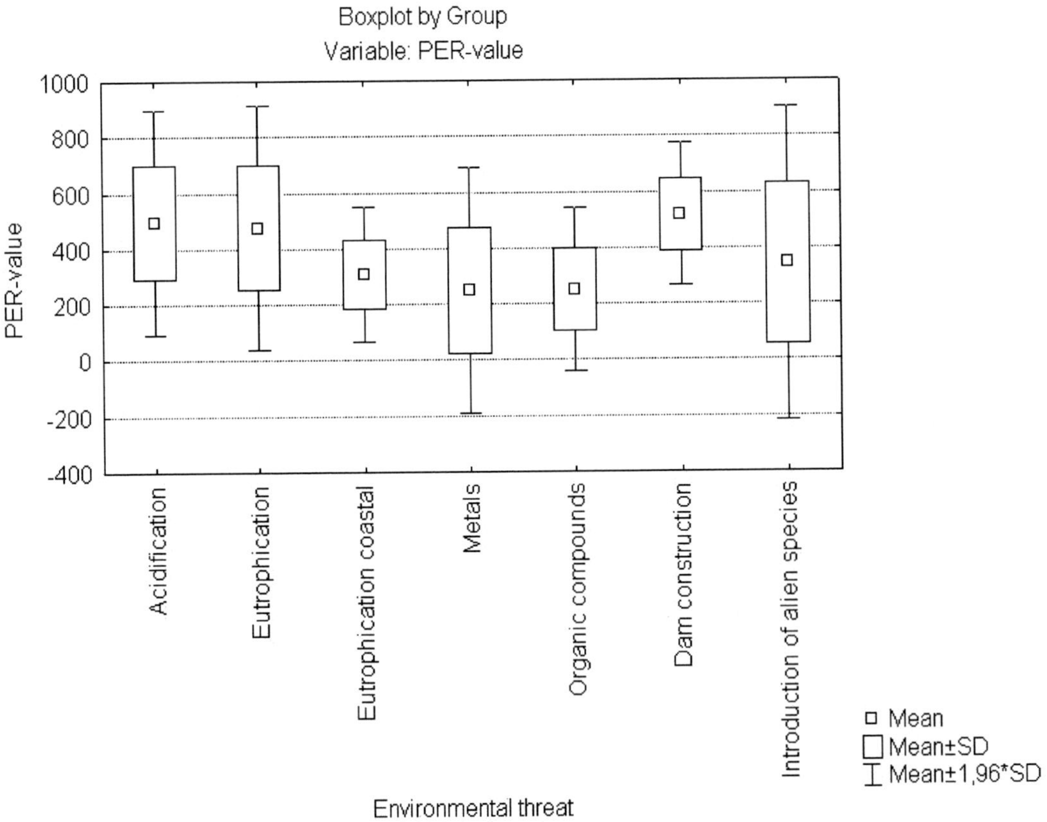

a)

Figure 1. Continued on next page.

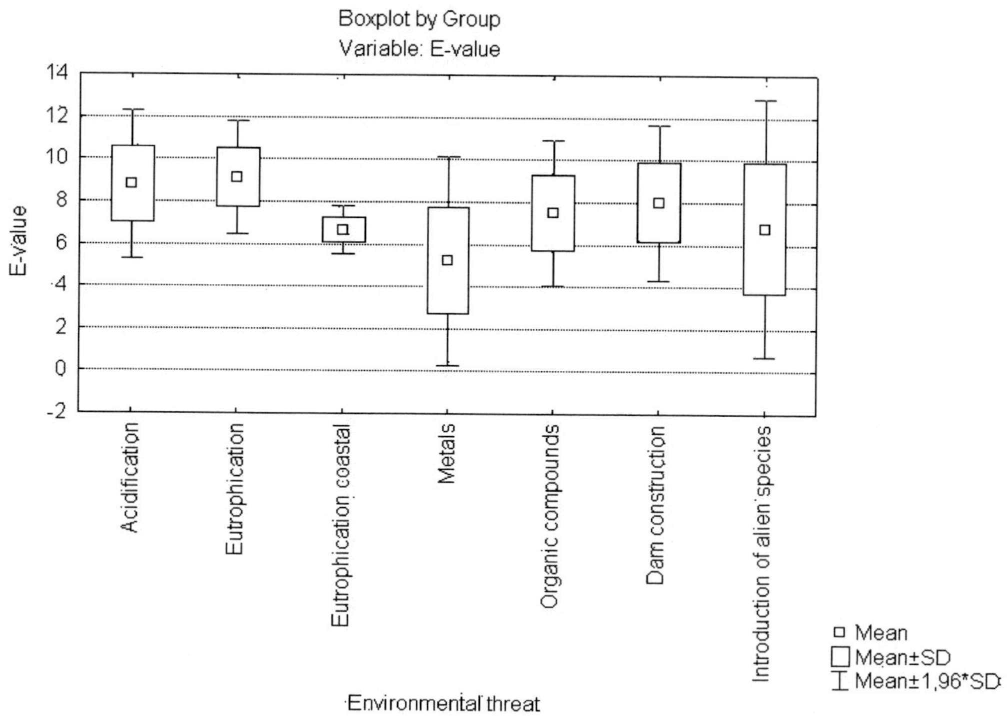

b)

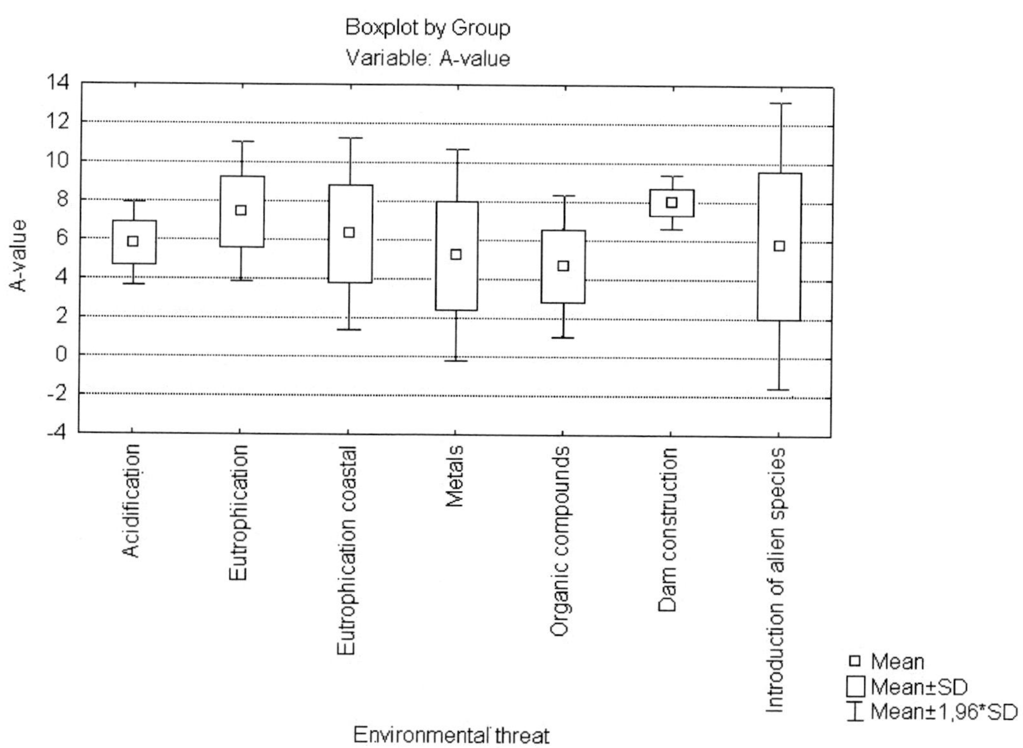

c)

Figure 1. Continued on next page.

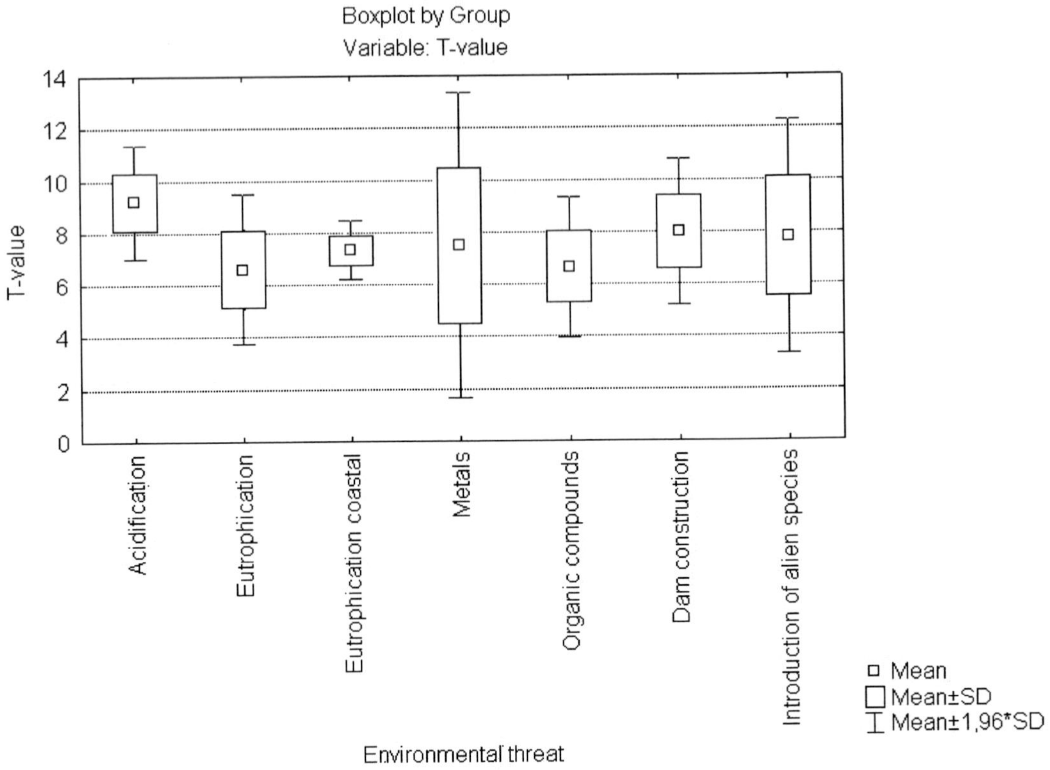

d)

Figure 1. The figures show Box & whisker plots from PER-analysis for the environmental threats acidification, eutrophication in inland water, eutrophication in coastal areas, metals, organic compounds, introduction of alien species and dam construction (a) PER-values, b) E-values, c) A-values and d) T-values).

Interesting and productive discussions centered on applying the methodology to other kinds of physical and biological threats,. There was strong agreement that water power-plants represent major ecological risks, also confirmed in the students PER-analysis in which dam construction was accorded the highest mean PER-analysis value (518) amongst all environmental threats. Since water power plants exploit the river environment, exerting diminishing effects both upstream and below their locations, the biggest challenge would be to establish the spatial distribution of the A-value for the chosen ecological effect-level. The student view was to first decide the effect-level and then find data that supported the chosen A-level (for example, water-level distributions in the river and within the dam before and after dam-construction). The student assignments found that dam construction for the purpose of water power plant energy-production is the highest-ranked ecological risk and hence a major threat to sustainable development. The observation is supported by the Swedish National Environmental Agency which tabulated varying threats to red-listed species in its publication "Biological Diversity"; it confirms that water power plants represent the significantly greatest threat to red-listed species in inland water systems.

In spite of the probable uncertainty in estimating PER-values for the introduction of alien species, some students successfully addressed this challenge. A wide variation in PER-, E- and A-values was found. The difficulty of establishing if the effect is large, large real, clear real, etc. was obvious. A comparison was made with alternative E-value estimation scale presented in this paper (Appendix). The students found that the alternative scale could better include these biological interactions and it was recommended to further test and develop the scale for inclusion in an ecological risk-assessment method. Since the introduced species are mobile and may interact with other animals, it was concluded that there was a large, inherent variation in the area dimension and time frame of such an environmental threat which could be very difficult to establish with high significance. Thus, there is a great need for population development data to permit the A- and T-values to be established

In general, estimates of the variables for same environmental threats were notably inconsistent. There may be a diversity of reasons, including psychological, for the variation in estimated E- and T-values. The psychological effect may have arisen from personal concern or worry. In everyday language, worry and risk are used interchangeably [Sjöberg, 1998]. It is recognised that negative or positive concern may considerably influence estimates of risk [Johnson & Tversky, 1983, Finucane et al. 2000]. However, it is also acknowledged that anger may bias blame-judgements but not risk assessments [Böhm, 2003]. A further psychological effect may be attributable to the teachers presentation-methodology which, in this case, encourages critical thinking, as advocated in the approach of Freire [1970]. The encouragement to critically read and evaluate course-information may have been too extensive – thereby disposing the students to minutely question all course-book and course leaders' evaluations. A stage of conflict may have arisen regarding the information presented. When a stage of conflict identifies inadequate or misleading information, it is a positive force, especially if it leads to searching for correct science-based information [Mitchell, 1997]. On the other hand, if the conflict leads only to offence towards, misuse or no use of science-based information, it is a negative force.

Information concerning the process-stage to be used in evaluations may also have been unclear to the students. This possibility is, to some extent, corroborated by the fact that some students did not estimate the E-value for either acidification or eutrophication as being 10. Wide variations in the T-value estimations additionally appear to support the possibility. Lastly, the students' knowledge and awareness of the evaluated threats might well have been inadequate. The finding of the widest discrepancies amongst estimations for the introduction of alien species lends support to the possibility of limited knowledge being an important factor in increasing variation. If viewed in the context of limited knowledge and awareness, the psychological influences of concern and/or worry on the evaluation may be significant.

Variation in the A-value is understandable as the area dimension of a threat indeed differs between regions. In ecosystems featuring diffuse borders, such as coastal areas, the inbuilt variation in size-estimations is high and, in common with E- and T-value estimations, subject to psychologically-induced influence in some measure.

Conclusion

The students found the PER-methodology simple as well as easy to understand and learn. It also advanced their own success in sourcing relevant data regarding environmental threats and their risk-evaluation. A strong motivation to compare and rank relevant environmental cases from different parts of the world was clearly evident as were efforts to corroborate thir PER-analyses using data in the scientific literature. Based on the manifest engagement in conducting the PER-analysis to rank two different environmental threats, it was clear that the students considered the methodology to be highly appropriate for environmental management issues. They maintained that using this methodology allowed them to comprehensively debate the issues of resource usage and environmental management plans and the use of incongruous words, for example "green energy", to describe different activities as more or less environmentally-friendly. It is likely that the justifications presented could be both understood and accepted also by individuals not having a background in natural science. This was borne out by the fact that those students who were awarded the highest examination grade were not natural scientists

Whilst the methodology was found to be applicable to a general evaluation of an environmental situation in a region, the students strongly suggested some adjustments to increase its application. In this context, they proposed further developments and/or clarifications of the E- and T-value scales. They also concluded that there was an urgent need to clarify the E-value using relevant examples. The alternative E-value scale introduced during the course, and also proposed in this paper, was considered worthy of further evaluation as it was more readily accessible. Also, it would be an interesting exercise to evaluate the proposed E scale with the original E-scale using environmental data. In relation to the T-value, since the exponential scale and the time frame were difficult to understand, a modification of the T-value scale was also suggested by the students

Although there were few assignments dealing with non-chemical environmental threats, their findings indicated that the methodology was appropriate for biological and physical threats and for comparing chemical, biological and physical threats. The potential of the methodology for ranking of all kinds of environmental threats was generally acknowledged. In the meantime, however, some developments in the methodology are desirable in the following areas:

- The PER-value
  - Is it reasonable to equally value effect, area distribution and time scale?
- The E-value.
  - Definition of the ecological effect (concentration levels, process stage etc.)
  - Determination of the strength of the ecological effect (new E-value scale?)
- The A-value
  - Determination of the spatial distribution of an effect for mobile animals?
  - Determination of the spatial distribution of a physical effect?
- The T-value
  - What is the most appropriate T-value scale?
  - How strong should the T-value be in the PER-analysis?

The most important conclusion reached by the students concerning this methodology is very interesting because

- it is a simple and rather straightforward method to compare environmental threats
- they do not need expert knowledge in natural science
- it is rather easy to explain the results of the analysis to common people
- using this method it is possible to evaluate the size of a environmental threat and to compare it with other threats
- they learn a method to optimize remedial management actions
- they realize that the spatial distribution of environmental threats is important
- they realize that the time distribution of environmental threats is important
- they learn how to present and defend evaluations of environmental threats

The students have been active and not hesitant to question evaluations in the course book or evaluations presented by the course leader not hesitant to new challenges in ecological risk assessment. This strongly supports that the intended methodology which was to follow the views of critical education [Freire, 1970; Greenall Gough and Robottom, 1993 ref in Diduck, 1999] or a democratic educational dialogue that shifts the centre of the learning process from the teacher to the learner was successful.

A final remark is that increased knowledge and awareness and increased access to scientific data would most certainly decrease uncertainty and variation in risk assessment. It could also clarify and make explicit opposing values, interests and actions at the heart of a conflict [Diduck, 1999]. This would be of utmost importance for education for and awareness of sustainable development as increased science based knowledge most certainly reduces conflicts. The world is though continuously evolving so there is no static level of knowledge facts. New knowledge facts is continuously presented of which some is of good quality and well scientifically verified while some is of poor quality supported by poor questionable science. For sustainable development it is important that presented knowledge is based on good science. It is extremely important that the educational institutes encourage the students to search for science based knowledge. Higher education has therefore the obligation to promote science based exploration, evaluation and critique of emerging ideas amongst students in all their study programmes.

Acknowledgements

We would like to thank all our students that have participated in courses with PER-analysis for all comments, criticism and scientific discussion on the method per se and our way to teach the methodology, you have all contributed to this paper. We also thank Dr Andrew Petersen for his magnificent work with English editing of the document. We would also like to thank our children Susanna and Stefan for patiently enduring our didactic discussions at the dinner table.

Appendix. Proposed New E-value Scale

Score	Effect
1	No known effect
2	Inhibitory effects found in laboratory studies (short time studies, for ex 48h) on single, not key functional, species for the ecosystem. Unlikely to find effects in the field. Not on any key functional species.
3	Inhibitory effects found in laboratory studies (long time studies) on single species (populations). Likely to find effects in the field. Not on any key functional species.
4	Inhibitory effects found in mesocosm studies on single species (resulting in small ecosystem changes). Effects will also be found in the field. Not on any key functional species.
5	Inhibitory effects on populations found in mesocosm studies (resulting in small ecosystem changes). Effects will also be found in the field. Not on any key functional groups of organisms.
6	Field studies have found loss of species resulting in ecosystem changes
7	Field studies have found loss of populations resulting in ecosystem changes
8	Several populations disappear resulting in large ecosystem changes
9	A large number of populations disappear resulting in very large ecosystem changes
10	A total change of key functional groups of organisms

Explanations to the proposed E-value levels;

1	No effect found
2	Effects in a 48 h Daphnia test
3	Effect levels for the most sensitive generation of Daphnia
4	it is found that some species is very sensitive and is reduced in numbers (significantly)
5	several species is found to be reduced in numbers (significantly)
6	A dominating species is found to disappear/has disappeared
7	field observations show a strong reduction of some dominating characteristic species
8	field observations show strong reductions of a number of dominating species (typical species), for example reduced littoral vegetation. From process point of view the step has turned to regular occurrence of the process state studied. For eutrophication occurrence of algal blooms is frequent but not dominating the system. For acidification long periods of heavy precipitation and/or snow melting uses the ecosystems buffering capacity resulting in a low pH. The ecosystem do though have internal buffering capacity resulting in recovery to “normal” pH when the acid inputs reduces
9	The system has changed to mainly a system typical for the end phase of the process studied. For example a lake with heavy ”algal blooms”, not occurring all years or a lake with minor remaining buffering capacity against acidification resulting in some natural recovery.
10	A totally changed ecosystem. For example an acid lake or a lake dominated by heavy blooms. An example of a biological effect could be a total dominance by the alien species resulting in a total change of the species with all previously common species lost.

References

Begon M., Townsend C.R. and Harper J.L. 2006. *Ecology : from individuals to ecosystems*. Blackwell Publishing, Oxford

Böhm, G. 2003. Emotional reactions to environmental risks: Consequentialist versus ethical evaluation. *Journal of Environmental Psychology* **23**:199-212.

Cunningham W.P., Cunningham M.A. and Woodworth Saigo B. 2006. *Environmental Science – a global concern*. 9th Edition, McGraw-Hill Higher Education

Diduck, A. 1999. Critical education in resource and environmental management: Learning and empowerment for a sustainable future. *Journal of Environmental Management* **57**:85-97.

Finkel, A.M. & Golding, D. (Eds). 1994. Worst things first?: The Debate over Risk-based Environmental Priorities. *Resources for the Future*, Washington.

Finucane, M.L., Alhakami, A., Slovic, P. & Johnson, S.M. 2000. The affect heuristic in judgements of risks and benefits. *Journal of Behavioural Decision Making* **13**:1-17.

Freire, P. 1970. Pedagogy of the Oppressed. New York:Seabury

Fricker, A. 1998. Measuring up sustainability. *Futures* **30**:367-375.

Greenall Gough, A. & Robottom, I. 1993. Towards a socially critical environmental education: water quality studies in a coastal school. *Journal of Curriculum Studies* **25**, 301-316.

Håkanson, L. 1999. *Water Pollution – methods and criteria to rank, model and remediate chemical threats to aquatic ecosystems*. Backhuys Publishers, Leiden

Johnson, E.J. & Tversky, A. 1983. Affect, generalization, and the perception of risk. Journal of Personality and Social Psychology **45**:20-31

Mitchell, B. 1997. *Resource and Environmental Management*. London: Longman.

Orr, D.W. 1994. *Earth in Mind*. Washington: Island Press.

Reinert, K.H., Jayjock, M.A. & Weiler, E.D. 2006. Using Risk Assessment to Facilitate and Enhance the Movement to Sustainability. *Journal of Environmental Quality Management* **15**(3): 1-8

Sjöberg, L. 1998. Worry and risk perception. *Risk analysis*. **18**:85-93.

Slovic, P. 1999. Trust, emotion, sex, politics, and science: Surveying the risk-assessment battle field. *Risk analysis* **19**:689-701

Swedish Environmental Protection Board. 1994. *Biological Diversity in Sweden – A Country Study*. Naturvårdsverkets förlag, Sweden.

Suter II, G.W. (ed.). 2007. *Ecological Risk Assessment* (2nd edition). CRC Press. Taylor & Francis Group, Boca Raton.

The World Commission on Environment and Development. 1987. *Our Common Future*. OUP, Oxford

United Nations Conference on Environment and Development. 1992. *Agenda* **21**. UN. Geneva.

US EPA. 1998. *Guidelines for Ecological Risk Assessment*. EPA/630/R-95/002F.

Waara, K-O, Petersen, A., Lanaras, T., Paulauskas, V., Kleiven, S., Crosa, G., Penttinen, O-P, Quesada, A., Kovats, N., Galinou-Mitsoudi, S., Lapinska, M., Newton, A., Balode, M., Hindak, F., Marsalek, B. 2006. Inland Water Quality Assessment – A Joint European Masters Programme. *Journal of Science Education and Technology* **15**(5):409-415.

Walker B. and Salt D. 2006. Sustaining Ecosystems and People in a Changing World. *How can landscapes and communities absorb disturbances and maintain function*? Island Press, London.

Wals and Jickling. 2002. "Sustainability" in higher education – From doublethink and newspeak to critical thinking and meaningful learning. *International Journal of Sustainability in Higher Education*, **3**:221-232.

In: Science Education in the 21st Century
Editor: Ingrid V. Eriksson, pp. 263-277

ISBN: 1-60021-951-9

Chapter 9

ADOPTING THE ABDUCTIVE INQUIRY MODEL (AIM) INTO UNDERGRADUATE EARTH SCIENCE LABORATORIES

Phil Seok Oh
Gyeongin National University of Education, Korea

Abstract

A subject matter with a long history of development, such as science and mathematics, has a cultural notion of what counts as the practice of the disciplinary inquiry. Earth science, integrating geology, atmospheric and oceanic sciences, and astronomy into a whole, has a distinctive methodology in its intellectual and cultural tradition. Many authors agree that abduction is the scientific method characterizing the backward reasoning process of earth science. They also contend that abduction should be applied to the school curriculum so that students can be familiar with the discipline-specific inquiry method. Considering this position, I have adopted the Abductive Inquiry Model (AIM) into my earth science laboratory course for prospective secondary science teachers. In this chapter, I will describe what abduction is and what the AIM looks like, and then present some features of the undergraduate laboratories in which the AIM was used to teach and learn earth science.

Introduction

A subject matter with a long history of development, such as science and mathematics, has a cultural notion of what counts as the practice of the disciplinary inquiry. This practice should be reflected into the school curriculum so that students can be familiar with the discipline-specific inquiry method (Bybee, 2004; National Research Council, 1996, 2000). Earth science, integrating geology, atmospheric and oceanic sciences, and astronomy into a whole, has a distinctive methodology in its intellectual and cultural tradition. The dominant pattern of reasoning employed in earth science proceeds backward "from effect (traces) to causes" rather than "from causes (test condition) to effects" (Cleland, 2002, p. 484). That is to say, the focus of earth scientific research is on explaining existing natural phenomena with long past events or unobservable causes at present while the major work of classical experimental sciences is making a prediction and testing it in a controlled laboratory setting (Ault, 1998; Cleland, 2002; Dodick & Orion, 2003; Kleinhans, Buskes, & de Regt, 2005). For example, earth scientists 'read' the structural and compositional characteristics of rock layers to make interpretations about the processes that shaped the strata through the geologic time. Retrodictive (or, more broadly, postdictive) tasks of this sort calls for a scientific method which differs from the hypothetico-deductive method of experimental sciences and also characterizes earth science as an *interpretive* and *historical* science (Frodeman, 1995).

The methodological uniqueness of earth science stems mainly from the nature of evidence available to earth scientists and phenomena addressed in earth scientific research. First of all, the evidence uncovered from the earth is mostly fragmented (Cleland, 2002; Kleinhans, Buskes, & de Regt, 2005; Selles-Martinez, 2004). Not every consequence of ever-happened processes remains in the earth environment, and even the recorded data disappears due to weathering, metamorphism, climate changes, and so on. Hence, the information gathered from field work is not enough to determine a single explanation conclusively. With such limited and partial evidence, earth scientists devote their efforts to formulating plausible hypotheses which consist of causative conditions or processes that could possibly result in the phenomena in question.

The fact that the natural phenomena explored in earth science are mostly inaccessible and uncontrollable affects how earth scientists perform their investigations as well (Cleland, 2002; Kim, 2002; Kleinhans, Buskes, & de Regt, 2005; Selles-Martinez, 2004). Just like events going on under the deep seafloor or in the earth's interior, there are many earth scientific processes which cannot be observed directly. Besides, no method has been established to restore particular natural activities such as volcanic eruptions, earthquakes, and atmospheric entry of meteorites. To study such unobservable and uncontrollable phenomena, earth scientists have to follow a postdictive pattern of reasoning which, like retrodiction, runs backwards to infer the hypothetical states of unknown, but not necessarily past, conditions.

Then, what kind of a scientific method distinguishes earth science from other disciplines? Many authors argue that *abduction* is the logic of reasoning which characterizes the inquiry practice of earth science (Baker, 1996, 2000; Engelhardt & Zimmermann, 1982; Frodeman, 1995; Kim, 2002, 2003; Kleinhans, Buskes, & de Regt, 2005; Oh & Kim, 2005; Rhoads & Thorn, 1996). They also suggest that this method should be employed when earth science is taught in schools, because the school curricula ought to be faithful to the nature of the disciplinary inquiry. Considering this position, I have made use of abduction in teaching earth science to prospective science teachers by adopting the *Abductive Inquiry Model* (AIM) in undergraduate laboratory courses. In what follows, I will discuss what abduction is and what the AIM looks like as well as describe some features of the undergraduate laboratories in which the AIM was utilized for the learners to practice solving retrodictive and postdictive problems of earth science.

Abduction

Charles S. Peirce (1839-1914), an American philosopher known as the founder of pragmatism, referred to abduction as a form of inference. It means that abduction, like deduction and induction, has a logical structure to draw a conclusion from two premises. Using the commonly-quoted example, the syllogistic forms of deduction, induction, and abduction can be presented for comparison as follows:

Deduction

(rule, P → Q)	All the beans from this bag are white.
(case/condition, P)	These beans are from this bag.
(result, Q)	These beans are white.

Induction

(case/condition, P)	These beans are from this bag.
(result, Q)	These beans are white.
(rule, P → Q)	All the beans from this bag are white.

Abduction

(result, Q)	These beans are white.
(rule, P → Q)	All the beans from this bag are white.
(case/condition, P)	These beans are from this bag.

Peirce used the terms ‘rule’, ‘case’, and ‘result’ for the major premise, the minor premise, and the conclusion of the syllogistic form, respectively. Deduction is then the argumentation which draws a result from a rule and a case; Induction is the process in which a rule is generated with a case and a result; Abduction is the inference to derive a case or condition from a result and a rule (Goudge, 1950; Magnani, 2001; Walton, 2004). Functionally, abduction goes backward from effects to causes so that it can provide a possible reason why a certain phenomenon occurs in such a way. In other words, abduction explains the result Q by taking it as a typical instance of the rule P → Q. To fulfill this abductive function, it is important to find a rule because the explaining statement (P) is already present in the premise (P → Q) and the conclusion can be drawn automatically following the structure of the inference (Bonfantini & Proni, 1983).

From the perspective of formal logic, however, abduction has a critical flaw. In deductive reasoning, the conclusion is necessarily true when all the premises are true and the reasoning process is valid. This is why deduction is known as the self-evident, truth-conservative inference. But, in the case of abduction, the truth of the conclusion cannot be guaranteed by the truth of the premises. This is because abduction is the inference which, given the necessary condition Q, draws the sufficient condition P out of the rule P → Q as the conclusion. Therefore, in abductive reasoning, the premises can only partially support the validity of the conclusion and the conclusion goes beyond the content which the antecedents can ensure logically. For this reason, abduction is considered an ampliative inference (Goudge, 1950; Magnani, 2001; Walton, 2004).

The ampliative nature of abduction appears to be a rather strong point in the practice of scientific inquiry. Induction is another type of ampliative reasoning, but it can only provide a statistical generalization by identifying the regularity among individual events. On the contrary, abductive reasoning goes beyond the simple observation and allows for generating explanatory hypotheses. The example below shows the use of abduction in the context of earth scientific research (Engelhardt & Zimmerman, 1982, p. 81-82).

First premise (resulting state of affairs): Diamonds were found in volcanic pipes in South Africa.

Second premise (law): Diamond is produced only from carbon and carbon compounds when the temperatures reach at least 1000℃ and when the pressures are at least 55 kbar.

Conclusion (controlling state of affairs): In the formation of the volcanic pipes, material was brought up from depths where pressures of at least 55 kbar obtain.

As illustrated above, abduction enables earth scientists to explain an observed phenomenon by yielding a statement about a causative process. The explanatory hypothesis conjectured by abduction is what is already presupposed in the rule – the second premise of abductive inference. Hence, it is crucial in the process of abductive reasoning to identify a suitable rule, and there are several ways for scientists to do this. When scientists encounter a puzzling observation, they may first utilize some elements of their background knowledge to explain it. The background knowledge here indicates a broad range of information which includes personal or cultural knowledge and formal or informal experiences, namely all the general information the scientists have (Harrowitz, 1983). Therefore, confronting with an unexpected phenomenon, scientists recall certain information among their background knowledge and then infer abductively by using the recalled information as a rule.

If the scientists cannot find something appropriate from their own background knowledge, they search for scientific facts and principles from various resources in order to discover a rule for abduction. Further, when no specific rule is established yet, the scientists attempt to explain a phenomenon by developing a new rule. It has been widely discussed that scientists employ a variety of thinking strategies to create a rule for abductive inference (Engelhardt & Zimmerman, 1982; Magnani, 2001; Shelley, 1996; Thagard, 1988, 1992). For example, Alfred Wegener (1880-1930) used analogical thinking to suggest his innovative idea of continental drift. Analogy is a cognitive strategy based on a similarity between two things. In the times of Wegener, it was generally known to scientists that the coastal lines of South America and Africa fit together as if the two continents had been somehow connected in the long past. The real issue was to explain how the two continents had been broken apart and located far away from each other as we see today (Oreskes, 2003). To solve this problem, Wegener, who had observed huge masses of ice during his excursion to Greenland, made an analogy: "This first split in Cretaceous time into two parts, which then, *like floating icebergs, drifted farther and further apart.*"(Giere, 1988, p. 230, italic originally). Thus, he analogically formed the rule which says, 'If the two continents were joined together and drifted away from each other, then the costal lines match well' (c.f., Thagard, 1992), and, from this rule, he abductively derived the hypothesis of continental drift.

As a matter of course, scientists can come up with a variety of rules by drawing on their background knowledge, retrieving information from some resources, or creating new ones by themselves. In other words, a large number of rules are at their disposal, and the scientists have to select the most plausible ones with well-defined evaluative criteria (Goudge, 1950). As soon as an appropriate rule is identified, a hypothesis can derive from the rule as the result of abductive inference. In this respect, abduction is often regarded as the process of reasoning to certain rules, formulating explanatory hypotheses with the rules, and evaluating the hypotheses in order to explain some phenomena (Goudge, 1950; Magnani, 2001; Walton, 2004).

According to Peirce, explanatory hypotheses inferred by abduction have many varieties (Goudge, 1950), one of which fits well to retrodictive or postdictive problems of earth science. That is, there are hypotheses which refer to facts physically incapable of being observed by investigators. This is the case with hypotheses about past events or current inaccessible conditions. For instance, "Fossils are found; say, remains like those fishes, but far in the interior of the country. To explain the phenomenon, we suppose the sea once washed over this land" (Goudge, 1950, p. 196; Walton, 2004, p. 6). This explanation is inherently ambiguous even though it sounds plausible. Such contingent and provisional nature of explanation, however, characterizes the abductive reasoning from which earth scientists gain benefits as they interpret the present fragmentary information to reconstruct the history of the earth environment (Ault, 1998). Likewise, earth scientists take advantage of abduction for their research on inaccessible targets, such as the structure of the earth's interior, which cannot be inquired by the inductive generalization or controlled experiment.

There is no doubt that, in science, it is quite significant to suggest plausible hypotheses to solve a problem. It is especially true to the earth scientific inquiry of which a goal is to provide accounts for the phenomena occurring in the situation that direct observations and controlled experiments are impossible. Abduction is the scientific method which earth scientists heavily rely on in order to fulfill this goal by developing explanatory hypotheses.

The Abductive Inquiry Model (AIM)

Earlier, Lawson (1995) recognized abduction as the creative process of generating a hypothesis and used it to construct the science instruction model called 'empirical-abductive learning cycle'. Learning by this model, students study the available observations and suggest hypothetical explanations for them before the teacher introduces scientific concepts related to the observed phenomena. That is, the empirical-abductive learning cycle is a general model of instruction which addresses the entire procedure of a science lesson.

The *Abductive Inquiry Model* (AIM) has been proposed recently with more emphasis on the nature of the earth scientific inquiry so that it can be utilized for teaching and learning earth science in schools (Kim, 2003; Oh & Kim, 2005). The AIM is framed in four interrelated steps of '*exploration – examination – selection – explanation*', as illustrated in Figure 1.

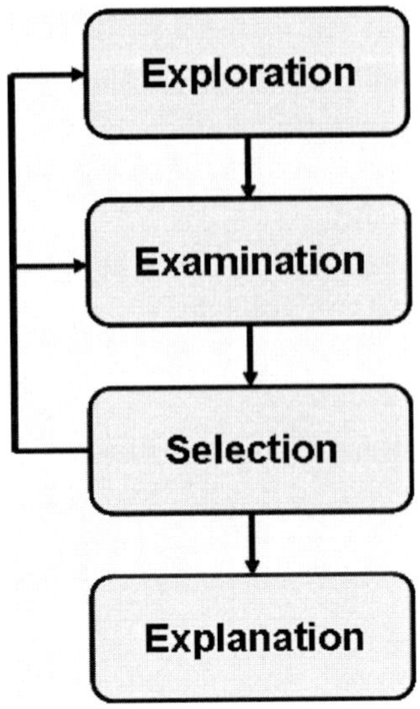

Figure 1. The Abductive Inquiry Model (AIM).

In the stage of *exploration*, students are engaged in a descriptive investigation of earth scientific phenomena. They are allowed to have hands-on experiences with earth materials, participate in geological fieldtrips, or observe objects and events in the sky. The phenomena explored by students can also be provided in the form of real time data or virtual images from the Internet. During exploration, the learners are supposed to describe the phenomena, discover empirical patterns embedded in the data, and find some aspects to be explained by abductive inference. For example, students may be interested in folded rock layers and encouraged to think about possible causes or processes that could form those shapes. Activities in this first step can be performed individually, in groups of several members, or through interactions with the teacher.

During the *examination* stage, students try to find appropriate rules and derive hypotheses from the rules in order to explain the problem identified in the previous step. They may recall some rules from their background knowledge or refer to various resources to discover scientific facts, principles, laws, or theories which they are going to use as rules for abductive inference. The learners can also create a new rule by reorganizing information from their background knowledge and other resources. In short, it is the major goal of this stage to examine all possible rules available to explain the phenomena in question. The pedagogical roles of the teacher are especially important in this process to help students come up with plausible explanations. The teacher may provide the learners with scaffolding by posing trigger questions to remind them of some suitable rules in their background knowledge, present scientific information associated with the problem, and become a model of using creative strategies which are necessary to produce novel rules.

In the next step, *selection*, students evaluate alternative rules and explanations they figured out previously and choose the most plausible ones. The teacher needs to mediate discussions among students so that they can reflect on the explanations from various perspectives. He or she should also provide appropriate criteria with which the learners can judge which hypotheses are valid and persuasive from the scientific viewpoint. If any new problems are found to be explained in this step, students may go back to the exploration and examination stages and go over the processes again.

In the *explanation* stage, students construct and suggest complete explanations of the phenomena. For finalizing their explanations, the learners make use of the rules selected and, if necessary, combine different rules to produce better one. Occasionally, they offer several different explanations together. Considering the nature of earth scientific problems that it is difficult to establish a single right answer, however, allowing different hypotheses would encourage students to think creatively to perform their inquiry tasks. The explanation provided by students in this stage often appears in the form of a causal scenario in which various events and processes are sequentially connected to result into the observed phenomena.

Earth Science Laboratories with the Abductive Inquiry Model (AIM)

The earth science laboratory is a one-semester-long course for prospective secondary science teachers. The goal of this lab course is to educate and empower students with expertise in teaching earth science through exercising the inquiry practice that characterizes important aspects of the discipline. The AIM as an inquiry learning model is well-suited to the problem-solving procedure in earth science, and I have adopted this model to develop the earth science laboratories during the past few years. Activities for this course came from various resources, including Brice and Levin (1981), Ford (1996), Slattery (1996), *KESS* (The Korean Earth Science Society, 1998), *USGS* (The U.S. Geological Survey, 1997), *ESCP* (Earth Science Curriculum Project, 1965), *Project WET* (The Watercourses and Council for Environmental Education, 1995). Among a number of activities in these books and programs, retrodictive or postdictive problems were selected and modified to some degree so that they could fit to the AIM. Table 1 lists the activities of my earth science laboratory course with brief descriptions.

It would be helpful for understanding the features of the abductive inquiry-based laboratories to provide more details of one of the activities presented in Table 1. *Sediment Analysis* could be an exemplar in that it involves a typical retrodictive problem of earth science. The problem in this activity requires students to carefully study sedimentary particles and use their properties as clues to infer where the sediments originated from, how they were transported and deposited, and what other processes they had undergone in the past. The *Sediment Analysis* activity was introduced originally by Brice and Levin (1981) in their earth history lab book, and it was revised for the earth science laboratory course in which the AIM was employed. The inquiry processes actually performed by the students enrolled in the course are described below according to the four steps of the AIM.

Table 1. Abductive Inquiry Activities for the Earth Science Laboratory Course

	Title	Description	Source
1.	Volcanoes and Plates	Students plot the locations of volcanoes, associate different rock types with them, and identify where the plate boundaries occur. They are then asked to explain why a particular type of volcanic rock is generally discovered in a particular plate boundary.	Ford, 1996
2.	Ring of Fire	Students mark the locations of different types of volcanoes on a world map and figure out why stratovolcanoes are mainly found in the Ring of Fire.	*USGS*, 1997
3.	Earthquakes in Real Time Data	Using the real time earthquake data from the USGS Web site, students explore the characteristics of earthquakes in California and Japan. They are then asked to explain the differences between the two areas with knowledge and ideas regarding the theory of plate tectonics.	Slattery, 1996
4.	Use of Earthquake Waves	Given data of earthquake waves, students construct time-distance graphs and analyze their graphs to interpret the structure of unobservable underground layers.	*KESS*, 1998
5.	Sediment Analysis	Students collect sediments, analyze their properties, and interpret the origins, transportations, and depository processes.	Brice & Levin, 1981
6.	Mapping an Ancient Stream Channel	Students are provided with fieldwork data of the thickness of rock layers and the direction of the ancient current. After retrieving the ancient stream channel with this data, students have to figure out where uranium deposits can be found and how the uranium ores have been formed.	*ESCP*, 1965
7.	An Ice Age Puzzle	Using rock descriptions of a glacier-formed area, students create geological columns and interpret a sequential events caused by the glacial activity.	*ESCP*, 1965
8.	Who is to Blame?	Students work on a true story-based problem in which they are asked to find the source of ground water contamination and explain how arsenic from an old cemetery polluted wells down far from the source.	*Project WET*, 1995

Exploration

From the foot of a nearby mountain, streamside, or several places on the campus, students collected proper amount of clastic sedimentary particles which were not solidified yet due to its exposure to the upper part of the surface layer. These sediments were made up of broken fragments of minerals and rocks which were in various sizes and shapes. In a laboratory, the preservice teachers in groups analyzed the sizes of grains in the sediments and summarized the size distribution with a table and histogram. They also investigated how many particles of quartz, feldspar, and limestone were included in their samples and computed the percentages of the grains. After producing such descriptive data, the students were asked to interpret the origins of their sediments and the ways they were transported and deposited. This task was a retrodictive one because it drove the learners to infer from the present evidence to the past causes, and it was quite sure that abduction should be employed to resolve the problem.

Examination

In order to interpret abductively the history of the sediments, students had to examine all possible rules which contained hypothetical explanations for the characteristics of the clastic particles. They could find some rules from their background knowledge but also were allowed to look up references if scientific information was needed for their thinking. To help the learners' rule-searching activity, several pedagogical questions were posed, such as, "What other evidence in the sediments looks useful to formulate your explanations?", "Are there some scientific theories and principles which you think can be applied to infer the sedimentary environment?", and "What kind of scientific information do you need more to solve the problem?"

Once they discovered an appropriate rule, students used the rule as a premise of abductive inference and derive a hypothesis from it in order to explain the descriptive data produced in the first stage. For example, identifying the fact that quartz grains were abundant in their samples, many groups of the preservice teachers recalled their background knowledge that quartz is quite durable to weathering. They then proposed that the sediment had been weathered for a long period of time. Those who had learned from a reference book that sediments from a beach often show the negatively skewed distribution of grain sizes paid their attention to their non-symmetrical histogram and concluded that their grain sample had originated from a shallow seawater environment.

In fact, the learners not only relied on well-known scientific facts and theories but also created themselves new rules for abductive reasoning. For instance, some students recognized that a large portion of their sediment grains were in thin-edged shape. To explain this distinctive feature, they constructed the new rule that if rocks were cut artificially, then the fragments appeared with sharp edges, and, using this rule, the learners suggested the hypothesis that their sediments originally came from a quarry where rocks were cut and elaborated by masons. There was also a group of preservice teachers who carried out additional explorations of their clastic sediment. They analyzed the sorting, abrasion rates, and roundness of the grains, and examined scientific laws concerning these sedimentary properties. In other words, the students went back to the first step of abductive inquiry, produced new descriptive data, and generated more explanations focusing on the newly-organized evidence.

As a consequence of such dynamic activities in this stage, the prospective science teachers came to suggest a number of explanations regarding the origins, transport systems, and depository environments of their sediments.

Selection

Working in the selection stage, students chose, among many alternatives, the most plausible accounts for the process of sediment transport and deposit. In other words, in this step of the abductive inquiry, the learners were supposed to evaluate different rules and hypotheses they figured out in the pervious stage. They should then determine what could be a possible explanation of the sediments and which statements should be eliminated from the list of candidate hypotheses.

It was clear that the preservice teachers used various evaluative criteria to assess, select, and eliminate competing explanations. For instance, a group of students first proposed that their sediment with the high portion of quartz and feldspar came from the erosion of granites exposed to a water environment. But, when they realized that limestone grains, which are weak to chemical weathering by water, were also found in their sample, they excluded this idea as an account for the formation of the sediment. That is to say, the learners evaluated the hypothesis in light of new evidence, and, in this case, the discovery of the limestone grains turned to be contradictory evidence with which the students eliminated their first hypothesis.

Sometimes students selected and combined different rules to yield a more complicated explanation. It was interesting to all of the students that there was sediment which included both large and small grains in great numbers and showed a bimodal size distribution. Facing such an unusual sample, a group of the preservice teachers chose two different rules simultaneously and integrated them into one. The first rule they selected was that sedimentary deposits result mainly from natural processes, and the other is that poorly sorted mixtures were often produced by men. Combining these two rules, they proposed it as the final explanation that at first a certain type of sediment had been formed through a natural process and then mixed with other sedimentary grains by human beings for the purpose of, for example, gardening. According to the students who proposed this explanation, their decision to select the rule of human-induced process was based on their own background knowledge that a bimodal distribution of grain sizes can hardly be discovered in the natural environment. Thus, the theoretical consistency was one of the important criteria with which the preservice teachers selected and eliminated particular hypotheses, no matter how scientifically sound and valid their theory was.

Explanation

The abductive inquiry activity described thus far was completed with the explanation stage. In this step, the learners presented their final accounts for the past earth scientific processes resulting in the current sediments by using the rules and hypotheses they selected previously. Actually, the places where the students collected their samples were diverse, and the geological information of circumstantial landscapes was limited. For this reason, it was not easy for the preservice teachers to perform precise analyses. In a positive side, however, such limitations made it possible that the students tried various ideas and discussed different opinions without feeling pressure to find a single right answer. Consequently, a variety of explanations were proposed by the prospective science teachers. It was also worth noting that these accounts were often in the genetic form of explanation in which a certain event affected the next process, which in turn influenced the following one, and, likewise, finally resulted into the present evidence.

Another exemplar of abductive inquiry, *Earthquakes in Real Time Data* is presented in Figure 2 in the form of student worksheet. This activity was offered to the students after they learned about volcanic activities and its relationship with the plate tectonic movement through two previous investigations. *Earthquakes in Real Time Data* deals with a postdictive problem for which the learners should employ the abductive method to explain the distinctive features in earthquakes between two different areas. The difference of this problem from that of the *Sediment Analysis* activity is that in this case students have to infer the hypothetical states of non-observable earth plates and their movements, rather than the events in the long past.

Earthquakes in Real Time Data

※ Visit the Web site, *http://earthquake.usgs.gov*

Exploration

1. Table 1 shows the information of the earthquakes which have recently occurred in California, United States. Mark the locations where the earthquakes took place on the map by using the latitude and longitude data given on the table. Think about what specific features characterize the recent earthquakes in California.

Table 1. Recent Earthquakes in the California Area (*partially presented here*)

	Date	Time	Latitude	Longitude	Depth (km)	Magnitude	Region
1	09/07/05	09:25	40.5 N	125.2 W	3	3.2	Offshore Northern, CA
2	09/06/05	11:35	36.4 N	118.0 W	2	3.2	Central CA
...	...	...	...	...	...	...	...
25	08/25/05	18:18	34.4 N	119.8 W	7	2.6	Santa Barbara Channel

2. Table 2 shows the information of the earthquakes which have recently occurred in Japan. Mark the locations where the earthquakes took place on the map by using the latitude and longitude data given on the table. Think about what specific features characterize the recent earthquakes in Japan.

Table 2. Recent Earthquakes in the Japan Area (*partially presented here*)

	Date	Time	Latitude	Longitude	Depth (km)	Magnitude	Region
1	09/06/05	09:13	38.3 N	141.7 E	71	4.5	East Cost, Honshu
2	09/05/05	16:18	29.9 N	129.1 E	205	4.4	Ryukyu Islands
...	...	...	...	...	...	...	...
27	08/26/05	06:09	36.7 N	142.5 E	38	4.4	East Cost of Honshu

3. Examine carefully Tables 1 and 2 and the maps on which you plotted the earthquake locations. Describe the similarities and differences between the earthquakes in California and those in Japan

※ "Why do the earthquakes which occur in California and Japan have different characteristics as you have discovered through Exploration?" Now find out an answer(s) to this question following the steps below.

Figure 2. Continued on next page.

Examination

1. Recall your earth science-related knowledge and experiences, and find out answers to the question above:
   (i) Describe what you consider to be useful for solving the problem among your background knowledge, including scientific facts, laws, principles, and theories as well as your own experiences.
   (ii) Using the knowledge that you think is useful, figure out answers to the given question. Based on your creative and candid ideas, try to come up with various possible accounts about the different features in earthquakes between California and Japan.

2. Think about what kind of information you need to provide better answers to the problem:
   (i) Identify, if any, other features in the earthquake data and maps which seem to be helpful for answering the question.
   (ii) Describe what kind of knowledge, including scientific facts, laws, principles, and theories, is needed to construct better explanations regarding the differences in earthquakes between California and Japan. Find the information you need from reference books, the Internet, and other resources.

3. Using the information you gathered, provide your answers once again. You may revise and modify the answers you first presented during activity 1. You can also make up new explanations based on the newly-obtained information.

Selection

4. Consider the explanations you offered in the previous step, and find out which one(s) is the most plausible and which one(s) is not. To select proper or improper answers, you may discuss with your peers or use the materials from the Internet or reference books. Describe the reasons why you judge a certain explanation(s) appropriate and others not.

Explanation

5. What is the explanation(s) you finally chose as an answer(s) to the question? On the basis of the answer(s) you selected, explain the earth scientific phenomena you explored, namely, the differences in earthquakes between the areas of California and Japan.

Figure 2. Student Worksheet for the *Earthquakes in Real Time Data* Activity.

Conclusion

Abduction is the process of reasoning in which explanations are formed, evaluated, and suggested for certain phenomena. This is also a scientific inquiry method used frequently by earth scientists. Earth science as an interpretive and historical science involves a task of retrodiction and postdiction which are cognitive simulations of long past events or inaccessible conditions at present. Abduction can guide such a backward-directed pattern of reasoning, enabling earth scientists to infer from evidence to underlying causes. This is the rationale with which I have adopted the AIM in my undergraduate earth science laboratory course. Success of the new program was evidenced by the active participation of the course enrollees in every activity, their understanding of the nature of earth science, and positive comments found in their lab notes such as the followings:

This class was very meaningful to me in terms of the process of solving problems, rather than the solutions. Forming a hypothesis by myself, not to speak of making a conclusion, ... it was more or less burdensome to me, but I felt that, doing step by step, I was developing.

Writing experimental reports, I thought that I had been doing a genuine job. ... I had to think more creatively ..., and the feeling of achievement was so great after I completed the reports.

In this lab, there seemed more time to think for myself and discuss with friends. It was good to try out various ideas freely with no pressure of getting the right answer even though what I thought might not be correct.

Student responses like those above imply that the earth science laboratories have provided a model for using the abductive method when the prospective teachers learn science in other courses, perform independently earth scientific research, or teach earth science in schools. I believe further that abduction is helpful for younger students to study earth science and that the development of abductive inquiry activities for secondary and even elementary school students should follow. From the perspective of science education research, it is needed to examine the effects of using the AIM on student learning as well as explore more qualitatively how students learn earth science when they are engaged in abductive inquiry.

References

Ault, C. A. Jr. (1998). Criteria of excellence for geological inquiry: The necessity of ambiguity. *Journal of Research in Science Teaching,* **35**, 189-212.

Baker, V. R. (1996). Hypotheses and geomorphological reasoning. In B. L. Rhoads & C. E. Thorn (Eds.), *The scientific nature of geomorphology* (pp. 57-85). New York: John Wiley & Sons.

Baker, V. R. (2000). Conversing with the earth: The geological approach to understanding. In R. Frodeman (Ed.), *Earth matters: The earth sciences, philosophy, and the claims of community* (pp. 2-10). Upper Saddle River, NJ: Prentice Hall.

Bonfantini, M. A., & Proni, G. (1983). To guess or not to guess? In U. Eco & T. A. Sebeok (Eds.), *The sign of three: Dupin, Holmes, Peirce* (pp. 119-134). Upper Saddle River, NJ: Prentice Hall.

Brice, J. C., & Levin, H. L. (1981). *Laboratory studies in earth history* (3rd ed.). Dubuque, IA: Wm. C. Brown.

Bybee, R. W. (2004). Scientific inquiry and science teaching. In L. B. Flick & N. G. Lederman (Eds.), *Scientific inquiry and nature of science* (pp. 1-14). Dordrecht, The Netherlands: Kluwer Academic Publishers.

Cleland, C. E. (2002). Methodological and epistemic differences between historical science and experimental science. *Philosophy of Science,* **69**, 474-496.

Dodick, J., & Orion, N. (2003). Geology as a historical science: Its perception within science and the educational system. *Science & Education,* **12**, 197-211.

Engelhardt, W. von, & Zimmermann, J. (1982). *Theory of earth science* (translated by L. Fisher). Cambridge, UK: Cambridge University Press.

ESCP (Earth Science Curriculum Project) (1965). *Investigating the Earth: Laboratory manual.* Boulder, CO: American Geological Institute.

Ford, B. A. (1996). *Project earth science: Geology.* Arlington, VA: National Science Teachers Association.

Frodeman, R. (1995). Geological reasoning: Geology as an interpretive and historical science. *GSA Bulletin,* **107**(8), 960-968.

Giere, R. N. (1988). *Explaining science: A cognitive approach.* Chicago, IL: University of Chicago Press.

Goudge, T. A. (1950). *The thought of C. S. Peirce.* New York: Dover Publications.

Harrowitz, N. (1983). The body of the detective model: Charles S. Peirce and Edgar Allan Poe. In U. Eco & T. A. Sebeok (Eds.), *The sign of three: Dupin, Holmes, Peirce* (pp. 179-197). Upper Saddle River, NJ: Prentice Hall.

KESS (The Korean Earth Science Society) (1998). *Earth science laboratory.* Seoul, Korea: Kyo-hak-youn-gu-sa. (in Korean)

Kim, C.-J. (2002). Inferences frequently used in earth science. *Journal of the Korean Earth Science Society,* **23**(2), 188-193.

Kim, C.-J. (2003). Preparing teachers for systems science methodology. In V. J. Mayer, (Ed.), *Implementing global science literacy* (pp. 255-266). Columbus, OH: The Ohio State University.

Kleinhans, M. G., Buskes, C. J. J., & de Regt, H. W. (2005). Terra Incognita: Explanation and reduction in earth science. *International Studies in the Philosophy of Science,* **19**(3), 289-317.

Lawson, A. E. (1995). *Science teaching and the development of thinking.* Belmont, CA: Wadsworth Publishing Company.

Magnani, L. (2001). *Abduction, reason, and science: Process of discovery and explanation.* New York: Kluwer Academic/Plenum Publishers.

National Research Council (1996). *National Science Education Standards.* Washington, DC: National Academy Press.

National Research Council (2000). *Inquiry and the National Science Education Standards: A guide for teaching and learning.* Washington, DC: National Academy Press.

Oh, P. S., & Kim, C.-J. (2005). A theoretical study on abduction as an inquiry method in earth science. *Journal of the Korean Association for Research in Science Education,* **25**(5), 610-623. (in Korean with an English abstract).

Oreskes, N. (2003). From continental drift to plate tectonics. In N. Oreskes (Ed.), *Plate tectonics: An insider's history of the modern theory of the earth* (pp. 3-27). Boulder, CO: Westview Press.

Rhoads, B. L., & Thorn, C. E. (1996). Observation in geomorphology. In B. L. Rhoads & C. E. Thorn (Eds.), *The scientific nature of geomorphology* (pp. 21-56). New York: John Wiley & Sons.

Selles-Martinez, J. (2004). International Earth Science Olympiad: What to test and how to do so. *Seoul Conference of the International Earth Science Olympiad: Conference Proceedings*, 136-142.

Shelley, C. (1996). Visual abductive reasoning in archaeology. *Philosophy of Science,* **63**, 278-301.

Slattery, W. (1996). Earthquakes, volcanoes, and the information superhighway. *Science Activities,* **33**(3), 8-12.

Thagard, P. (1988). *Computational philosophy of science*. Cambridge, MA: The MIT Press.

Thagard, P. (1992). *Conceptual revolutions*. Princeton, NJ: Princeton University Press.

The Watercourses and Council for Environmental Education (1995). *Project WET: Curriculum and activity guide*. Bozeman, MT: Author.

USGS (The U. S. Geological Survey) (1997). *Volcanoes! Teaching guide*. Reston, VA: Author.

Walton, D. (2004). *Abductive reasoning*. Tuscaloosa, AL: The University of Alabama Press.

In: Science Education in the 21st Century
Editor: Ingrid V. Eriksson, pp. 279-290
ISBN: 1-60021-951-8

Chapter 10

USE OF COPIES OF OLD SCIENTIFIC INSTRUMENTS FOR PHYSICS TEACHING

José Alberto Maroto Centeno[1a], Manuel Quesada-Pérez[1b] and Juan Esteban Bueno-González Tejero[2c]

[1]Group of Physics and Chemistry of Linares, Departament of Physics, EPS of Linares (University of Jaén) C/ Alfonso X el Sabio, 28, 23700 Linares, Jaén, Spain

[2]Group of Physics and Chemistry of Linares (University of Jaén)

Abstract

By means of the review of the old physics books, we can have a look into the world of the physics education of a century ago. We know that the teachers preferred to use qualitative explanations and experimental exhibitions rather than theoretical demonstrations as nowadays. In fact, the physics texts show a lot of illustrations of physics instruments that could be found in the physics laboratory of every educational centre. This way of physics teaching was widely used in educational centres of different levels (primary school, secondary school, university) up to the end of the second World War. Nevertheless, in the course of the last years, a number of different authors have proposed using copies or redesigns of these old scientific instruments in order to make the process of learning physics more attractive. In this chapter we carry out a review of some of the more interesting researches and experiences published in the last years. In some cases, new experimental devices have been designed in order to carry out qualitative exhibits (for example, colour disks based on Maxwell disks are used to show the colour vision mechanism). On the other hand, some old scientific instruments commonly used in qualitative exhibits have been redesigned in order to carry out quantitative evaluations (for example, an experimental device which permits the analysis of the transition from laminar to turbulent flow regime has been designed by means of the use of a Mariotte bottle). Also, we carry out a novel and introductory analysis of a quantitative intermittent fountain that is based on the well-known qualitative intermittent fountain. As a conclusion, an intermittent fountain can be used as an attractive and useful laboratory experiment in which the students can control the number and speed of the intermittences. In this chapter, it can be seen that some fields of experimental physics education (as mechanics,

[a] E-mail address: jamaroto@ujaen.es
[b] E-mail address: mquesada@ujaen.es
[c] E-mail address: jebutesa@ujaen.es

electromagnetism and optics) have been extended by means of the use of copies and devices based on old scientific instruments.

Introduction

By means of the review of old physics books we can have a look into the world of the physics education of a century ago. We know that the teachers preferred to use qualitative explanations and experimental exhibitions rather than theoretical demonstrations as nowadays. In fact, the physics texts show a lot of illustrations of physics instruments that could be found in the physics laboratory of every educational centre. On the other hand, both the theory shown in the books and the illustrations of the scientific instruments are useful tools that permit us to analyze the historical context in which these books were written: physics theories used at this age, technological applications carried out with the instruments, knowledge of prestigious makers, books authors and so on.

This way of teaching physics was widely used in educational centres of different levels (primary school, secondary school, university) up to the end of the second World War. Márquez [1] has identified the principal causes of the decline of physics teaching based on experimental exhibitions. In short, the overcrowding of the universities, the huge growth of scientific knowledge, and the nature of the new physics advances (quantum theory and relativity theory), which do not permit easy and intuitive experimental exhibitions, seem to be the main causes that led to a radical change in the way of teaching physics. From then, physics, like mathematics, had been taught by formulating principles or laws (axioms) from which can be obtained a number of physics properties (theorems and corollaries). On the other hand, the physics laboratory is mainly used for both the learning of methods for data treatment and for the checking of quantitative physics laws. As a consequence, the appeal of physics has highly decreased with time from the students' point of view.

Nevertheless, in the course of the last years, a number of different authors have proposed using copies or redesigns of old scientific instruments in order to make the learning process more attractive. In this chapter, we carry out a review of some of the more interesting researches and experiences published in the last years. In some cases, new experimental devices have been designed in order to carry out qualitative exhibits (for example, colour disks based on Maxwell disks are used to show the colour vision mechanism). On the other hand, some old scientific instruments commonly used in qualitative exhibits have been redesigned in order to carry out quantitative evaluations (for example, an experimental device which permits the analysis of the transition from laminar to turbulent flow regime has been designed by means of the use of a Mariotte bottle). Also, we carry out a novel and introductory analysis of a quantitative intermittent fountain that is based on the well-known qualitative intermittent fountain. As a conclusion, an intermittent fountain can be used as an attractive and useful laboratory experiment in which the students can control the number and speed of the intermittences. In this chapter it can be seen that some fields of experimental physics education (as mechanics, electromagnetism and optics) have been extended by means of the use of copies and devices based on old scientific instruments.

Historical Aspects: The Foundation of the Experimental Physics

A number of different factors explain the arrival of didactic programs of experimental physics in the European Universities. First of all, the growing of the so-called "Experimental Philosophy", that is, the teaching of Natural Philosophy by means of experiments. The experimental philosophy or experimental physics was a direct consequence of the movement that gave place to modern science: the Scientific Revolution. Secondly, we must keep in mind that experimental science was highly supported by the illustration ideology. Finally, a new social class specially interested in the new science, the middle class, emerged in the XVIII century.

Oxford was the first university to give courses of experimental physics. J. Keill started the first course in 1700. In 1710 J.T. Desaguliers relieved Keill and wrote the text titled "A Course of Experimental Phylosophy". The role played by Desaguliers in the spreading of experimental physics was very important. The programs of the new experimental physics became so famous in British universities (mainly Oxford and Cambridge) that they were exported to other countries like Holland. W. J. 'sGravesande travelled to Great Britain where he met in Newton and Desaguliers. 'sGravesande became a mathematics professor of the Leiden University in 1717 and, later, a physics professor. Then he started a course of experimental physics based on the methods and books learned during his visit to Great Britain. 'sGravesande wrote a text entitled "Physices Elementa Matematica, Experimentis Confirmata. Sive, Introductio ad Philosophiam Newtonianam" which clarified the areas that were included in the physics programs since then. P. Musschenbroek was the continuer of the work of 'sGravesande and his books were widely studied in a lot of European countries. Nollet, after visiting to Holland where he learned the new didactic methods of Musschenbroek, started a course of experimental physics in 1735 which was taught in the Collége de Navarre. Nollet published in 1748 the first three volumes of a text titled "Leçons de Physique" which included his own course of experimental physics. Nollet decided to avoid the use of equations in his course. Due probably to this fact, the Nollet course was the most popular one in France.

In short, the previous four authors were the most relevant figures that contributed to the birthing and spreading of experimental physics in the XVIII century. With regard to the XIX century, it must be said that not only the society but also the science changed rapidly. The middle class grew and an increase of the cultural level of the population became evident. On the other hand, different branches of the science were independently developed (Chemistry, Astronomy, Geology and, finally, Physics). In this ambit, the experimental physics, which was founded in the XVIII century, reached its golden age at the XIX century. It must be underlined the special role played by Faraday, who started in 1826 his "Friday Evening Discourses" and "Christmas Lectures" in the Royal Institution. These public conferences contributed highly to the divulgation of physics, they were continued by some famous scientists (Helmholtz, Dejar, Lissajous, Millar, Tyndall, etc.) and they have been taught until now.

After this brief historical review, we are going to analyze in this chapter some of the most interesting and recent works based on old physics instruments.

New Additive Colour Mixing Exhibits

One of the simplest but most effective tools for studying additive colour mixing and complementary colours is the colour top, which is a rapid spinning disk whose visible surface has been coloured. The spinning of the disk visually mixes the light reflected by the different colours on the disk to produce the illusion of a single pure hue. This device was used in several famous exhibits carried out by James Clerk Maxwell in the middle of the nineteen century in order to prove that any natural colour could be produced from the mixing of three primary colours (red, green and blue). Theoretically speaking, almost any colour of red, green or blue will produce acceptable colour mixing demonstrations, in practice, it is very difficult to predict the relative proportions of each colour required to produce a specific hue, which implies important difficulties in order to carry out these colour exhibits successfully.

In this recent work the use of a commercial PC program (Corel Photo-Paint 9) has facilitated the design of a collection of coloured disks [2]. The characteristics of this program has also permitted to select the most suitable primary colours and the development of an analytical method to calculate the relative proportions of each primary colour necessary to produce a wide variety of colours (that is, the trichromatic coordinates of each colour). The collection of coloured disks showed in this work (which includes primary colours and six test colours) can be downloaded free from internet [1]. In order to carry out the colour exhibits the researchers and teachers only require to buy (there are commercial ones available relatively cheap) or manufacture the top. We recommend these spectacular visual exhibits for educational purposes in order to show both the additive colour mixing and the colour vision mechanism.

Use of a Mariotte Bottle for the Experimental Study of the Transition from Laminar to Turbulent Flow

Although the theory of laminar flow is well developed and many solutions are known [3,4], there is no theoretical analysis nor computer solution that can simulate the small scale random fluctuations of turbulent flow. Because of this lack, the theory for turbulent flow is semi-empirical and is based on dimensional analysis and physical reasoning that permits us only to evaluate the average properties and the variance of the fluctuations. Nevertheless, the theory of turbulent flow is surprisingly effective.

It is of much practical interest for engineers and physicists to have prior knowledge of the flow regime (laminar or turbulent) that will govern the movement of a fluid. In 1883, Osborne Reynolds demonstrated that the nature of the flow regime depends on the value of a certain dimensionless parameter that represents the ratio between the inertial force and the viscosity [5]. This parameter, known as the Reynolds number, R, has been widely used. It has been experimentally demonstrated that for the flow of fluids through circular tubes, there is a critical value of R of approximately 2300 that marks the transition from one flow regime to another.

---

[1] A compressed file named COLLECTION.zip, which is available on the web page www4.ujaen.es/~jamarotoprograms.html, includes some files with the collection of coloured discs shown in this work.

In the work reviewed in this section, a Mariotte bottle was designed and constructed in order to analyze the transition regime from laminar to turbulent flow [6]. A Mariotte bottle is a device that provides a constant effusion velocity for liquids. Historically, the Mariotte bottle was frequently used in the 19th century in oil lamps for domestic illumination [7]. It consists of a closed bottle whose cover is crossed by a vertical tube. To achieve a constant effusion velocity, the lower end of the bottle must be constantly submerged in the liquid that fills the bottle.

In this work several straight, smooth, circular glass tubes with different lengths and diameters were inserted in a hole in the lower part of the bottle's wall, and the flow regime of water was investigated by means of simple flow rate measurements. The velocity of the liquid that pours out of the tube versus the distance between the lower end of the vertical tube and the level of the horizontal tube was determined for each tube. The experimental Reynolds numbers were then determined and compared with the theoretical values calculated by means of the formulae for laminar and turbulent flows.

The use of glass tubes of different but similar lengths and diameters allowed to check that the transition from laminar to turbulent flow is independent of these parameters, and depends, for a given material (glass) and liquid (water), only on the Reynolds number.

In short, the use of a Mariotte bottle to empty a container by means of straight, smooth, circular glass tubes has permitted the design of a useful laboratory experiment that provides experimental evidence of the transition from laminar to turbulent flow. Because the device is very simple and the results obtained are clear and unambiguous, we recommend this device for use as a laboratory experiment for students of fluid mechanics.

A Barlow Wheel Used for the Evaluation of the Lorentz Law

Peter Barlow was born in Norwich, U.K., in 1780. The son of a modest family, he became an eminent scientist in his age. A member of the Royal Society, he was decorated with the Cowley medal in 1825 as reward for his work with the compass placed in big ships. In fact, he was able to eliminate the disturbances due to big iron blocks. He also invented an achromatic telescope (by using carbon sulfur instead of flint glass) and realized noteworthy improvement in the field of railroad construction. Finally, in 1828 he invented the commonly know Barlow wheel, that is, a primitive dynamo-electric machine. Peter Barlow died in 1862.

The Barlow wheel used in this work [8] belongs to an ancient instrument collection which is in the General Physics Laboratory of the Polytechnics School of Linares (University of Jaén, Spain) which dates form the ending of the XIX century. This apparatus has a historical value, and this inscription in its wood plate basis: "INS$^{\underline{TS}}$ DE CHIMIE ET DE PHYSIQUE, Produits Chimiques, G. FOUNTAINAINE 18, Rue Monsieur-le-Prince (PARIS)". The figure 1 shows a recent photo of this old scientific instrument.

A Barlow Wheel consists of a metallic cogwheel, which moves around a horizontal axis. During the rotational wheel movement, one or two cogs take contact with the mercury contained in a bucket and located between the magnetic poles. When the current circulates across the wheel support, the wheel axis, one of the four arms wheel and the mercury, a force moment that leads to the rotational motion of the wheel appears due to the magnetic influence. The Barlow wheel has a reversible operation form: if the direct-current power supply is replaced by a voltmeter and makes the wheel rotate, a measurable current intensity can be obtained.

Figure 1. A recent photo of a Barlow wheel dated from the ending of the XIX century.

The Lorentz Law is a fundamental formula that explains the action of an electric and/or magnetic magnetic field on a charge. In some cases the point of interest is the action of only a magnetic field on a moving charge or current element. Some practical applications [9] of this analysis are the velocity selector, the cyclotron, the Hall effect, the mass spectrometer and the direct-current motor. The velocity selector, the cyclotron and the mass spectrometer, however, are not suitable for a basic University laboratory, while the direct-current motor is nearer to the engineering field and the Hall effect is commonly used as a Hall drill for magnetic fields measurements. On the other hand, the theoretical analysis of the Barlow wheel includes the Lorentz and the Petroff laws, which characterize the friction between the axis and the bearing. In the stationary state (when the angular velocity is constant), a simple formula can be obtained which relates the current intensity and the angular velocity. In practice, with the use of some complements such as a light barrier connected to a counter, a voltmeter, and a direct-current power supply, a simple evaluation of the Lorentz law can be carried out. So, the Barlow wheel, originally designed for qualitative exhibits, is converted into a quantitative and useful assembly in a laboratory of General Physics.

Copyes of Old Scientific Instruments and the Experimental Teaching of the Mechanics

n this section we have selected two works based on old scientific instrument which we consider very interesting in order to innovate the experimental teaching of the mechanics.

First of all we can have a look to the Escriche pendulum, an apparatus invented by the Spanish physicist Tomás Escriche in 1876. We know that Escriche studied in Madrid at the Faculty of Sciences and was a teacher at the private University and Institute of Oñate (Spain) from 1869 to 1873. After developing his expertise in teaching at several institutes of secondary education, he became a professor of physics and chemistry at the institute of Barcelona and was named director there in 1913.

During his years of public teaching, he was looking for a simple experiment to demonstrate to his pupils that the period of a pendulum is proportional to the square root of the inverse of g, the gravitational field strength. This is complicated because g varies by only a small amount at different places on Earth. Escriche used an ingenious trick to give a wider range of g values: he made the motion of the pendulum depend on a component of g. Escriche had the idea of an inclined pendulum, or "pendulum of variable acceleration" which is analyzed in an interesting work [10]. We know that his pendulum was used for many years in physics classes in Spanish secondary education. The Escriche pendulum was invented more than 100 years ago. Following the same line of reasoning as used by Escriche, a similar apparatus has been described and built [11]. Because of its simplicity and versatility, the idea can still be useful to physics students in the laboratory.

On the other hand, we know that measurements of the gravitational field strength, g, has made scientist to invent a huge variety of instruments. One of these inventions is the Morin Machine, built by A. Morin (1795-1880) in the XIX century. In a very suggestive paper [12] the authors show a simple model of the Morin Machine built for didactics purposes. They carried out an easy and precise evaluation of g. Furthermore, the simplicity of the machine makes it a tool which can be well used in other experiments with a high physical meaning. Some of these experiments are related in the text of this paper.

Introductory Analysis of a Quantitative Intermittent Fountain

The University of Jaén, Spain, carried out recently a program of didactics innovation in order to review the actual didactics methods. The principal objectives of this program were to improve the active participation of the students in the learning process and, on the other hand, to improve the autonomous learning. The authors of this chapter proposed a project which is based on an old scientific instrument called intermittent fountain (code PID06A). This project received funding.

An intermittent fountain is basically a pipette which generates intermittences during the empty process of a container. This instrument was widely used during the XIX century with didactics purposes. The classical appearance of this useful and beautiful instrument is shown in the figure 2 which is an image scanned from an old physics book [13]. Nevertheless, we carried out a redesign of this instrument in order to obtain quantitative measurements of different physical parameters. The figure 3.a show a photo of this manufactured intermittent fountain which permits to analyze the different parts of it. We can see that a large container (with a radius of 4.5 cm and a height of 12 cm) can be filled with a liquid (water) through a tube which is hermetically closed by a thread (A). The level of water inside of the container can be checked by using a transparent scale (B). The emptying process of the container is carried out through two tubes of radius 1 mm and a tube of radius 1.4 mm. These tubes can be individually closed by inserting rubber stoppers. A key (D) controls a tube (E) that penetrates the large container and permits the air exit during the fill process. On the other hand, the emptying process of the container begins just when the container is full of water and the key (D) is opened.

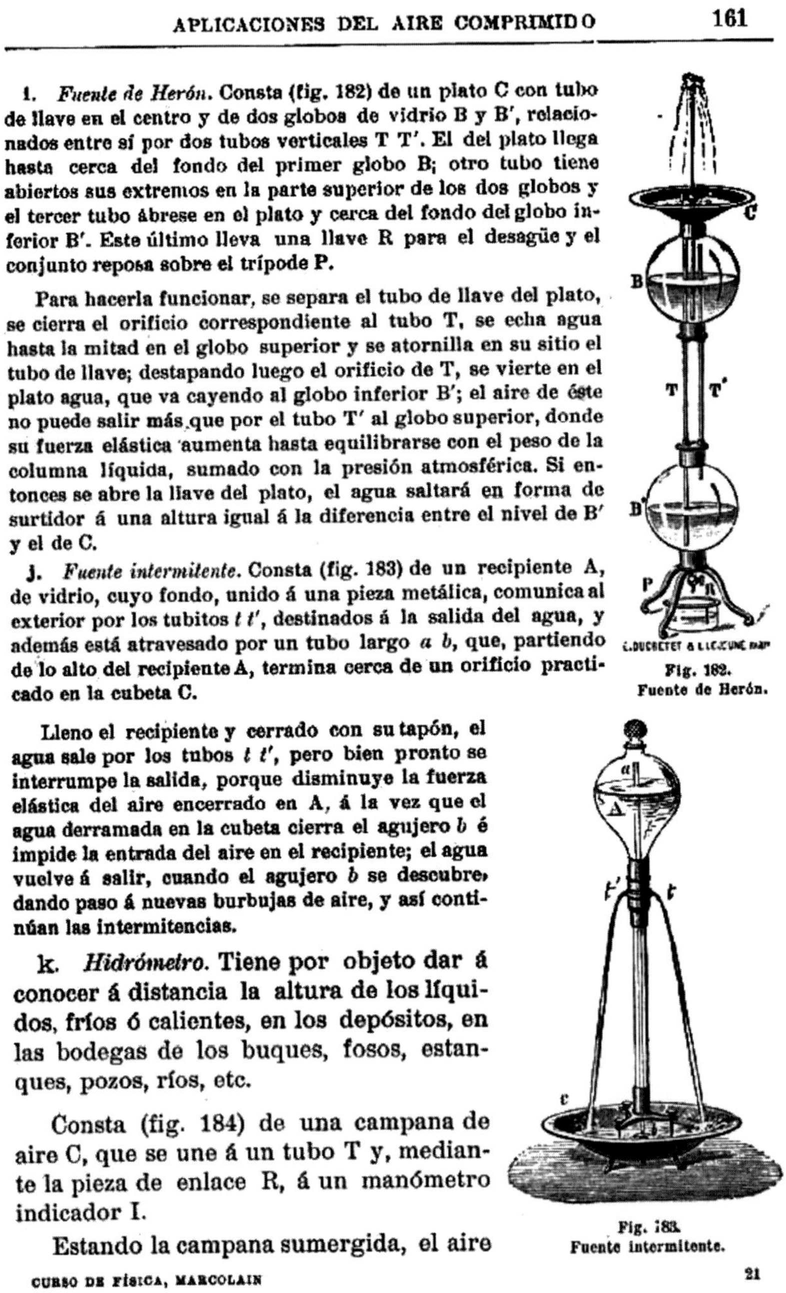

APLICACIONES DEL AIRE COMPRIMIDO 161

i. *Fuente de Herón.* Consta (fig. 182) de un plato C con tubo de llave en el centro y de dos globos de vidrio B y B', relacionados entre sí por dos tubos verticales T T'. El del plato llega hasta cerca del fondo del primer globo B; otro tubo tiene abiertos sus extremos en la parte superior de los dos globos y el tercer tubo ábrese en el plato y cerca del fondo del globo inferior B'. Este último lleva una llave R para el desagüe y el conjunto reposa sobre el trípode P.

Para hacerla funcionar, se separa el tubo de llave del plato, se cierra el orificio correspondiente al tubo T, se echa agua hasta la mitad en el globo superior y se atornilla en su sitio el tubo de llave; destapando luego el orificio de T, se vierte en el plato agua, que va cayendo al globo inferior B'; el aire de éste no puede salir más que por el tubo T' al globo superior, donde su fuerza elástica aumenta hasta equilibrarse con el peso de la columna líquida, sumado con la presión atmosférica. Si entonces se abre la llave del plato, el agua saltará en forma de surtidor á una altura igual á la diferencia entre el nivel de B' y el de C.

j. *Fuente intermitente.* Consta (fig. 183) de un recipiente A, de vidrio, cuyo fondo, unido á una pieza metálica, comunica al exterior por los tubitos *t t'*, destinados á la salida del agua, y además está atravesado por un tubo largo *a b*, que, partiendo de lo alto del recipiente A, termina cerca de un orificio practicado en la cubeta C.

Fig. 182.
Fuente de Herón.

Lleno el recipiente y cerrado con su tapón, el agua sale por los tubos *t t'*, pero bien pronto se interrumpe la salida, porque disminuye la fuerza elástica del aire encerrado en A, á la vez que el agua derramada en la cubeta cierra el agujero *b* é impide la entrada del aire en el recipiente; el agua vuelve á salir, cuando el agujero *b* se descubre, dando paso á nuevas burbujas de aire, y así continúan las intermitencias.

k. *Hidrómetro.* Tiene por objeto dar á conocer á distancia la altura de los líquidos, fríos ó calientes, en los depósitos, en las bodegas de los buques, fosos, estanques, pozos, ríos, etc.

Consta (fig. 184) de una campana de aire C, que se une á un tubo T y, mediante la pieza de enlace R, á un manómetro indicador I.

Estando la campana sumergida, el aire

Fig. 183.
Fuente intermitente.

CURSO DE FÍSICA, MARCOLAIN 21

Figure 2. Intermittent fountain. Image Scanned from a physics book dated from 1900.

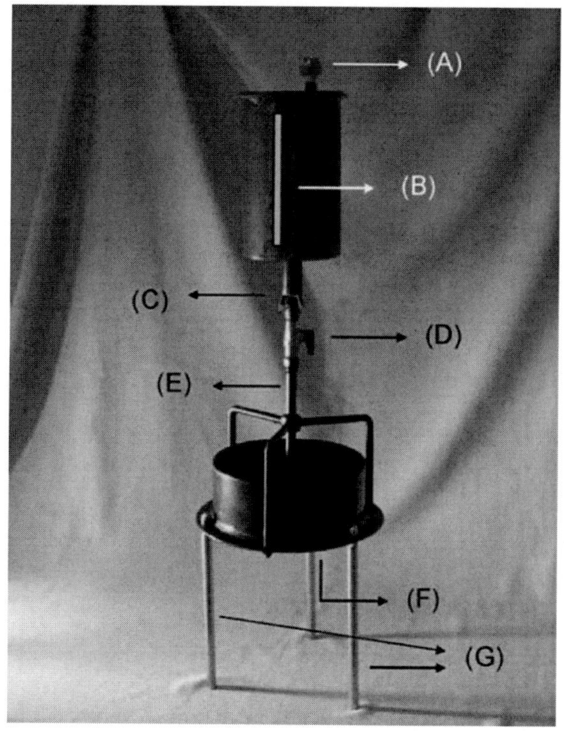

Figure 3.a. Photo of the manufactured intermittent fountain analyzed in this chapter.

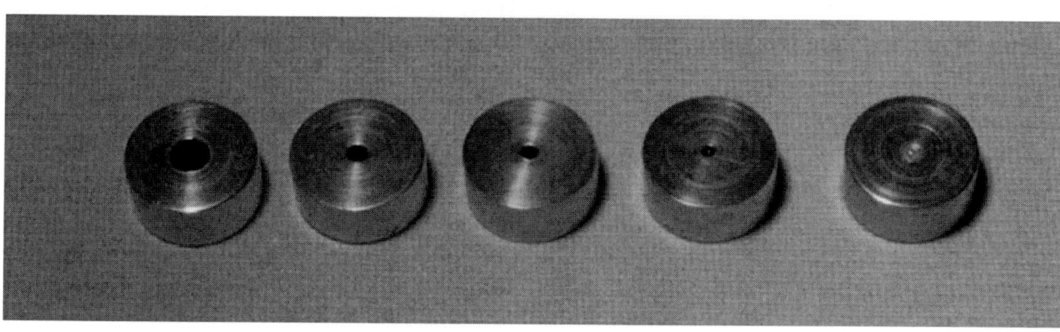

Figure 3.b. Photo of five different threads which can be inserted in the hole (F).

In order to understand the working of an intermittence, we must take into account that when the low end of the tube (E) is obstructed then the atmospheric pressure stops the empty process of the large container. In the low part of the instruments there is an open-air container (with a radius of 7.5 cm) which collects the water emptied by the tubes (C). This container has a hole (F) in the base in which can be inserted five different thread that are shown in the figure 3.b. These threads are drilled and have circular holes with radius of 0, 1.1, 1.3, 1.5 and 3 mm. Also, it is must be said that there is a distance of 1cm between the low end of the tube (E) and the basis of the container. Therefore, when the liquid level in this container reached 1 cm, an intermittence takes place. Three large legs (G) permits both the leveling of the instrument and the use of an additional container (placed in the room inside of them but not shown in the figure) in order to collect the water emptied by the low container.

A theoretical study of the intermittent fountain shows that there is not a simple formula that fits the working of it. Nevertheless, the working of this fountain is governed by the basic laws of the fluid mechanics (that is, the continuity equation and the Bernoulli law). Anyway, it is evident that an essential condition for the running of an intermittence is that the emptying flow rate of the upper container must be major or equal than the maximum emptying flow rate of the lower container (that takes place when the liquid height reaches 1 cm). As consequence, the next relationship must be fulfilled:

$$A_{low} v_{low} \leq v_{up} N_L A_L + v_{up} N_S A_S \tag{1}$$

where A_{low} is the section of the hole opened in the thread used in the experiment, v_{low} is the maximum exit speed of the water through the lower container, v_{up} is the exit speed of the water through the tubes (C), A_L and A_S are, respectively, the two different sections of the tubes (C) (subscript L refers to large and subscript S to small). N_S is the number of small tubes used (0, 1 or 2) and N_L the number of large tubes used (0 or 1).

Finally, the speeds in equation (1) can be evaluated by using the Torricelly law:

$$v = (2gh)^{1/2} \tag{2}$$

where h is the height of the column of liquid and g is the gravitational field strength.

The intermittent fountain permits to carry out both qualitative and quantitative experiences of fluid mechanics. From the qualitative point of view, the students use to find very interesting to study different selections of both emptying tubes (C) and threads used during the emptying process in order to check different working modes of the intermittent fountain. To sum up, they seek to fulfill diverse aims as, for example, to generate the major number of intermittences as possible or to generate intermittences as much time as possible.

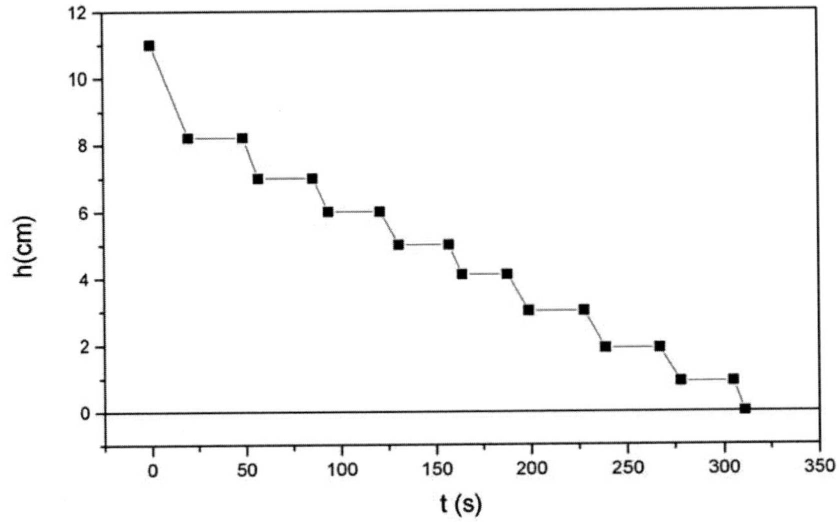

Figure 4. Height of the water in the upper container versus the time. hi = 11 cm; $N_L = 1$; $N_S = 2$.

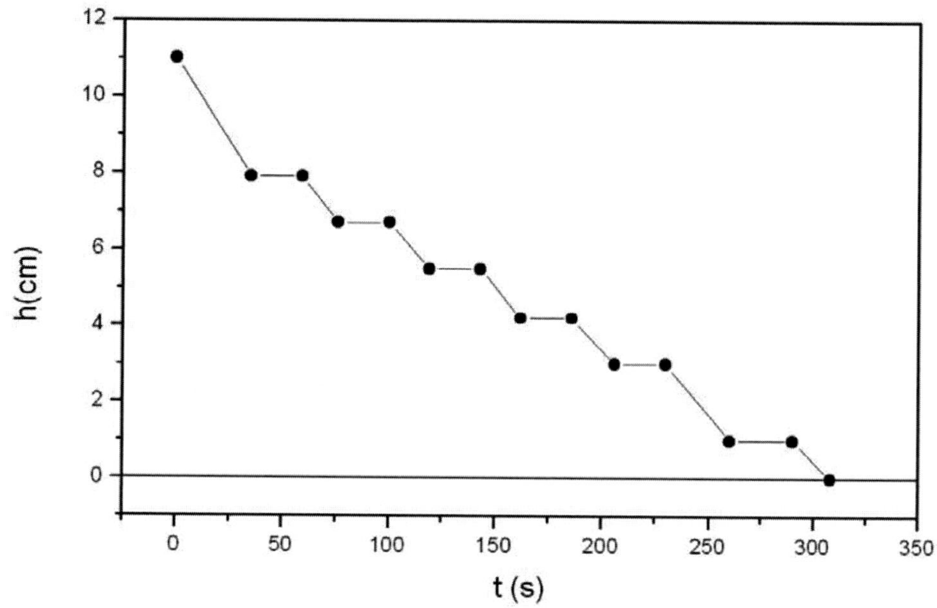

Figure 5. Height of the water in the upper container versus the time. hi = 11 cm; $N_L = 0$; $N_S = 2$.

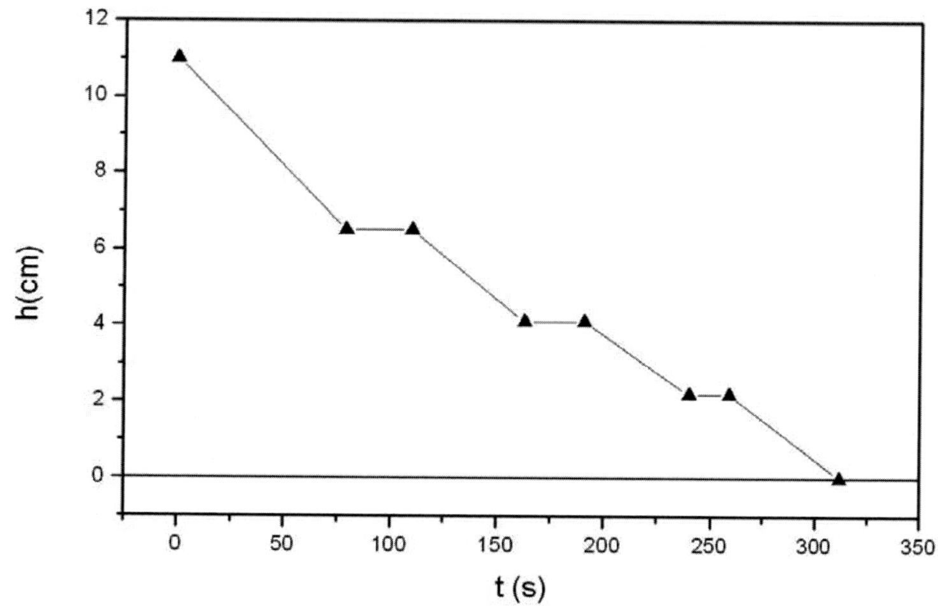

Figure 6. Height of the water in the upper container versus the time. hi = 11 cm; $N_L = 0$; $N_S = 1$.

The following three figures show some interesting results obtained by using a thread with a radius of 1.1 mm (the smaller of them) and an initial height of water in the upper container, *h*, of 11 cm. The data shown in figure 4 were obtained by using one large tube and two small tubes (that is, $N_L = 1$, $N_S = 2$). On the other hand, the data shown in figures 5 and 6 were obtained by using the following combination of tubes: $N_L = 0$ and $N_P = 2$ in figure 5, and N_L

= 0, $N_P = 1$ in figure 6. Figures 4, 5 and 6 show the height of the water in the upper container versus the time. In these figures, every horizontal segment means an intermittence. By reviewing these figures it can be easily seen that the total number of intermittences increases with the increment of the flow rate thorough the tubes (C). This has a very simple explanation: the number of intermittences depends on the balance between the incoming and the exit flow rate in the low container and when the incoming flow rate is quite larger than the exit flow rate then the tube (E) is frequently plugged. This explanation can be qualitatively proved by analyzing the time for which the first intermittence takes place. It can be seen that this time takes a value less than 25 s in figure 4, a value between 25 and 50 s in figure 5 and close to 75 s in figure 6. As conclusion, the time for which the first intermittence takes place decrease with the increment of the incoming flow rate.

Nevertheless, the total time used in the emptying process of the upper container seems to be independent of the incoming flow rate (a total time slightly above 300 s). Finally, from the artistic point of view, the different working modes of the intermittent fountain can be easily controlled. So, when the intermittent fountain is running as in figure 4, then there are a great number of intermittences and, therefore, there is not flux of water during the majority of time. In the opposite case, when the intermittent fountain is running as in figure 6, then there are a low number of intermittences and, therefore, there is a flux of water during the majority of time.

This was only an introductory analysis of a quantitative intermittent fountain. A wide theoretical treatment of it will be shown in future works.

References

[1] Márquez Delgado, R. *Rev. Esp. Fis.* 1996, 10 (1), 36-40.

[2] Maroto, J.A.; Peñalver Valverde, C.C.; Bueno-González Tejero, J.E. *Phys. Educ.* 2006, 41 (5), 448-452

[3] Schlichting, H. *Boundary Layer Theory*; 6th ed.; McGraw-Hill: New York, NY, 1968.

[4] White, F.M. *Viscous Fluid Flow*; McGraw-Hill: New York, NY, 1974.

[5] Reynolds, O. *Philos. Trans. R. Soc.* 1883, 184, 935-982.

[6] Maroto, J.A.; de Dios, J.; de las Nieves, F.J. *Am. J. Phys.* 2002, 70 (7), 698-701.

[7] Fernández de Fígares, M. *Manual de Física y Nociones de Química*; Librería de la Señora Viuda e Hijos de Zamora: Granada, 1873.

[8] Maroto, J.A.; de Dios, J.; de las Nieves, F.J. *IEEE Trans. Educ.* 2000, 43 (3), 316-320.

[9] Tipler, P.A.; Mosca, G. *Physics for Scientists and Engineers*; 5th ed; W.H. Freeman and Company: New York and Basingstoke, 2003.

[10] Vaquero, J.M.; *Phys. Teach.*, 2000, 38, 424-425.

[11] Feliciano, J.; *Phys. Teach.*, 1998, 36, 51.

[12] Granados Ruíz, D.; Sánchez, C.; *Rev. Esp. Fis.*, 1999, 13 (1), 40-44.

[13] Marcolain San Juan, P. *Curso Elemental de Física Moderna*; Tipografía de Emilio Casañal: Zaragoza, 1900.

In: Science Education in the 21st Century
Editor: Ingrid V. Eriksson, pp. 291-316
ISBN: 1-60021-951-9

Chapter 11

RESEARCH ON LEARNING AND TEACHING OF STUDENTS' CONCEPTIONS IN SCIENCE

Mei-Hung Chiu[1] and Jing-Wen Lin[2]
National Taiwan Normal University

Abstract

Research in science learning and teaching continues to show the difficulties in learning science and the challenges in changing/restructuring students' conceptions about scientific phenomena at various age levels. For the past decades, we have moved from behaviorism to constructivism, from Piaget and Bruner to cognitive scientists, from unitary representation to multiple representations, from a single explanation framework to multiple frameworks (Chiu, Guo, & Treagust, 2007). Such changes make us wonder what other researchers have achieved and what challenges face us in science learning. This article reviews the large body of literature on students' conceptions in science education over the past 30 years and summarizes what teachers and schools work effectively on developing their students' cognitive structures. We cover the following perspectives: (1) the trajectory of the development of students' conceptions in science, (2) the sources of students' scientific conceptions, (3) the theories of conceptual change in science learning, (4) ways of assessing students' learning, and (5) the learning and teaching strategies for promoting science learning. A particular model proposed by Chiu (2007) was also introduced. Finally, the implications for science education are discussed.

Introduction

The past few decades have seen a tremendous amount of research investigating how children perceive the world, acquire scientific knowledge, construct their understanding, and explain the function of the physical world. Much research has revealed that children construct their mental models through intuition and presuppositions about phenomenological events. However, the inconclusive findings concerning learners' learning in understanding the

[1] E-mail address: mhchiu@ntnu.edu.tw
[2] E-mail address: jwlin@mail.nict.gov.tw

enterprise of science concerning how students learn and come to understand scientific concepts leave open many questions. For example, there are many points of view about the developmental stages of scientific conceptions, theories of conceptual change, methodological issues for conducting research, and empirical results from different age levels and contents of learning materials in the area of science. This paper discusses these issues from the existing literature in order to shed some light on our understanding of learning and to reconsider the practice of science education.

The Trajectory of the Development of Students' Conceptions in Science

Three Important Approaches to Investigating the Development of Students' Conceptions in Science

There are three important approaches to investigating the development of students' conceptions. The first one is a survey across grades. This large-scale developmental study provides insight into the characteristic ways in which scientific ideas progress during childhood. Using this, science teachers could follow the development of concepts to introduce their pupils to the accepted scientific viewpoint at the appropriate time.

The second approach that has gained more and more acceptance in science education studies consists of observing students throughout the entire learning pathway. Compared to the first approach, this approach is a micro-scale or medium-scale investigation. It focuses on one student or specific target students to explore his/her cognitive itineraries in the course of several weeks. This approach requires detailed analyses of students' learning pathways in order to argue the effectiveness of a specific learning situation. In addition to the overall evaluation of a sequence, researchers could also test hypotheses that underlies the design of the learning situations and then to improve them.

The last approach is a novel one; its aim is to promote the investigation of students' conceptual developmental pathways and learning difficulties using a more systematic and scientific methodology.

Investigating by a Survey Across Grades

Some selected topics that have been well studied in an across-grades survey are: living things in biology, gas in chemistry, and electricity in physics. The results in these topics are almost repeated across different countries.

For many pupils, animals are restricted to large land mammals, especially those with four legs, a land habitat, fur, and noise production (Bell, 1981). Therefore, worms and spiders are not well recognized as animals by young pupils. And with respect to the example of a person, the age at which over 50% of students recognize a person as an animal is about 10 to 12 years. Tema's (1989) work is similar to Bell's investigation.

Gas is around us at anytime, anyplace. However, several studies show that pupils are not aware that air and other gases possess material characteristics (Brook & Driver, 1989; Séré, 1986; Stavy, 1988). Stavy (1988) conducted a study to investigate nine- to 15-year-old students' concepts of the substantive nature of air. She asked students to predict what would happen to the weight of a carbon dioxide cartridge before and after it was used to make soda water. She also asked students to predict the weight of a cup of soda water before and after the bubbles had escaped from it. The results showed that the development of the "air is light" idea appears between ages nine and 11. Brook and Driver (1989) report the same results.

With respect to the topic of electricity, four models appear to be popular with different age groups (Shipstone, 1985; Psillos, Koumaras, & Tiberghien, 1988; Shepardson & Moje, 1994). The unipolar model regards only one wire as active. The second model, the clashing currents model, says that current is flowing from both ends of a battery to a lightbulb and clash to produce light. Adherents of the third model, the current-used-up model, already have a correct concept of circuit. However, they think the current is used up by a bulb and, so, the returning current is less than the one going out. The fourth model is the scientific model. Here, the current is identical from beginning to end. Shipstone (1985) finds that 50% of 12-year-old students, 60% of 14 year olds, and less than 40% of 17 year olds hold the current-used-up model. Less than 10% of 12 year olds, less than 40% of 15 year olds, and 60% of 17 year olds hold the scientific model.

It is worth mentioning that, in 2003, the National Science Concept Learning Study (NSCLS) conducted a national across-grades survey of students' science concepts in Taiwan to develop a databank of concepts held by students (Chiu, Treagust, & Guo, 2007). A lot of time, effort, and human power and financial resources went into this survey, which included three subjects in science: physics, chemistry, and biology. NSCLS researchers adopted and then modified Treagust (1988, 1995)'s two-tier diagnostic format design to investigate the distribution of students' conceptions and to explore the trajectory development of 4th, 8th, 9th, and 11th grade students. For example, for the chemistry phase, Chiu (2007a) found that elementary school students' ideas were more intuitive and phenomenon oriented than older students. Although high school students showed cognitive development, their similar conceptions about chemical concepts were similar to junior high students' ideas. With respect to the development of the concept of nature and the behavior of particles, we found that both junior and senior high school students have a good understanding of the nature of particles, such as the size of the gas particle, though enior high school students' cognitive development in this area outperforms the junior high school students'. However, with respect to the behavior of particles, such as the movement of particles inside a balloon when the apparatus is placed into a hot water tank, 38% of junior high and 23% of senior high students believed that the particles moved to the edge of the balloon to expand the balloon. Such a discrepancy could be considered as evidence of cognitive development and performance based on intuitive thinking.

Investigating via the Learning Pathways Approach

The underlying belief of this approach is that learning is self- development of a cognitive system; therefore, learning is seen as constructing new cognitive elements and changing the properties of cognitive elements. Researchers who adopt this approach usually describe students' cognitive states before instruction, sometimes after instruction. They investigate the effects of curricula, media, and certain types of instruction on learning, as well as look at those teacher approaches that were involved in a change of students' cognitive states due to instruction and processes. In addition, much of the current research seems to devote to describe the intermediate concepts and the temporal processes of changing such students' conceptions (Niedderer, 1997). For example, Schwedes and Schmidt (1992) discuss the reality of expected cognitive conflicts and uncover some unexpected difficulties encountered by students in developing an analogy between hydraulic and electric circuits. Psillos and Kariotoglou (1999) investigate several different learning pathways of students who were engaged in a specific teaching sequence on fluids and, thus, take into account their differential reaction to a conflict. Similar results can be found in Arnold and Millar's work (1996) about the use of an analogy in the teaching of heat, temperature, and thermal equilibrium.

As for describing the intermediate states, Galili, Bendall, and Goldberg (1993) and Niedderer and Golberg (1995) come up with some clear research results. Galili et al. (1993) clearly indicate that there is an intermediate concept of optical image—a "relevant ray model" between the "holistic model" and the scientific model. Niedderer and Goldberg (1995) point to students' four cognitive states in terms of core statements and complementary statements as well as iconic representations. The four cognitive states are: "everyday life current" prior conception, the first intermediate conception—"positive and negative current", the second one—"electron current" and the third one—"current with electron gas pressure". All these conceptions were referred to the intermediate concept that students developed. In summary, such research data analysis makes it possible to discuss and improve the effectiveness of teaching and learning strategies.

A Novel Approach to Investigating the Developmental Pathways of Students' Mental Models

As seen above, the investigations of students' developmental pathways not only take a great deal of time and effort, but they also face the problem of research results validation (Méheut, 2004 a). Lin and Chiu (Lin, 2006; Lin & Chiu, 2006, 2007a) propose a novel approach, the cladistical approach, to solve these problems. This approach's construction of the evolutionary pathways of pupils' mental models for a specific topic draws on evolutionary epistemology (e.g., Toulmin, 1972; Hull, 1988), biological cladistics, and cognitive psychology (e.g., e.g., Chi, 1992, Chi and Roscoe, 2002 & Vosniadou, 1994). Using computers and a correct coding of pupils' mental models and cognitive characters, this approach can construct in a very short time a clear representation of a hypothetical evolutionary pathway that represents the learning difficulties for a specific topic. Furthermore, because we have a clear representation to show how students' mental models evolve, it is possible to set up different experiments to support or reject this hypothetical pathway. As such, this approach uses a more systematic and scientific methodology.

Because it is the first cross-domain study on the use of cladistics in science education, Lin and Chiu (2006) deliberately broke it down into several stages to explore its feasibility. First, they selected a topic, electricity, to show their hypothetical mental model's evolutionary pathway, which is reconstructed by computer software PAUP 4.0. They named it Tree 70 because it is the 70th tree that is auto-coded by software (Lin, 2006; Lin & Chiu, 2006). Second, they adopted the first approach (the across-grades survey) to validate the conceptual evolutionary pathway of Tree 70 (Lin, 2006; Lin & Chiu, 2007a). Then, they exhaustively searched databases to find support in literature. Finally, they designed relevant curricula and found that they helped students learn more effectively.

According to the analysis of Tree 70, the collective tendency of students' developmental pathways in a series circuit system begins from the alternative conceptions of bipolar and bidirectional of current, and follows the concepts of closed loop and incorrect concepts of electricity resistance. It also indicates that the source-consumer model, strength of current, consistent current, sequential inference, and electric energy transfer in a bulb are the most difficult concepts for students to grasp. In summary, this approach suggests that a possible sequence, based on students' preconceptions, is helpful for science educators in designing related curricula.

The Sources of Students' Scientific Conceptions

Warren, Ballenger, Ogonowski, and Rosebery (2001) state that understanding the interactions between life-world sources and scientific knowledge could help teachers to understand the gap between students' success in science learning and teachers' efforts at teaching science. Although research into the sources of students' conceptual frameworks is still developing, the main sources have been explored. The following enumerates some theoretical and empirical studies on these sources.

The Important Sources of Students' Conceptions

Formal instruction. Without a doubt, the most important source of students' conceptions is formal instruction and the interaction between teachers and students. This source not only results in students' correct conceptions but also in students' alternative conceptions. Most previous studies focus on how inefficient teachers can bring about alternative conceptions for students (Duit, 1991; Griffiths & Preston, 1989). However, Taber (2001) argues that even with a keen, capable, and subject-knowledgeable teacher, motivated, capable pupils could fail to learn effectively. The critical reason is mainly a mismatch between the students' cognitive structures and the teacher's expectations.

Problematic representations in textbooks are the other influencing factor in students' learning in science classes. It isn't always the case that the representations are incorrect—sometimes they are just oversimplified. Oversby (2000) analyzed textbooks and argued that although some representations, such as models, have their strengths and limitations, they simply provide explanations of phenomena. In an attempt to prevent textbooks from misleading students, teachers should think carefully about choosing appropriate models or multiple models, make their important features manifest, and assist students by comparing and elaborating on every specific example in each model.

Culture and linguistic effects. Halliday (1978) claims that cultures have different patterns of meaning that not only determine ways in which members of a community perceive the world, but also determine what they attend to. For instance, in Brewer, Hendrich, and Vosniadou's (1987) cross-cultural study of children's development of cosmological models, they found that Indian pupils believe the shape of earth to be like a round dish, American pupils think it is like a hollow ball, and Samoan pupils like a ring. There are many cultural factors that influence students' science learning. However, language plays the most manifest role among these cultural factors. Herron (1996) argues that the meanings of the same words in chemistry and everyday life are very different. Lin and Chiu's (2007b) study shows that Taiwanese students usually see the meaning of "中和" as "mean, middle, neutrality, or the golden mean in Confucianism" in Taiwanese daily culture. Therefore, they usually think that mixing any kind of acid and base could produce a neutral solution. In this situation, we suggest teachers could help students realize not only the superficial level of scientific symbols and words, but also their links to the real world.

The experiences of everyday life. Everyday life abounds with sense impressions, such as the feel of heat, voice, light, motion, and oxidization, and the growth of plants and animals, and so on. Undoubtedly, these experiences are all sources of concepts (Driver, 1981). Claxton (1993) concludes that everyday experiences have two obvious characteristics. The first one is piecemeal, fragmentary, and local. They are like many "minitheories" that are developed in response to particular experiences, predicaments, or needs. They work just well enough on their home ground, but their applicability may be rather circumscribed. For instance, our ideas about wood in the context of fuel might be quite different from those about wood in the context of plant growth (Barker & Carr, 1989).

The second characteristic is that these minitheories make up a whole package. They are not just comprised of content, but also of a process that we could adopt to predict and interpret phenomenon. Claxton (1993) claims that this approach is content focused, and it provides a complementary point to illustrate the dynamics of knowledge change.

Social environment. The time students expose themselves to the social environment is greater than the limited learning time in science classes. Therefore, its influence is far greater than formal instruction's (Driver, 1981). In daily impromptu conversation—chatting with friends, parents, or even experts—students easily pick up fragmentary scientific knowledge and alternative concepts anytime and anywhere. In addition, a variety of media, such as television (Lin & Chiu, 2007b; Lucas, 1993), newspapers (Wellington, 1991), and museums (Lucas, 1993) are possible avenues of information. For example, the TV news that "someone splashed sulfuric acid on students at a school entrance" can give students the misconception that all acids are corrosive and harmful (Lin & Chiu, 2007b).

Causal effect and intuition. When explaining natural phenomena, it is widely accepted that intuition has a major impact on students' thinking. With respect to electricity, Closset (1983) shows that students always use "sequential-inference" to interpret electric current; for heat and temperature, Kesidou and Duit (1993) find that students often hold a direct proportional relationship between calorific capacity and temperature; and Stavy and Tirosh (2000) identify three intuition types of students' responses: "more A – more B," "same A – same B," and "everything can be divided." Taber and Watts (1996) also point out that students usually use intuition statements—such as "this is its way," "it is normal," and "this is common sense"—to illustrate physical phenomena. These studies indicate that students always use the schema of linear casual effect and ignore other relevant variables in tackling problems.

The Weights of the Sources of Students' Conceptions

As mentioned above, the categories of sources have gradually become clear. However, the ways in which to explore how these sources influence students' understanding are still fuzzy. Lucas (1993) claims that using a heuristic typology, which considers both the learner and the source, helps us think about accounts of learning from diverse sources. Therefore, the matching of sources and students' mental models is important and valuable. Lin and Chiu's (2007b) study is to investigate the weights of the diverse sources of each student's mental models in acids and bases. In their study, they clearly state the percentages of each source contributes to each student's mental models. They found evidence that the weight of sources is not consistent with the findings reported in previous studies. Their research results show how students amalgamate knowledge from the systems of life-world and science. Furthermore, it reminds researchers and teachers to re-examine the diverse sources previously ignored.

Theories of Conceptual Change in Science Learning

Research in conceptual change has received a lot of attention in past decades. In an effort to provide an explanation for human mental behavior—how human beings or learners construct and restructure their mental representations—various areas of study generated theories of conceptual change. These areas include psychology, philosophy, computer science, linguistics, cognitive psychology, and social science. In the following sections, we draw some theories from these areas to discuss how we can improve our science learning and teaching based on research theories and empirical findings.

Researchers have different viewpoints about conceptual change: for instance, weak vs. strong (Carey, 1985), within an ontological tree (conceptual change) vs. across an ontological tree (radical conceptual change) (Chi, 1992, 2005; Chi, Slotta, & deLeeuw, 1994; Chi & Roscoe, 2002), fragmental vs. coherent (diSessa, 1993; diSessa & Sherin, 1998; diSessa, Gillespie, & Esterly, 2004; Thagard, 1992, 2001), enrichment vs. restructuring (Vosniadou, 1994), and semantic conceptual change vs. theoretical conceptual change (Tiberghien, 1994, p. 86). These theories or viewpoints can be categorized into two types: one corresponds to a deep modification in the structuring of the field of objects and events. In other words, it requires a new organization of knowledge insofar as there are new semantic relations between the model and the experimental fields (events, Tiberghien, 1994, p. 85) but in a way that the theory is not radically changed (p. 87). According to the other type, theoretical conceptual change corresponds to a change at the level of theory (for instance, scientific revolution in the history of science, Kuhn, 1970) and, in particular, with respect to causality. In other words, it requires a restructuring in the syntax of the model, in semantics, and in the theory that coheres with science. Given the different degrees of conceptual change that researchers refer to, the explanations of students' difficulties of learning sciences are inconclusive.

Conceptual development involves the enrichment and restructuring of an existing knowledge structure (see Carey, 1985; Vosniadou & Brewer, 1992). It has been argued by some researchers that the process of conceptual change is a slow and incremental enterprise rather than a sudden shift of theory. The two components involved in the enterprise of conceptual change that warrant careful investigation are product and process. Apparently, past research has focused mostly on the investigation of the changes of students' conceptions before, as opposed to after, intervention with conceptual change approaches (such as Guzutti & Glass, 1992). A limited amount of research describes how and why the conceptual change occurs.

Product

The discussion about the product of conceptual change is centered on naïve conceptions, piecemeal knowledge coherence, and consistency. Vosniadou (1999, p. 12) mentions that we not only have a restructured system, in the sense of developing radically different representations of reality that were not available before but, as the proponents of the multiple representations view would argue, also a more flexible system, a system that makes it easier to tolerate different perspectives and different points of view.

One limitation of conceptual change research is that it pays little attention to the development of cognitive flexibility. Furthermore, Vosniadou (1999) argues that learners do not have an available representation for restructuring and they might not have the type of flexible system that makes it easy to assume different perspectives and different points of view of a phenomenon and the nature of concepts. Chi (2005) claims that without a suitable category for locating a concept, a learner might have difficulty conceptualizing or acquiring correct scientific concepts. Duit (1999) claims that conceptual change is viewed by leading approaches with a constructivist orientation in science education, not as an exchange but as the development of new conceptions that are initially fruitful in particular contexts (the old conceptions may continue to be valuable in other contexts). Science instruction should not, then, strive to replace everyday ideas totally with scientific concepts, which provide a much more fruitful framework than students' own conceptions (Jung, 1986, cited from Duit, 1999, p. 270). DiSessa and Sherin (1998) propose a coordination class that claims a systematic collection of strategies for reading certain types of information that come from the world.

Process

Researchers argue that the process of conceptual change is either a case of incremental development or one of rapid development of conceptions. Such research requires a detailed analysis of a learning trajectory or a longitudinal investigation that provides some reliable evidence. Duit (1999, p. 270) elucidates that conceptual change in science education is not simply a matter of concept exchange but also the development of new conceptions that are fruitful, intelligible, and plausible. He further points out that the old conceptions developed initially in a particular context may continue to be valuable in other contexts.

Different Approaches for the Theories of Conceptual Change

Ontological approach. Chi and her colleagues (Chi, 1992; Chi, Slotta, & deLeeuw, 1994) propose a theory of conceptual change—the Incompatibility Hypothesis—that relies on three suppositions: an ontological one concerning the nature of entities in the world; a metaphysical one concerning the nature of certain science concepts; and a psychological one concerning students' naïve conceptions.

The first supposition is that entities in the world may be viewed as belonging to different ontological categories. The three primary ontological categories (or trees) are called Matter, Processes, and Mental States (Chi, 1992, p. 28). Each tree has its hierarchical subcategories and attributes to show its existence and connection with major categories. Categories within a given tree differ ontologically from categories on other trees because they do not share any ontological attributes. Therefore, conceptual change occurs when a concept has to be reassigned to an ontologically distinct category (across trees). Chi and her colleagues (1994, p. 34) claim that the difficulty of learning science concepts stems from a *mismatch* or *incompatibility* between the categorical representations students bring to an instructional context and the ontological category to which the science concept truly belongs.

The second supposition, a metaphysical assumption about the nature of scientific concepts, is quite abstract (Chi et al., 1994, p. 31) and perhaps only an approximation of the scientific definition of conceptual change. Emergent (Chi & Roscoe, 2002; it was called a constraint-based interaction in 1994) is a subcategory of Processes. It is determined by a known or knowable set of constraints (1994, p. 31). For instance, a current exists only when electrically charged particles are moving, usually because of an electric field. This implies an interaction of components belonging to a scientific phenomenon. However, the students did not take the emergent nature of the concept into account. On contrary, the students considered the current concept as a matter. That was the major reason of creating difficulty of ontological shift that is necessary for conceptual change. Similar concepts that are difficult to learn might be attributed to the same characteristics, such as the concept of heat, force, and light.

The third supposition, a psychological assumption about the nature of misconceptions, deals with the ontological status assigned by students to certain scientific concepts. For instance, students often think of force as a kind of impetus imparted to a body or as an intensive property that a body possesses. As Chi and others (Chi et al., 1994) point out, misconceptions about certain biological and physical science concepts arise because students have initially placed them in a category to which they do not belong. This miscategorization of their conceptions explains learning or the failure of learning science. Under the Incompatibility Hypothesis, it is the ontological status of a student's initial ideas about such concepts that renders them difficult to learn, because acquiring the true concept requires a shift in ontological status (radical conceptual change, Chi et al. 1994, p. 35). They further argue that misconceptions are not robust (could be removed), not consistent, not persistent, not homogeneous, and not recapitulated (Ibid., p. 36) with some empirical evidence in their research laboratory. For instance, electrical current, heat transfer, and light are misrepresented as processes-as-substance (see Slotta, Chi, & Joram, 1995; Slotta & Chi, 2006).

Epistemological approach. Vosniadou and her colleagues (Vosniadou, 1994; Vosniadou & Brewer, 1992) adopt a semantic rather than a syntactic description of conceptual change. They argue that the reorganizations of knowledge taking place during the knowledge acquisition process occur via reinterpretation of a hierarchy of constraints that differ in their degree of entrenchment. The deepest and most difficult constraints to change are those they term *presuppositions.* Presuppositions (Vosniadou & Brewer, 1992,p. 180) form the basis of an individual's ontology and epistemology and exert an enormous influence on the knowledge acquisition process. Vosniadou and Brewer also include a number of additional constructs, such as beliefs and mental models, to act as second-order constraints on the knowledge acquisition process. Therefore, constraints come to exert a powerful role in children's understanding of complex physical phenomena.

Vosniadou and Brewer applied three criteria (or values) that scientists use to evaluate the adequacy of a theory: accuracy, ontological consistency, and simplicity (adopted from Kuhn, 1977) to judge children's mental models of the day/night cycle. Their study finds, first, that the mental models of day/night constructed by the young children were for the most part empirically accurate and consistent with their observations. Second, results showed logically consistent responses by the young children to questions, which is inconsistent with other research (e.g., diSessa, 1993). Finally, with respect to simplicity, which refers to the use of the same mechanism (Ibid., p. 176) to account for different phenomena, their study finds that majority of the children used the same mechanism to explain the disappearance of the sun and the disappearance of the moon.

Vosniadou and Brewer (1992, p. 180) criticize Thagard's (1992) and Chi's et al. (1994) approaches because they do not explain why the reorganizations of knowledge that are characterized as branch jumping or crossing ontological trees are so difficult for scientists who are developing science or for children who are acquiring new knowledge.

Affective/social approach. Many researchers no longer believe that conceptual change takes place solely from a cognitive perspective but that it is also involved in intentional, social, and cultural contexts that influence a learner's understanding of science. Sinatra (2000) defines "intentional learning" as the kind of learning that is goal directed and deliberate, internally initiated rather than initiated by the environment, and under the conscious control of the learners, who can initiate, redirect, or cease learning at will (Vosniadou, 2003). As several researchers state, conceptual change is a process that can be facilitated by socio-cultural factors and educational setting (e.g., Strike & Posner, 1992; Pintrich et al., 1993; Southerland & Sinatra, 2003). For instance, Southerland and Sinatra (2003) found that students' epistemological beliefs and cognitive dispositions related to their degree of acceptance of human evolution. Similarly, students who enjoy critical, open-minded thinking were also more likely to accept human evolution. Southerland and Sinatra claim that intentional-level cognitive processes are essential for allowing students to do the deep processing necessary to compare rival knowledge and belief systems and to choose which explanation for biological change they will accept. Intentional learning can play an essential role in conceptual change about evolution.

Longitudinal conceptual change study. Clark (2006) interviewed 50 8th graders for their understanding of thermal equilibrium five times during the semester and then again preceding their 10th and 12th grade years to follow their subsequent progress. He found that the students evolved from having disjointed sets of context-dependent ideas to having integrated cohesive perspectives.

Multidimensional perspectives of conceptual change theories. Duit and Treagust (2003) propose in their paper a multiperspective framework to delineate a conceptual change process. They primarily take an epistemological, an ontological, or a social/affective position to understand the nature of conceptual change. From the epistemological perspective, the best-known approach in science education is based on the conceptual change model (CCM) proposed by Posner, Strike, Hewson, and Gertzog (1982) in which if a learner is dissatisfied with his or her prior conception and an available replacement conception is intelligible, plausible, and/or fruitful, accommodation of the new conception may follow. These conditions are not only necessary for conceptual change but other components (such as anomalies, analogies, metaphors, exemplars and images, past experiences, and others) addressed in Strike and Posner (1992) also influence a learner in the construction of his or her knowledge structure. Vosniadou and Ioannides (1998) argue that the conceptual change approach requires from the student meta-conceptual awareness of learning and pre-supposition of the phenomenon—both of which have been underestimated in the gradual process that's been taken in understanding students' conceptual structure.

The second perspective looks at ontological changes, from intuitive or common sense conceptions to scientific concepts of science entities. Chi (1992) and her colleagues (1994) claim that radical conceptual change cannot happen without a shift from one ontological category to another during learning. Vosniadou (1994) also emphasizes in her framework theories the importance of ontological change for restructuring knowledge. Chi and colleagues claim that that some scientific concepts that are held as Process concepts by scientists are often misinterpreted by students as being Matter entities. Or some Emergence concepts are mis-categorized into Event concepts. For instance, students considered the concept of electricity to be Matter because it could be used up—as a Matter of the electricity that requires the interaction between electrons.

The third perspective is to consider the impact of affection and social cognition which expand the attributes of the conceptual change model to a rather fruitful and powerful one. Pintrich, Marx, and Boyle (1993) argue that to make conceptual change happen requires not only rationality but also motivation. Pintrich and Sinatra (2003) call this process an intentional conceptual change that takes the learning environment and an individual's characteristics into account for meaningful learning.

In addition to the three major perspectives discussed above, alternative approaches for conceptual change are still blooming. For instance, Grosslight et al. (1991) state that students' modeling ability is mainly stuck on the one-to-one correspondence between reality and the model, which limits their usage and explanations of the models for understanding the scientific principles and phenomena in learning science. In their study, they found that experts tend to consider the models as tools for describing, thinking, simulating, and explaining, whereas secondary school students tend to consider the models to be copies of actual objects rather than representations of entities for communication, explanation, or idea exploration.

Integrative Approach—Promoting chemical education via Research And InstructioN-Based/Oriented Work (RAINBOW). The framework of RAINBOW (Chiu, 2007b, 2007c) dealt with developmental research, science education research, cognitive science, and school environment orientation. It has seven perspectives (developmental, ontological, epistemological, affective, evolutional, instructional, and integrative). The following section describes RAINBOW's theoretical background and then provides an example to illustrate how an empirical study fulfilled the goals of the use of different perspectives to investigate students' conceptions in science. Some theories discussed in the previous sections are summarized in a brief statement to link to the related studies.

The perspectives of the RAINBOW model are depicted in Figure 1. The components are as follows:

1. **Developmental approach (DA)**: Toulmin (1972) proposes an evolutionary point of view about the development of knowledge concepts to explain how humans developed knowledge gradually in each conceptual population. This opposes Kuhn's (1970) revolutionary viewpoint. Toulmin's approach is able to explain why certain concepts have survived over time.
2. **Ontological approach (OA):** This approach adopts the incompatible theory about students' conceptions of Chi and her colleagues (e.g., Chi, 1992, 2005; Chi & Roscoe, 2002; Chi, Slotta, & deLeeuw, 1994) in which conceptual change is the process of shifting concepts across ontological boundaries. This reassignment process—like perspective shifts and accommodation—requires learners to reassign a concept to a completely different ontological category. The other types of reconceptualization are better described as "conceptual reorganization," which simply involves hierarchical conceptual shifts or incremental changes, but no ontological shifts.
3. **Epistemological approach (EA):** This approach involves two main theories. One is Posner, Strike, Hewson, and Gertzog's (1982) epistemological theory, which explains how conceptual change happens when students conceptualize science entities. The other is Vosniadou's (2002) framework theory, which refers to the conceptual system that young children form to interpret their observations of the physical world, as well as their interpretations of the information provided by their culture. One more consideration is students' awareness of their conceptual learning.
4. **Affective/Social approach (AA):** Several researchers noticed that learning cannot be simply dealt with by considering only rational and cognitive factors (Strike & Posner, 1992; Pintrich, Marx, & Boyle, 1993; Pintrich & Sinatra, 2003). They claim that other factors, such as self-efficacy, classroom social contexts, intentions, expectations, and motivations, are also needed for conceptual learning.
5. **Evolutional approach (VA):** This approach adopts many evolutionary theories of science learning, including an interdisciplinary technique called cladistics in biology in which a tree is a graphic representation and a hypothesis of the genealogic relationships between taxa (referred to as students' mental models in Lin & Chiu, 2007a). The technique of cladistics is to provide a taxonomy—building a hypothetical tree—to investigate which characteristics of a concept have evolved during learning, in particular, to investigate students' conceptual developmental pathways and learning difficulties (Lin & Chiu, 2006) that is later validated by empirical research (Lin & Chiu, 2007a).
6. **Instructional approach (IA):** Although it is quite difficult to make a research claim simply from the perspective of instructional research, this paper intends to bring science educators' attention to Meheut's Teaching-Learning Sequences (TLS) (Meheut, 2004 a and b) as a foundation for discussing the instructional approach. TLS includes a gradual, research-based evolutionary process to build a bridge between science and a student perspective. A related approach, a notion of educational reconstruction proposed by Komorek and Duit (2004), also takes into account the content structure of science, students' understanding of science content, and the construction of instructional modules. Other instructional techniques include modeling activities (e.g., Gilbert, 2005) and Prediction-Observation- Explanation-Comparison (POEC, modified from White & Gunstone, 1992).

7. **Approach with Integration (AI)**: Various combinations of the approaches have been discussed above that can help us to achieve research purposes and should be encouraged. In order to understand better learners' learning in science, a multiple approach is needed. The RAINBOW model attempts to show how to integrate various perspectives for promoting science learning. Not only does the RAINBOW model encourage science educators to integrate five approaches (DA, OA, EA, AA, and VA) in their research, it also allows their use in different combinations to accomplish different research purposes—and then with the consideration of these approaches, to design the instructional approach (IA). This approach mirrors the current trend of cognitive science that encourages learning and teaching via multiple representations as well as by considering social and contextual factors that play central roles in science learning. The circle in the center of Figure 1, which takes the shape of the chemical structure of benzene as an illustration represents a hybrid approach and interaction between and among the six dimensions of the RAINBOW model.

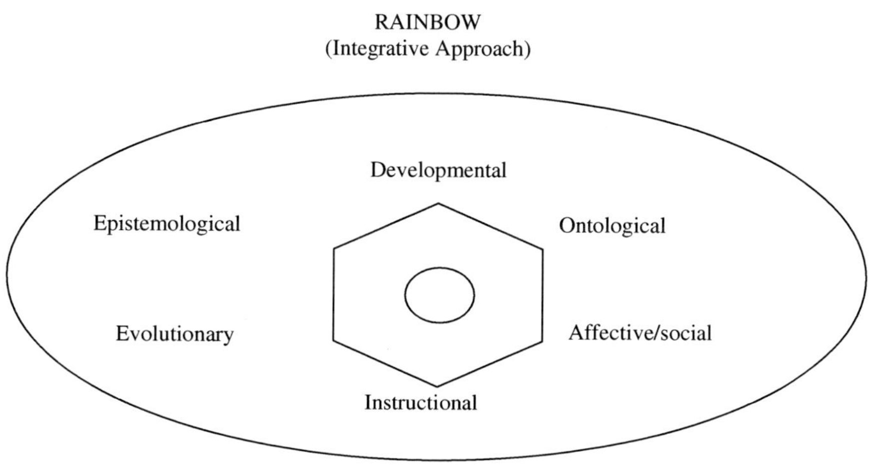

Figure 1. Components of the RAINBOW Model.

Chiu (2007b, 2007c) integrated the approaches discussed above to examine how students developed their conceptions of the particulate nature of gases. The results showed that multiple approaches not only did they promote students' learning of abstract concepts but they also elicited students' motivation in learning science. This study has also revealed that from the epistemological perspective, the students were able to repair their incomplete and flaw mental models of scientific models via learning with multiple representations of gas particle models. Also, the results reveal that in both the treatment and control groups, students' mental structures contain consistent models. This result echoes the claims of Chi and Roscoe (2002) and Vosniadou (Vosniadou & Brewer, 1992; Ioannides & Vosniadou, 2001), who advocate that students hold consistent mental models that influence their learning of new information. In addition, from the affective and social perspectives, the students expressed their motivation and positive attitude toward the modeling activities during instruction. From the analyses of the completeness of students' mental models, we found the improvement of students' performance was mainly on the ontological nature of movement of gas particles.

Furthermore, along the same lines, Wu (2007), a member of the first author's research group, investigated conceptions of characteristics of gas particles held by the students at various levels, ranging from 6th, 8th, 9th, 11th, 12th to juniors in the university. She found that the young students incrementally developed their structure of gas particles. Misunderstanding of the expansion of gas particles' volume as an increase in temperature was developed at 6th grade, but it was gone at the 9th grade and replaced and well-developed at the 12th grade. This further provides evidence that our design was able to reflect the RAINBOW's perspective, which promotes and elicits students' learning in science and leads to a better profile of our knowledge of conceptual development and change with respect to the complex concepts in science.

Ways of Assessing Students' Learning

Science education reform is now a worldwide trend. Most curriculum reform reports point out that assessment influences teaching is widely accepted in most countries. However, assessments nowadays narrow the nature of teaching, display the characteristics of subjects incorrectly, and underestimate the injustice of accessing education. Moreover, assessments do not provide any useful information for educators about what students already know (Treagust, 1995). In light of these limitations, several alternative assessment techniques were developed that consider the real nature of an assessment.

Wandersee, Mintzes, and Novak's (1994) review of related literature finds that there are more than 14 methods that explore students' existing knowledge. Researchers adopt the interview technique the most, with multiple-choice test next. Using interviews to explore students' ideas began with Piaget. Since then, several interview techniques have been developed, the most widely adopted types perhaps being interview-about-instance (IAI) and interview-about-event (IAE).

Interviewing is a valuable method. However, aware of the limits posed by teachers' heavy teaching loads and time constraints, Treagust (1988, 1995) developed a two-tier diagnostic test to improve the quality of assessments. Concept mapping and portfolio are the other two important alternative assessments. The former is valuable for representing students' knowledge frameworks, the later helps to document strengths and weaknesses in performance under realistic circumstances. The following introduces these important methods, which help us explore students' prior ideas in an effective way.

Portfolio

The portfolio has been viewed as an alternative assessment instrument characterized by its particular construction, implementation, and evaluation processes for collecting evidence about performance. It has also been described as an educational assessment that encourages students to assume responsibility for their own learning as well as a means for monitoring progress and for anticipating students' learning needs in the course of their development (Wiggins, 1993; Orland-Barak, 2005). Based on a review of the literature, three types of portfolios are identified: dossier type, reflective, and course-related learning (Smith & Tillema, 2001; Winsor, Butt, & Reeves, 1999).

Dossier portfolios document work performance that is related to external appraisal or review (Graves & Sunstein, 1992; Wiggins, 1993; Delandshere & Petrovsky, 1998). This type of portfolio focuses on remarkable achievements that may be evaluated according to specific external standards. The key feature in identifying this type of portfolio is the collection of evidence of prototypical performance in fixed formats (Peterson, 1995). Together with related instruments, such as work samples (Wood, 1994), this type of portfolio has enjoyed widespread popularity as a tool for standardized performance testing (Bennett & Ward, 1993).

Portfolios used for reflective learning document and illuminate the cyclical process of professional growth in understanding work experiences through continuous monitoring over an extended period of time (Smith, 1998). This type of portfolio is guided by students' goals and learning needs and is used as a document in and alongside professional practice to integrate evolving thoughts and actions (Smith & Tillema, 2001).

Course-related learning portfolios usually serve to inform students of information about redirecting and regulating their program-related development in competence. They place a high value on the recognition, interpretation, and utilization of relevant learning experiences related to course standards.

In all, studies show that elementary and middle school portfolio assessments produce many desired instructional benefits, such as positive changes in curriculum and instruction. However, especially after the first two or three years of implementation, problems appear. They include the additional demands placed on teachers' and students' time that is required in maintaining portfolios and preparing explanatory documentation (Stecher, 1999).

Concept Mapping

Concept mapping is an efficient way to represent students' knowledge structures and changes in knowledge structures. This method is rooted in a constructivist epistemology that assumes human beings construct meanings of the world from their past experiences (Novak, 1996; Novak & Musonda, 1991).

A concept map represents the interwoven relationships between concepts (Novak, 1996). Usually, a concept in a concept map is perceived as an event or an object, and it is designated by a label. This method is a powerful tool for assessing high cognitive-level abilities, such as synthesis and evaluation. Students are usually asked to utilize several given concepts and then build relationships between these concepts. To complete this task, students must evaluate which concepts are the most significant superordinate concepts and which are the subordinate ones as well as provide the appropriate linking words to describe concept relationships. If students are also asked to add several more concepts that are related to the concepts provided, the challenge of recall, synthesis, and evaluation is further strengthened. This method can provide highly specific information to teachers for the purpose of identifying specific places where the instructional design has failed to attain the intended goal of teaching important concepts or propositions (Novak, 1996).

Although a concept map is a valuable tool for researchers and teachers to represent students' knowledge structures, it is found to be useful in many more applications as well, such as facilitating meaningful learning, designing teaching materials, identifying alternative conceptions, evaluating learning, and improving science teachers' and students' understanding of the constructed nature of knowledge (Novak & Gowin, 1984; Novak, 1990; Novak & Wandersee, 1990). With the growing attention in schools toward effective teaching and learning, concept mapping is also found to promote cooperative learning and other kinds of group work (Novak, 1996).

Interview-about-Instances and Interview-about-Events

The techniques of interview-about-instances (IAI) and interview-about-events (IAE) are grounded in Osborne and Gilbert's work (Osborne & Gilbert, 1979; Osborne, 1980). The IAI technique is usually based on discussions of cards containing pictures of familiar objects or phenomena in pupils' worlds. The conversation that takes place between researchers and students always focuses on whether the card portrays an instance or a non-instance of the concept being explored. The conversation of the IAE technique also focuses on a concept, but the typical stimulus is an activity carried out with the pupil. Students' ideas are constructed from a variety of prior experiences, and instances cards and event activities are capable of capturing the richness of the variety. They provide a shared external focus (pictures or activities) for students to talk about their rich prior experiences (Carr, 1996).

To explore the idea of "living," for example, Carr (1996) chose some cards featuring animals and plants to elicit students' ideas. He also chose the sun and a fire as further examples. The fire card prompted many students to use the metaphoric sense of this word: Students described how a fire "breathed" and "fed." The focus pictures or choice of events in the examples seek to engage students about their viewpoints of the science concepts. All of them are necessary to move beyond the narrow focus of the typical examples represented in schools and many science definitions.

IAI and IAE are good methods to explore students' ideas deeply. However, those students who are shy might be uncomfortable with this approach. In such a case, one of the solutions might be to escape from the method gracefully and, without recrimination, thank the student for the time spent.

Two-Tier Diagnostic Test

Wolf et al. (1991) suggest that assessments should assess how students think rather than what knowledge they possess. This alternative assessment emphasizes integrating students' learning context into the assessment items. Moreover, it calls for including the learning processes involved in students' reading, writing, and problem solving. Treagust (1988, 1995), a pioneer who followed this viewpoint, developed the two-tier diagnostic test technique.

According to Treagust (1988, 1995), a two-tier diagnostic instrument should have three stages and 10 steps. First, define the concepts in a specific topic and build the concepts framework according to a concept map and proposition statements (related to steps 1 to 4 below). Second, review the literature exhaustively and develop open-ended questions that include multiple items for interviews (related to steps 5 to 7 below). Finally, develop a two-tier diagnostic instrument and modify it to an appropriate degree (related to steps 8 to 10 below). Following are the detailed contents of the 10 steps.

1. Examine proposition statements to check the related concepts in a specific topic.
2. Develop a concept map to represent the links of key concepts in a specific teaching context.
3. Utilize a concept map to examine propositions. Propositions are directly related to a concept map. (The aim of this step is to check the internal validity of examined content.)
4. Examine content validity by involving science education experts, science teachers, and science experts.
5. Exhaustively review related literature about students' alternative conceptions. (This step develops a preliminary two-tier instrument.)
6. Collect students' alternative concepts using ill-structured interviews. (After reviewing the literature, researchers always get to students' deep ideas by conducting interviews in order to design each item in a two-tier diagnostic test.)
7. Utilize free response to develop the second-tier items in a two-tier diagnostic test. (In this step, researchers always provide blanks for students to fill in to understand their response reasons.)
8. Develop a two-tier diagnostic instrument. The first tier of the test is mainly for statements of fact, which include scientific answers and students' alternative conceptions; the second tier is to obtain students' reasons for their answers in the first tier. It includes students' responses and the research results in the literature review.
9. Check all items, and cover all propositions in a specific topic.
10. Continue to modify and elaborate the test items to ensure that the instrument can diagnose students' alternative conceptions.

A two-tier diagnostic test is quite suitable for large-scale samples to investigate the frequency of students' common alternative conceptions. For example, the American Chemical Society has used it as recommended examples for conceptual questions (Voska & Heikkinen, 2000). In addition, researchers in Taiwan have conducted a national two-tier diagnostic survey to investigate the patterns and frequencies of their students' science knowledge. More details about the procedures of developing and conducting this type of research were depicted in Chiu (2007a) and Chiu, Guo, & Treagust (2007). However, some limitations are also reported. Griffard (2001) argues that the interview data in Treagust's (1988; 1995) study is used to support the development of two-tier test items rather than focus on the interview content. Moreover, he questions the very nature of two-tier diagnostic tests: According to Griffard, the aim of a two-tier diagnostic test is not to identify which alternative conceptions students hold but the distributions and frequencies of students' alternative conceptions. In addition, Odom & Barrow (1995) claim that the context and terminologies in items influence students' real understanding. Finally, Griffard and his colleagues claim that students always adopt test-taking strategies to choose answers rather than retrieve the answers from memory (Griffard & Wandersee, 2001; Griffard, 2001).

Learning and Teaching Strategies for Promoting Science Learning

The following sections draw attention to instructional strategies from ontological, epistemological, psycho-cognitive, and RAINBOW perspectives that are related to the theories of conceptual change discussed earlier.

Ontological Approach

As mentioned earlier, Chi et al. (1994, p. 28) propose a theory of conceptual change that relies on three suppositions: ontological, epistemological, and psychological. Based on these suppositions, Slotta and Chi (2007) taught students about particles movement by a simulation program first and then introduced electricity. Results showed that the teaching of the ontological nature of the particles promoted the generation of conceptual change in the students' knowledge of electric circuit. This approach makes evident the importance of explicitly explaining scientific concepts from the point of view of ontological categories.

Epistemological Approach

Several researchers have claimed that some often-referred-to strategies might not be as useful as previously thought. For instance, Vosniadou (1999) argues that it is not possible to remove a learner's conceptions about science without changing a learner's beliefs or presuppositions about the relevant scientific concepts. Vosniadou (1999) argues that if students' beliefs are fragmented, then instructional interventions aimed at producing cognitive conflict do not make much sense in terms of helping students construct appropriate explanations of a phenomenon. Vosniadou's argument is that students usually regard the counterexamples (or cognitive conflicts) inconsistently and superficially, therefore working at this level does not guarantee that the learners make the radical kinds of conceptual change that are usually necessary for science learning. In other words, cognitive conflict might not be an effective way to explain phenomena. This is consistent with Chi's claim that the cognitive conflict or analogy strategy does not emphasize the ontological nature of phenomenon, and that, therefore, the cognitive conflict approach does not offer an opportunity for radical conceptual change.

Tiberghien (1994) generates a theory via an epistemological analysis of physics learning that contains the explanatory system for scientific phenomenon and the meaning of interpretation and prediction of a theory. She found that modeling allows researchers to establish and compare the analyses of the acquisition of the learner's personal knowledge per se, the content of physics teaching, and physics knowledge in order to differentiate between the different types and characteristics of students' learning in science. Echoing Tiberghien's research, Alzatea, Tamayo, & Puig (2007) found that the students who used models came up with more coherent and cohesive explanations about respiration then those who did not and that those models were used more specifically according to the various contexts. They also found in some subjects a coexistence of various stable and useful explicative models with respect to different situations.

Psycho-Cognitive Approach

Méheut (2004b) depicts an "integrated constructivist" approach that takes into account the psycho-cognitive characteristics of learners (such as difficulties and misconceptions of students) for examining problems and didactic situations in which students can develop new knowledge. This approach was titled "teaching-learning sequence" (TLS) to denote the close linkage between proposed teaching and expected student learning as a distinguishing feature of a research-inspired, topic-oriented sequence (Méheut & Psillos, 2000; Psillos & Méheut, 2001). In other words, a TLS is both an interventional research activity and a product that includes well-researched teaching-learning activities empirically adapted to student reasoning (Méheut, 2004a, p. 516). Some researchers have experimented with TLS on various levels of students, from elementary school students (Lin, 2006) to secondary school students (Méheut, 2004b). The results from Méheut (2004b) show that by taking into account students' difficulties, conceptions, and ways of reasoning, those students who attended the TLS were able to improve particle model building—developing explanations, manipulating the different variables of the model, and correctly predicting experiments that dealt with pressure and temperature concepts.

Concluding Remarks and Implications

Science learning has received a lot of attention from researchers in many disciplines, such as cognitive science, philosophy of science, science education, cognitive psychology, study of science, and pedagogy. For the past decades, researchers sought to improve on previous explanations for learners' developmental trajectories, mental representations and inter-relationships among them, conceptual change, and learning paths in science. For instance, many science educators believe that students rely on explanations that are based on everyday experiences and inconsistent logical thinking, which are considered to be limited and fragmented. However, Vosniadou (p. 283) shows that children's thinking does not appear to be quite as limited as previously suggested. Vosniadou and Brewer (1992) find that elementary school children can provide well-defined explanations of the day and night cycles. It seems that the children show their sensitivity to issues of logical consistency and simplicity in their explanations. Besides, Vosniadou (2007) points out, students must learn how to restructure their naïve, intuitive theories based on everyday experience and lay culture. In other words, they must undergo profound conceptual change. This type of conceptual change cannot be achieved without systematic instruction that takes into consideration both individual, constructivist, and socio-cultural factors.

Hatano and Inagaki (2000, p. 424) claim that when students recognize that their comprehension is inadequate, they are motivated to pursue understanding, and their motivation for comprehension can be amplified socially. Therefore, dialogue-type interactions and support for enduring comprehension activity by the group members or teachers are required for repairing or revising beliefs during the conceptual change process. A teacher's long-term endeavor to shape the classroom into a community of learners who share the goal of understanding serves as important groundwork for the success of strategies that bring about prolonged comprehension activities and conceptual change. Combining cognitive and socio-cultural approaches can effectively bring about instruction-based conceptual change. Beyond these approaches, Chiu's (2007b and 2007c) RAINBOW approach offers a wider range of perspectives to be considered when we intend to scaffold students' learning in science domain.

In summary, by merging the findings of theoretical research and instruction, in practice, we hope to bridge students' understanding of science with a meaningful context that bring profound improvements to science learning. In this paper, we have discussed the developmental path, sources of students' scientific conceptions, ways of assessing students' learning, and the learning and teaching strategies for promoting science learning. We hope the studies reviewed will shed some light of pedagogical importance on the design of instructional activities and materials for science learning. We call for a serious examination and future exploration of the importance of a framework for explaining students' learning and conceptual changes in science.

References

Alzatea, O. E. T., & Puig, N. S. (2007). High-school students' conceptual evolution of the respiration concept from the perspective of Giere's cognitive science model. *International Journal of Science Education*, **29**(2), 215-248.

Arnold, M., & Millar, R. (1996). Learning the scientific "story": A case study in the teaching and learning of elementary thermodynamics. *Science Education,* **80**(3), 249-281.

Barker, M., & Carr, M. D. (1989). Photosynthesis: Can our pupils see the wood for the trees? *Journal of Biological Education,* **23**(1), 41-44.

Bell, B. F. (1981). *Animal, plant, living: Notes for teachers.* (LISP Working Paper 30). Hamilton, New Zealand: Science Education Research Unit, University of Waikato.

Bennett, R. E., & Ward, W. C. (1993). *Construction vs. choice in cognitive measurement: Issues in performance testing and portfolio assessment.* Hillsdale, NJ: Lawrence Erlbaum.

Boulter, C. J., & Buckley, B. C. (2000). Constructing a typology of models for science education. In J. K. Gilbert & C. J. Boulter (Eds.), *Developing models in science education* (pp. 41-57). Netherlands: Kluwer Academic Publishers.

Brewer, W. F., Hendrich, D. J., & Vosniadou, S. (1987). *A cross-cultural study of children's development of cosmological models: Samoan and American data.* Paper presented at the Third International Conference on Thinking, Honolulu, HI.

Brook, A., & Driver, R. (1989). *Progression in science: The development of pupils' understanding of physical characteristics of air across the age range 5-16 years.* Leeds: Centre for Studies in Science and Mathematics Education, University of Leeds.

Carey, S. (1985). *Conceptual change in childhood.* Cambridge, Massachusetts: The MIT Press.

Carr, M. (1996). Interviews about instances and interviews about events. In D. F. Treagust, R. Duit, & B. J. Fraser (Eds.), *Improving teaching and learning in science and mathematics* (pp. 32-43). New York: Teachers College Press.

Chi, M. T. H. (1992). Conceptual change within and across ontological categories: Implications for learning and discovery in sciences. In R. Giere (Ed.), *Cognitive models of science: Minnesota studies in the philosophy of science* (pp.129-186). Minneapolis: University of Minnesota Press.

Chi, M. T. H. (2005). Commonsense conceptions of emergent processes: Why some misconceptions are robust. *Journal of the Learning Sciences*, **14**(2), 161-199.

Chi, M. T. H., & Roscoe, R. D. (2002). The processes and challenges of conceptual change. In M. Limon & L. Mason (Eds.), *Reconsidering conceptual change: Issues in theory and practice* (pp. 3-27). Netherlands: Kluwer Academic Publishers.

Chi, M. T. H., Slotta, J. D., & deLeeuw, N. (1994). From things to processes: A theory of conceptual change for learning science concepts. *Learning and Instruction [special* issue], **4**, 27-43.

Chiu, M. H. (2007a). A national survey of students' conceptions of chemistry in Taiwan. *International Journal of Science Education* [special issue], **29**(4), 421-452.

Chiu, M. H. (2007b). *Research and instruction-based/oriented work (RAINBOW) for conceptual change in science learning*. Paper presented at the second NICE symposium, July 30-31, Taipei, Taiwan.

Chiu, M. H. (2007c). *Promoting chemical education via research and instruction-based/oriented work (RAINBOW).* Paper presented at the IUPAC 2007 Congress, August 3-11,Torino, Italy.

Chiu, M. H. & Chung, S. L. (2007). *Investigating correctness, consistency, and completeness of students' mental models and paths of conceptual change in learning the nature of gas particles via multiple modeling activities.* Paper presented at the biannual meeting of European Science Education Research Association (ESERA), August 20-25, Molmo, Sweden.

Chiu, M. H., Guo, C. J., & Treagust, D. (2007). Assessing students' conceptual understanding in science: An introduction about a national project in Taiwan. *International Journal of Science Education* [special issue], **29**(4), 379-390.

Clark, D. (2006). Longitudinal conceptual change in students' understanding of thermal equilibrium: An examination of the process of conceptual restructuring. *Cognition and Instruction*, **24**(4), 467-563.

Claxton, G. L. (1993). Minitheories: A preliminary model for learning science. In P. J. Black & A. M. Lucas (Eds.), *Children's informal ideas in science* (pp.45-61). London: Routledge.

Closset, J. L. (1983). Sequential reasoning in electricity. *Research on Physics Education* (pp. 313-319). Proceedings of the First International Workshop, 23 June-13 July, La Londe les Maures, France, Editions du Centre National de la Recherche Scientifique, Paris.

Delandshere, G., & Petrovsky, A. (1998). Assessment of complex performances: Limitations of key measurements assumptions. *Educational Researcher,* **27**(2), 14–24.

diSessa, A. (1993). Towards an epistemology of physics. *Cognition and Instruction,* **10**(2 & 3), 105-225.

diSessa, A. A., Gillespie, N. M., & Esterly, J. B. (2004). Coherence versus fragmentation in the development of the concept of force. *Cognitive Science,* **28**, 843-900.

diSessa, A. A., & Sherin, B. L. (1998). What changes in conceptual change? *International Journal of Science Education*, **20**(10), 1155-1191.

Driver, R. (1981). Pupils' alternative frameworks in science. *European Journal in Science Education,* **3**, 93-101.

Driver, R., & Oldham, V. (1986). A constructivist approach to curriculum development in science. *Studies in Science Education,* **13**, 105-122.

Duit, R. (1991). Student's conceptual frameworks: Consequences for learning in science. In S. M. Glynn, R. H. Young, & B. K. Britton (Eds.), *The Psychology of Learning Science* (pp. 65-85). Hillsdale, NJ: Erlbaum.

Duit, R. (1999). Conceptual change approaches in science education. In W. Schnotz, S. Vosniadou, & M. Carretero (Eds.), *New perspectives on conceptual change* (pp. 263-282). Oxford: Elsevier Science Ltd.

Duit, R., & Treagust, D. (2003). Conceptual change: A powerful framework for improving science teaching and learning. *International Journal of Science Education*, **25**(6), 671-688.

Galili, I., Bendall, S., & Goldberg, F. (1993). The effect of prior knowledge and instruction on understanding image formation. *Journal of Research in Science Teaching,* **30**(3), 271-301.

Gilbert, J. K. (2005). Visualization: A metacognitive skill in science and science education. In J. K. Gilbert, J. K. (ed.), *Visualization in science education,* pp.9-27. The Netherlands: Springer.

Graves, D. H., & Sunstein, B. S. (1992). *Portfolio portraits.* Portsmouth: Heinemann Books.

Griffard, P. B. (2001). The two-tier instrument on photosynthesis: What does it diagnose? *International Journal of Science Education*, **23**(10), 1039-1052.

Griffard, P. B., & Wandersee, J. H. (2001), *A qualitative look at a quantitative approach to alternative conceptions research: The two-tier instrument.* NARST 2001, St. Louis, MO.

Griffiths, A. K., & Preston, K. R. (1989). *An investigation of grade* **12** *students' misconceptions relating to fundamental characteristics of molecules and atoms.* ERIC Document No. ED 286757.

Grosslight, L., Unger, C., Jay, E., & Smith, C. (1991). Understanding models and their use in science: Conceptions of middle and high school students and experts. *Journal of Research in Science Teaching*, **28**(9), 799-822.

Guzutti, B. J., & Glass, G. V. (1992, April). *Promoting conceptual change in science: A comparative meta-analysis of instructional interventions from reading education and science education.* Paper presented at the annual meeting of the American Education Research Association, San Francisco, CA.

Halliday, M. A. K. (1978). *Language as social semiotic.* London: Edward Arnold.

Hatano, G., & Inagaki, K. (2000). When is conceptual change intended? A cognitive-sociocultural view. In S. Vosniadou (Ed.), In G. M. Sinatra & P. R. Pintrich (Eds.) why two editor entries listed here?, *Intentional conceptual change* (pp. 407-427). London: Lawrence Erlbaum Associates need comma? Publishers.

Herron, J. D. (1996). *The chemistry classroom: Formulas for successful teaching.* Washington, DC: American Chemical Society.

Hull, D. L. (1988). *Science as a process: An evolutionary account of the social and conceptual development of science*. Chicago: The University of Chicago Press.

Ioannides, C., & Vosniadou, S. (2001). The changing meanings of force. *Cognitive Science Quarterly,* **2**(1), 5-62.

Jung, W. (1986). Alltagsvorstellungen und das Lernen von Physik und Chemie. *Naturwissenschaften im Unterricht—Physil/Chemie,* **34**(April), 2-6.

Kesidou, S., & Duit, R. (1993). Students' conceptions of the second law of thermodynamics: An interpretive study. *Journal of Research on Science Teaching,* **30**, 85-106.

Komorek, M., & Duit, R. (2004). The teaching experiment as a powerful method to develop and evaluate teaching and learning sequences in the domain of non-linear systems. *International Journal of Science Education*, **26**(5), 619-633.

Kuhn, T. S. (1970). *The structure of scientific revolutions*. Chicago: University of Chicago Press.

Kuhn, T. S. (1977). *The essential tension*. Chicago: University of Chicago Press.

Lin, J. W. (2006). *Investigating the influences of different teaching-learning sequences in textbooks on students with different mental models of electricity from the perspective of conceptual evolution*. Unpublished doctoral dissertation, National Taiwan Normal University, Taipei (in Chinese).

Lin, J. W., & Chiu, M. H. (2006). *Students' conceptual evolution in electricity—The cladistical perspective*. Paper presented at NARST 2006, April, 3-6, 2006, San Francisco, CA.

Lin, J. W., & Chiu, M. H. (2007a). *Students' conceptual evolution in electricity—An empirical evaluation of cladistical perspective*. Paper presented at NARST 2007, April 15-18, New Orleans, LA.

Lin, J. W., & Chiu, M. H. (2007b). Exploring characteristics and diverse sources of students' mental models in acids and bases. *International Journal of Science Education*. **29**(6), 771-803.

Lucas, A. (1993). Constructing knowledge from fragments of learning. In P. J. Black & A. M. Lucas (Eds.), *Children's informal ideas in science*, (pp. 134-147). London: Routledge.

Méheut, M. (2004a). Teaching-learning sequences: Aims and tools for science education research. *International Journal of Science Education*, **26**(5), 515-535.

Méheut, M. (2004b). Designing and validating two teaching-learning sequences about particle models. *International Journal of Science Education*, **26**(5), 605-618.

Méheut, M., & Psillos, D. (2000). *Designing and validating teaching-learning sequences in a research perspective*. Paper presented at the International Symposium, Paris.

Niedderer, H. (1997). *Learning process studies in physics: A review of concepts and results*. Paper presented at the American Educational Research Association, Chicago.

Niedderer, H., & Goldberg, F. (1995). Lernprozesse beim elektrischen Stromkreis. *Zeitschrift fur Didaktik der Naturwissenschaften (ZfDN)* **1**(1995), 73-86.

Novak, J. D. (1990). Concept mapping: A useful tool for science education. *Journal of Research in Science Teaching*, **27**, 937-949.

Novak, J. D. (1996). Concept mapping: A tool for improving science teaching and learning. In D. F. Treagust, R. Duit, & B. J. Fraser (Eds.), *Improving teaching and learning in science and mathematics* (pp. 32-43). New York and London: Teachers College Columbia University.

Novak, J. D., & Gowin, D. B. (1984). *Learning how to learn*. New York: Cambridge University Press.

Novak, J. D., & Musonda, D. (1991). A twelve-year longitudinal study of science concept learning. *American Educational Research Journal*, **28**, 117-153.

Novak, J. D., & Wandersee, J. M. (Guest Eds.). (1990). Concept mapping. *Journal of Research in Science Teaching* [special issue], **27**(10), 923-1075 (whole issue).

Odom, A. L., & Barrow, L. H. (1995). Development and application of a two-tier diagnostic test measuring college biology students' understanding of diffusion and osmosis after a course of instruction. *Journal of Research in Science Teaching*, **32**(1), 45-61.

Orland-Barak, L. (2005). Portfolios as evidence of reflective practice: What remains "untold." *Educational Research,* **47**(1), 25-44.

Osborne, R. J. (1980). Some aspects of the students' view of the world. *Research in Science Education,* **10**, 11-18.

Osborne, R. J., & Gilbert J. K. (1979). Investigation of students' understanding of basic physics concepts using an interview about-instances approach. *Research in Science Education,* **9**, 85-93.

Oversby, J. (2000). Models in explanations of chemistry: The case of acidity. In J. K. Gilbert & C. J. Boulter (Eds.), *Developing models in science education* (pp. 227-251). Dordrecht/Boston/London: Kluwer Academic.

Peterson, K. D. (1995). *Teacher evaluation.* Thousand Oaks, CA: Corwin Press.

Pintrich, P. R., Marx, R. W., & Boyle, R. A. (1993). Beyond cold conceptual change: The role of motivational beliefs and classroom contextual factors in the process of conceptual change. *Review of Educational Research,* **63**(2), 167-199.

Pintrich, P. R., & Sinatra, G. M. (2003). Future directions for theory and research on intentional conceptual change. In G. M. Sinatra & P. R. Pintrich (Eds.), *Intentional conceptual change* (pp. 429-441). London: Lawrence Erlbaum Associates.

Posner, G. J., Strike, K. A., Hewson, P. W., & Gertzog, W. A. (1982). Accommodation of a scientific conception: Toward a theory of conceptual change. *Science Education,* **66**(2), 211-227.

Psillos, D., & Kariotoglou, P. (1999). Teaching fluids: Intended knowledge and students' actual conceptual evolution. *International Journal of Science Education,* **21**, 17-38.

Psillos, D., Koumaras, P., & Tiberghien, A. (1988). Voltage presented as a primary concept on DC circuits. *International Journal of Science Education,* **10**(1), 29-43.

Psillos, D., & Méheut, M. (2001). *Teaching-learning sequences as a means for linking research to development.* Paper presented at the Proceedings of the Third International Conference on Science Education Research in the Knowledge Based Society, Thessaloniki .

Schwedes, H., & Schmidt, D. (1992). *Conceptual change and theoretical comments.* Paper presented at the Research in Physics Learning: Theoretical Issues and Empirical Studies, Kiel. Germany.

Séré, M. G. (1986). Children's conceptions of the gaseous state, prior to teaching. *European Journal of Science Education,* **8**(4), 413-425.

Shepardson, D. P., & Moje, E. B. (1994). The nature of fourth graders' understandings of electric circuits. *Science Education,* **78**(5), 489-514.

Shipstone, D. M. (1985). Electricity in simple circuits. In R. Driver, E. Guesne, & A. Tiberghien (Ed.), *Children's ideas in science* (pp. 33-51). Milton Keynes: Open University Press.

Sinatra, G. M. (2000). From passive to active to intentional: Changing conceptions of the learner. In G. M. Sinatra (Chair), *What does it mean to be an intentional learner? Alternative perspective.* Symposium presented at the American Educational Research Association Annual Meeting, New Orleans.

Slotta, J. D., & Chi, M. T. H. (2006). The impact of ontology training on conceptual change: Helping students understand the challenging topics in science. *Cognition and Instruction,* **24**, 261-289.

Slotta, J. D., Chi, M. T. H., & Joram, E. (1995). Assessing students' misclassifications of physics concepts: An ontological basis for conceptual change. *Cognition and Instruction*, **13**, 373-400.

Smith, K. (1998). Portfolios as an alternative assessment practice in higher education. In J. E. Forster (Ed.), *University teaching* (pp. 137-160). New York: Garland Publishing.

Smith, K., & Tillema, H. (2001). Long-term influences of portfolios on professional development. *Scandinavian Journal of Educational Research,* **45**(2), 183-203.

Southerland, S. A., & Sinatra, G. M. (2003). Learning about biological evolution: A special case of intentional conceptual change. In G. M. Sinatra & P. R. Pintrich (Eds.), *Intentional conceptual change* (pp. 317-345), London: Lawrence Erlbaum Associates, Publishers.

Stavy, R. (1988). Children's conception of gas. *International Journal of Science Education,* **10**(5), 552-560.

Stavy, R., & Tirosh, D. (2000). *How students (mis-)understand science and mathematics: Intuitive rules.* New York: Teachers College Press.

Stecher, B. (1999). The local benefits and burdens of large-scale portfolio assessment. *Assessment in Education,* **5**(3), 335-351.

Strike, K. A., & Posner, G. J. (1992). A revisionist theory of conceptual change. In R. Duschl & R. Hamilton (Eds.), *Philosophy of science, cognitive psychology, and educational theory and practice* (pp. 147-176). Albany, NY: SUNY.

Taber, K. S. (2001). The mismatch between assumed prior knowledge and the learner's conceptions: A typology of learning impediments. *Educational Studies*, **27**(2), 159-169.

Taber, K. S., & Watts, M. (1996). The secret life of the chemical bond: Students' anthropomorphic and animistic references to bonding. *International Journal of Science Education,* **18**, 557-568.

Tema, B. O. (1989). Rural and urban African pupils' alternative conceptions of animal classification. *School Science and Mathematics,* **85**(4), 304-316.

Thagard, P. (1992). *Conceptual revolution.* New Jersey, Princeton: University Princeton Press.

Thagard, P. (2001). How to make decisions: Coherence, emotion, and practical inference. In E. Millgram (Ed.), *Varieties of practical inference* (pp. 355-371). Cambridge, MA: MIT Press.

Tiberghien, A. (1994). Modeling as a basis for analyzing teaching-learning situation. *Learning and Instruction*, **4**, 71-87.

Toulmin, S. (1972). *Human understanding: The collective use and evolution of concepts.* Princeton, NJ: Princeton University Press.

Treagust, D. F. (1988). Development and use of diagnostic tests to evaluate students' misconceptions in science. *International Journal of Science Education*, **10**(2), 159-169.

Treagust, D. F. (1995). Diagnostic assessment of students' science knowledge. In S. M. Glynn & R. Duit. (Eds.), *Learning science in the schools: Research reforming practice* (pp. 327-346). Mahwah, New Jersey: Lawrence Erlbaum Associates.

Toulmin, S. (1972). *Human understanding: The collective use and evolution of concepts.* Princeton, New Jersey: Princeton University Press.

Voska, K. W., & Heikkinen, H. W. (2000). Identification and analysis of student conceptions used to solve chemical equilibrium. *Journal of Research in Science Teaching,* **37**(2), 160-176.

Vosniadou, S. (1994). Capturing and modeling the process of conceptual change. *Learning and Instruction* [special issue], **4**, 45-69.

Vosniadou, S. (1999). Conceptual change research: State of the art and future directions. In W. Schnotz, S. Vosniadou, & M. Carretero (Eds.), *New perspectives on conceptual change* (pp. 3-13), Oxford: Elsevier Science Ltd.

Vosniadou, S. (2002). Mental models in conceptual development. In L. Magnani & N. Nersessian (Eds.), *Model-based reasoning: Science, technology, values* (pp. 1-16), New York: Kluwer Academic Press.

Vosniadou, S. (2003). Exploring the relationships between conceptual change and intentional learning. In G. M. Sinatra & P. R. Pintrich (Eds.), *Intentional conceptual change* (pp. 377-406), London: Lawrence Erlbaum Associates.

Vosniadou, S. (2007). Conceptual change and education. *Human Development*, **50**(1), 47-54.

Vosniadou, S., & Brewer, W. (1992). Mental models of the Earth: A study of conceptual change in children. *Cognitive Psychology*, **24**, 535-585.

Vosniadou, S., & Ioannides, C. (1998). From conceptual development to science education: A psychological point of view. *International Journal of Science Education,* **20**(10), 1213-1230.

Wandersee, J. H., Mintzes, J. J., & Novak, J. D. (1994). Research on alternative conceptions in science. In D. Gabel (Ed.), *The handbook of research in science teaching and learning* (pp. 177-210). New York: MacMillan.

Warren, B., Ballenger, C., Ogonowski, M., & Rosebery, J. (2001). Rethinking diversity in learning science: The logic of everyday sense-making. *Journal of Research in Science Teaching*, **38**(5), 529-552.

Wellington, J. (1991). Newspaper science, school science: Friends or enemies? *International Journal of Science Education,* **13**(4), 363-372.

White, R. T., & Gunstone, R. (1992). *Probing understanding*. Philadelphia, PA: The Falmer Press.

Wiggins, G. (1993). *Assessing student performance: Exploring the purpose and limits of testing*. San Francisco, CA: Jossey Bass.

Winsor, P., Butt, R. L., & Reeves, H. (1999). Portraying professional development in pre-service teacher education. *Teachers & Teaching,* **5**(1), 59–33.

Wolf, D., Bixby, J., Glenn, J., III, & Gardner, H. (1991). To use their minds well: Investigating new forms of student assessment. In G. Grant (Ed.), *Review of research in education* (pp. 31-74). Washington, DC: American Educational Research Association.

Wood, R. (1994). Work samples should be more used (and will be). *International Journal of Selection and Assessment,* **2**(3), 166–172.

Wu, Y. S. (2007). *A cross-age investigation of students' conceptions and mental models of the particulate model of gas.* Unpublished master's thesis, National Taiwan Normal University, Taipei (in Chinese).

INDEX

A

B

C

D

E

F

I

J

K

L

M

N

O

P

Q

R

T

U

V

W

Y